TOURISM AND BORDERS

New Directions in Tourism Analysis

Series Editors: Kevin Meethan, University of Plymouth
                    Dimitri Ioannides, Southwest Missouri State University

Although tourism is becoming increasingly popular as both a taught subject and an area for empirical investigation, the theoretical underpinnings of many approaches have tended to be eclectic and somewhat underdeveloped. However, recent developments indicate that the field of tourism studies is beginning to develop in a more theoretically informed manner, but this has not yet been matched by current publications.

The aim of this series is to fill this gap with high quality monographs or edited collections that seek to develop tourism analysis at both theoretical and substantive levels using approaches which are broadly derived from allied social science disciplines such as Sociology, Social Anthropology, Human and Social Geography, and Cultural Studies. As tourism studies covers a wide range of activities and sub fields, certain areas such as Hospitality Management and Business, which are already well provided for, would be excluded. The series will therefore fill a gap in the current overall pattern of publication.

Suggested themes to be covered by the series, either singly or in combination, include – consumption; cultural change; development; gender; globalization; political economy; social theory and sustainability.

Also in the series

Christian Tourism to the Holy Land
Pilgrimage during Security Crisis
Noga Collins-Kreiner, Nurit Kliot, Yoel Mansfeld and Keren Sagi
ISBN 0 7546 4703 X

Urban Tourism and Development in the Socialist State
Havana during the 'Special Period'
Andrea Colantonio and Robert B. Potter
ISBN 0 7546 4739 0

Tourism and Borders
Contemporary Issues, Policies and International Research

Edited by

HELMUT WACHOWIAK
International University of Applied Sciences Bad Honnef • Bonn, Germany

Routledge
Taylor & Francis Group
LONDON AND NEW YORK

First published 2006 by Ashgate Publishing

2 Park Square, Milton Park, Abingdon, Oxon OX14 4RN
711 Third Avenue, New York, NY 10017, USA

Routledge is an imprint of the Taylor & Francis Group, an informa business

First issued in paperback 2017

British Library Cataloguing in Publication Data
Tourism and borders : contemporary issues, policies and
    international research. - (New directions in tourism
    analysis)
    1.Tourism - Political aspects 2.Interregionalism 3.Place
    marketing
    I.Wachowiak, Helmut
    338.4'971

Library of Congress Cataloging-in-Publication Data
Tourism and borders : contemporary issues, policies, and international research /
edited by Helmut Wachowiak.
        p. cm. -- (New directions in tourism analysis)
    Includes bibliographical references and index.
    ISBN 0-7546-4775-7
    1. Tourism. 2. Visitors, Foreign. 3. Boundaries. I. Wachowiak, Helmut. II.
Series.

    G155.A1T589336 2006
    338.4'791--dc22

                                                                    2006012522

ISBN 978-0-7546-4775-1 (hbk)
ISBN 978-1-138-27392-4 (pbk)

Contents

PART III   Communication and Information

PART IV   International Research

List of Figures

List of Maps

List of Tables

List of Authors

Maha Doppelfeld, B.A., Dipl.-Betriebswirtin (FH)
Friends of Hwange National Park Trust
Zimbabwe
Email: maha@yoafrica.com

Daniel Engels, B.A.
International University of Applied Sciences Bad Honnef • Bonn
Email: daniel.engels@fh-bad-honnef.de

Holger Faby, Dr.
European Institute of Tourism (ETI)
Germany
Email: faby@eti.de

Karin Hartmann, M.A.
Mobility Consulting - Swissenergy
Switzerland
Email: karin.ht@gmx.de

Dimitri Ioannides, Ph.D., Professor
Missouri State University
USA
Email: dii608f@missouristate.edu

Stephan Kroll, Ph.D., Assistant Professor
California State University
USA
Email: skroll@csus.edu

Heinz-Dieter Quack, Dr., Professor
European Institute of Tourism (ETI)
Germany
Email: hdquack@eti.de

Christian Rast, Dipl.-Geograph
ift Freizeit- und Tourismusberatung GmbH
Germany
Email: rast@ift-consulting.de

Dallen J. Timothy, Ph.D., Professor
Arizona State University
USA
Email: dtimothy@asu.edu

Helmut Wachowiak, Dr., Professor
International University of Applied Sciences Bad Honnef • Bonn
Germany
Email: h.wachowiak@fh-bad-honnef.de

Preface

Based on the forecasts of the World Tourism Organisation (UNWTO), a continuous growth of international tourist arrivals is expected over the next decades. Consequently, in addition to an understanding of 'international tourism', an understanding of 'border issues' is necessary. Whenever persons cross a political boundary, phenomena of the environmental, socio-cultural, and economical world arise. Consequently, research from scholars of different disciplines has included the relationships between tourism and borders since as early as the 1930s. But, as so often in academia and, in particular, in the geography of leisure and tourism, as well as the economic disciplines focused on tourism, it is a fact, that many scholars contribute to important issues but only little knowledge exchange between international research groups exists.

Who really knows the innumerable contributions besides the many valuable books and articles of Northern-American and Central-European scholars? Who is able to evaluate the academic value of new experience and knowledge that has been created from several thousands of projects to support cross-border development worldwide? Who has the interdisciplinary expertise to understand the many facets of tourism development in cross-border areas, in so far as they involve technical planning issues on different administrative levels, international and national law issues, financial issues, natural and man-made environments, the management of public and private stakeholder interests, destination marketing, and other issues?

Furthermore, the subject *tourism and borders* still seems to be a niche within the academic disciplines that deal with leisure, recreation, and tourism as social, ecological, and economical phenomena. It is definitely observed by many but maintained by few. This fact is contradictory in a world of changing borders, the growth of regional importance, and growing self-administration on the regional level that sometimes even lead to autonomy developments. Furthermore, changing tourists' attitudes and a new sensibility towards borders as attractions, as well as non-acceptance of them as barriers in a world of mobility, should foster more intensive research in this field.

Given the importance of the relationships between tourism and boundaries, it is astonishing that this subject only met with intensive academic interest in the 1970s (with the IGU Tourism and Recreation working group meeting in Ljubljana/Trieste in September 1978 as one highlight). It only rejuvenated significantly in the early 1990s (especially in Europe, due to the greater harmonisation between the member countries of the European Union).

For these reasons, this compilation of works from several scholars and practitioners in the field of the geography of tourism and recreation, and tourism management aims to jointly motivate the academic community to continue the important discussion on tourism and boundaries. In contrast to many given and still beneficial contributions, this book does not aim to reflect on the subject with new theoretical discussions nor does it provide best-case reports from around the world. It follows the idea of reflecting on the relationship between tourism and boundaries from the major contemporary facets concerned. First, a conceptual discussion of tourism and international boundaries creates a framework of the key subject. This is followed by discussions about new policies and planning debates, destination management and stakeholder collaboration issues in border regions, as well as contemporary concepts of communication strategies to achieve trans-boundary development. This also includes recent developments in computer based information- and reservation systems in border regions.

However, apart from discussing selected contemporary discussions, this book follows a second approach. In the form of an individual chapter, the reader will find a commentary literature overview that provides information about most contributions in this field since the 1930s. Structured and analysed by geographical realms as well as by sub-topics, this bibliography aims to support research activities of (under-) graduates and scholars in the fields of geography, management, sociology, and policy and planning.

Certainly, the literature overview cannot include all published material that exists and, indeed, many are not listed. This is true also for the selected facets discussed by each chapter. If more time, and pages, were available, more chapters with additional facets of the phenomena arising out of tourism as a cross-border activity would have been included. Like every book, this one has also had to cope with the gap between academic desires and feasibility. However, all gaps and missing contents will be justified, if this book succeeds in motivating more persons to contribute to the subject, which will increase in importance again in a world of new regions and trans-boundary regionalism.

Helmut Wachowiak

International University of Applied Sciences
Bad Honnef • Bonn, Germany

Acknowledgements

As early supporters of my academic life, it was Christoph Becker and Albrecht Steinecke that confronted me first with the subject of cross-border issues in tourism, and I am still grateful for this. Many people have been influential in the last few years, and therefore I should like to take the opportunity to thank Antoinette Klute-Wetterauer and Peter Thuy, as well as Claudia Schild-Franken, Philip Sloan, and Georg Ummenhofer for memorable years during the foundation stage of the International University of Applied Sciences Bad Honnef • Bonn. Wolfgang Haensch, for the many ideas, support, and new horizons he gave me in different areas of my profession. In addition, I will never stop developing new ideas together with Heinz-Dieter Quack, whose feedback and enthusiasm I have always perceived as being very beneficial. I was fortunate to attract the interest of Dallen J. Timothy, without whose encouragement, I should certainly have stopped working on this book at a very early stage. In addition, many thanks go to Dimitri Ioannidis, who supported the project by finding a publisher, for his valuable suggestions concerning structure and contents, and for his personal contribution parallel to many other obligations. In the same way, many thanks go to the team of Ashgate Publishing Ltd for transforming an idea onto real paper. To Rudi Hartmann, who I regard as a perfect supporter of international collaboration in the field of the geography of leisure, tourism, and recreation. Without his engagement before, during, and after the annual meeting of the AAG in Denver/Colorado in Spring 2005, I should never have had the opportunity to meet so many outstanding colleagues from around the world. I have asked many people for their opinion about this book and none of their suggestions were wasted. To the chapter contributors, I have to pass on my many thanks for their expertise, their time, and especially for their patience. A final thanks goes to Daniel Engels, who accompanied this book from the first stages until delivery. Major parts of the commentary bibliography chapter are based on his thoughts and are linked with innumerable hours he invested in studying the literature on tourism and borders worldwide.

To Birte, Kara and Tobin

Introduction

Helmut Wachowiak

International University of Applied Sciences Bad Honnef • Bonn, Germany

In modern day societies, tourism has become a major aspect of leisure and recreational behaviour (Franklin, 2003) and, as such, has spurred numerous academic discussions on the scope and shape of tourism and related activities throughout the world. The different forms of tourist activities and tourist spaces have been thoroughly researched so that scholars, students and researchers today can easily get access to information about tourism and choose among a variety of very detailed definitions highlighting specific facets of this social, economical, and physical phenomenon.

In recent years, tourism has been especially recognised for its tremendous economic values and implications. This is reflected in the numerous publications with special foci on economic issues implicated by recreational activities, the fast growing number of new management schools and business universities around the world with academic programmes in tourism, sports, events, hospitality, and aviation, all of which add evidence to this development. This recognition led to amplified interest in the relationships pertinent between tourism and other economic activities. Especially ongoing globalisation and technological advances that facilitate travelling were consequently acknowledged to have spurred participation in tourist activities and subsequently the academic interest in the economics of tourism. It was soon identified as an activity relating to people exchanging and consuming international goods and services thereby economically and socially influencing the generating regions and destinations alike.

Academic discussions in the field of tourism have thoroughly examined the various social, economic, and environmental aspects and defined the phenomenon of tourism as seen from numerous different angles. It should not be surprising then, that there is currently a vast amount of academic literature discussing the economic scope of tourism, its implications, opportunities and threats for established social and economic systems, its chances for enhancing or even creating new economies and the political and environmental requirements and implications of tourism planning, development and control.

However, one aspect has so far been entirely neglected within these various definitions. According to Timothy (2001) and Timothy and Tosun (2003), tourism necessarily involves crossing borders of some type (international, subnational, regional, natural or only perceived ones). It could be argued then, that all borders, in turn, also influence tourism since they are a means of controlling the flow of people and enabling officials to enforce restrictions on either desired or undesired people coming into or leaving a specific country. With ever increasing tourist diversification and the emerging desire to collect destinations through visa stamps in tourists' passports (Franklin, 2003; Davis and Guma, 1992; Urry, 1992), the ritual of crossing a border, as an activity in itself, has grown in attractiveness. Additionally, historic border manifestations have become popular tourist attractions, and relicts and remnants of former borders, such as the Berlin Wall, have reached heritage status.

These aspects highlight the importance of the relationship between borders and tourism. But it is this elementary relationship that has largely been ignored and taken for granted in earlier tourism research. As a consequence of the poor definition in current academia, this important aspect of tourism studies has gained momentum in recent years and come into focus for future tourism research activities. This fact is also taken into consideration in this book, which carefully administers and presents contemporary study findings and research issues related to explorations in the dynamic relationship between tourism and borders.

As mentioned above, ongoing globalisation has spurred tourism participation through opening countries to the outside world, facilitating border crossings and creating economic free-trade and open border areas, thereby indicating the scope of the relationship between tourism and borders. The increasing social perception of the world as a 'global village' in recent years has brought up new questions for tourism managers and indicated the importance of investigating such relationships in more detail. The fact that numerous amounts of tourists all over the world, sometimes even unawares, cross borders every day, either because the border regions are their preferred destination, or because they pass through them on their way to a final destination, might explain why this aspect of tourism research has recently been recognised for its importance and is starting to be considered a major aspect of future tourism research.

Tourism planners and academics, as well as local stakeholders who have tasted 'the sweet juice of tourist cash' (Smith, 1996:xiv) have just begun to realise that the tourism industry, in particular, is highly perceptible to border issues (entry visa regulations, foreign investments, and border crossing restrictions, to name just a few issues) and that such issues need to be resolved satisfactorily before they can positively influence the economic outcome of tourism ventures. Former peripheral borderlands find themselves suddenly prospering from valuable tourism revenues and managers and politicians subsequently face questions on whether a border region could be considered as a destination itself, how to market it as a competitive destination unit, and what to consider in managing it in a sustainable way. The relationship between tourism and borders, therefore, is about the relationship between international neighbours and the way they co-operate across different types of borders in bringing regional, local and national policies in line with those of their neighbours.

Besides the fact, that many contributions are available as case studies, project documentation and governmental reports, only few teaching and research monographies discuss the variety of facets related to the subject. Language barriers are also responsible for the existing weakness in knowledge transfer between academic communities around the world. As a consequence, more publications focussing on the subject 'tourism and borders' with a more holistic and international understanding, would support academic teaching (such as in courses of 'Tourism policy', 'Tourism planning', 'Sustainable tourism development', 'Geography of leisure and tourism', and more) as well as further research. Consequently, this book addresses undergraduate and post-graduate students (especially within Master-Courses), PhD candidates, scholars, and private and public tourism organisations (including libraries) in the fields of Tourism Management, Tourism Marketing, Tourism Policy and Planning, and the Geography of Leisure and Tourism.

In addition to the enlargement of the 'New Europe', the increase in research, as well as planning activities concerning destination development and sustainable development and policies, form the pillars of the topical structure of this book, which intends to communicate a truly international perspective on the subject.

The first part ('Conceptual issues and policies') discusses central up-to-date issues of the topic 'tourism and borders'. As a conceptual framework, *Dallen J. Timothy* summarises the implications of existing and changing borders on tourism, and vice versa. He highlights, that because borders are places where political entities collide, economies converge, and cultures blend, they are perhaps one of the best laboratories for studying the globalisation process. An examination of the supply and demand side of tourism in border areas, examples of opportunities (such as employment opportunities in weakly developed border regions), as well as threats (such as prostitution, traffic congestion due to different price levels of products and services, etc.) for destination stakeholders from a local to international level, will be provided.

Because the opportunities for tourism development in border areas are most relevant for Europe with its many political entities, *Holger Faby* describes the policies applied by the European Union to support cross-border development. Special emphasis is given to the initiative INTERREG, the most important and requested planning tool for developing, managing, and financing cross-border projects in leisure and tourism. He differentiates the three strands of this initiative, in line with the different types of border constellations between countries: cross-border co-operation, transnational co-operation, and inter-regional co-operation.

As a current example of the last type of geographic borders, *Dimitri Ioannides* provides insight into an INTERREG IIIC project that aims to improve the effectiveness of regional development policies through large scale information exchange and co-operation networks. The GEDERI project (Gestion et Développement des Regions Insulaires) focuses on the development and management of Europe's island regions. It recognises that European islands share problems and concerns including economic marginality, a limited resource base, poor accessibility, chronic depopulation and a brain drain, plus environmental and/or socio-cultural threats brought on, among other things, by excessive tourist development. Finally, suggestions are provided on how the project aims to contribute to the sustainable development of Europe's islands.

The second part ('Destination marketing and management') takes into consideration, that the knowledge of modern management strategies gained in other non-peripheral destinations could be beneficial for joint destination management policies in cross-border areas in the future. However, due to the multiple differences between regions, cross-border destination marketing and development has to cope with many problems as well.

The first contribution to this section concentrates on the incoming tourism segment as a link between both the international tourism market, and the understanding of tourism and political boundaries. With special focus on the German incoming market, *Christian Rast* and *Stephan Kroll* analyse the specific problems associated with the marketing of inbound tourism and explain marketing issues of cross-border tourism from the point of view of the host destination. While inbound tourism is necessarily concerned with 'boundaries', the discussion of cross-border issues within this context has so far been sparsely covered by the existing literature. Furthermore, this chapter proposes an integral and systematic approach to analysing inbound tourism and its international components in order to fill the research gap. With reference to the German inbound market, differences in the tourist behaviour between tourists from different countries will be pointed out and appropriate suggestions for future destination marketing policies will be given.

The organisation of destination marketing depends heavily on the supporting administrative and financial structure in different countries. *Heinz-Dieter Quack*, who compares the destination organisations in France and Germany, will demonstrate this. Despite the fact that both countries play an important role in international tourism, the national borders mark a significant change in the organisation of tourism marketing. Different administrative structures of local, regional, and national authorities result in different decision making processes. This is especially important if it comes to cross-border co-operation, especially on a local and regional level.

As one of few successful tri-national co-operations, *Katrin Hartmann* discusses this problem in depth. She provides insight into the destination marketing efforts of the Lake Constance area that defines the border triangle between Austria, Germany, and Switzerland. She emphasises, that the implementation of a destination management process can help to make cross-border co-operation more effective, more professional, more sustainable, binding, and closer to the market. To achieve this result, professional organisational structures need to be implemented, the cross-border core businesses must be defined, and electronic cross-border marketing networks must be established. Her analysis also takes into consideration the conditions necessary for the implementation of a destination management process in cross-border areas and reflects on the associated opportunities and problems.

The third part ('Communication and Information') brings two selected facets of contemporary development in cross-border tourism to our attention: '*How to communicate joint sustainable development strategies in border regions effectively?*' and '*How to organise tourist oriented information systems across borders*'. Both questions are regarded as being the most relevant issues related to cross-border communication between destination stakeholders. Inter-cultural differences frequently add further barriers to joint cross-border development in tourism.

Maha Doppelfeld, in her detailed analysis of collaborative stakeholder planning in cross-border regions, exemplifies the problems as well as the opportunities of stakeholder collaborative planning to achieve sustainable tourism development. Based on research conducted in the Great Limpopo Transfrontier Park in Southern Africa, she notes that the formation of political boundaries has caused many ecosystems to be separated and disrupted. This problem has been identified, and today many countries with neighbouring protected areas, have decided to join these areas in order to reverse adverse impacts and improve the protected areas. The formation of trans-boundary parks not only benefits the environment, but also encourages friendly and co-operative relations between countries.

Another, and most relevant contemporary facet of communication issues is also dealt with by *Holger Faby*'s analysis of tourism information and communication systems in border areas. This chapter focuses on the technical issues, restrictions, and outlook. Here, Faby manages to highlight the pre-requisites of building up cross-border co-operation and network-based amenities, such as services on the Internet and mobile information systems: necessary *organisational measures* of especially small and medium sized enterprises in the hospitality sector in border regions, and necessary *technical measures* of destination marketing organisations and enterprises in border regions. This presents a vast field of research for tourism, market research, and cartography. Cartographic-scientific aspects, such as the scientific findings drawn from the observations made on the reaction of the users, their expectations and of the users' technical affinities in the examined field, should be highly rated. This is a reason for regarding cartography as both the science and the technique of the graphic, communicative, cognitive, and technological treatment of geospatially relevant information.

The last but also biggest section of this book ('International research') mainly consists of a commentary literature bibliography of 1,280 items written on the field of tourism and borders. The bibliography intends to support the conduction of further research in this field, as well as to support marketing and planning activities of organisations actively engaged in cross-border collaboration. To enable the reader to quickly access the desired information and easily skip through passages of minor interest, the chapter is further divided into six sections. The first section in this chapter contains literature that enables the reader to gain a general insight into types, scales, scope and functions of different borders, the social, economic and environmental importance and characteristics of border regions, the emergence of regionalism, and the politics of cross-border co-operation. The following five sections introduce literature on Europe, the Middle East, Asia, America, and Africa. Each section is further split into subsections introducing specific fields of interest, such as politics or environmental co-operation for example. In addition, a complete bibliography, listed alphabetically by the author's surname, is available at http://www.ashgate.com/subject_area/downloads/sample_chapters/Tourism_and_Borders_Bibliography.pdf, enabling the reader to quickly locate a title written by a specific author. In order to keep the international character of this scientific field, each title is listed in the language in which it was originally published.

The selected contemporary issues of the subject 'tourism and borders' presented in this book contain many remarks on future research on tourism and political borders. These have to be acknowledged with respect to the complexity of the subject and are, therefore, not formulated as recommendations. However, if the remarks made by the chapter contributors meet the interest of academics around the globe they could be further discussed and developed as a roadmap for future debates and knowledge exchange within the growing field of tourism. In addition, of course, language borders still exist between academic groups around the world dealing with issues in the wider field of leisure, recreation, hospitality, and tourism.

References

Davis, G. and Guma, G. (1992), *Passport to Freedom: A Guide for World Citizens*, Washington: Seven Locks Press.

Franklin, A. (2003), *Tourism – An Introduction*, London: SAGE Publications.

Smith, V.L. (1996), Foreword. In: Price, M.F. (ed.) (1996), *People and Tourism in Fragile Environments*, Wiley: Chichester.

Timothy, D.J. (2001), *Tourism and Political Boundaries*, London: Routledge.

Timothy, D.J. and Tosun, C. (2003), 'Tourists' perceptions of the Canada – USA border as a barrier to tourism at the International Peace Garden', *Tourism Management*, Vol. 24, pp. 411–421.

Urry, J. (1992), 'The Tourist Gaze and the "Environment"', *Exploration in Critical Social Science*. Vol. 9, No. 3, pp. 1–26.

PART I
Conceptual Issues
and Policies

Chapter 1

Relationships between Tourism and International Boundaries

Dallen J. Timothy
Arizona State University, USA

Introduction

Every day millions of people cross international borders for a variety of reasons. Primary among these is to work in a neighboring country. In addition to this, however, millions more people cross each day for purposes directly opposite of work–leisure. In most cases, people cross some form of political boundary every time they leave home for a weekend, go on an extended holiday, and in many cases they cross municipal, or city boundaries to dine out, watch a movie, or go skating. While these lower level boundaries, such as municipalities, counties and states or provinces, appear to have little effect on tourism and recreation, this is an unfounded assumption. In most cases, these lower administrative boundaries have significant bearings on property and sales taxes, education, law enforcement, public utilities, and social services. While these issues are closely linked to tourism in many ways, the most notable boundaries from a tourism perspective are found at the international level.

Because borders are places where political entities collide, economies converge, and cultures blend, they are perhaps one of the best laboratories for studying the globalization process. Likewise, tourism, one of the most globalized of all industries, has many unique characteristics. When the two, borders and tourism, run together, several interesting and unique relationships become evident: boundaries as tourist attractions and destinations, borders as barriers to travel and the growth of tourism, boundaries as lines of transit, and the growth of supranationalism, to name but a few. This chapter describes and examines some of these relationships between international boundaries and tourism. The author has written extensively in the past about many of these issues, so much of this chapter will be a re-evaluation process based on his previous work and more recent observations and empirical material.

The appeal of borders

Crossing international boundaries has fascinated people for centuries. Even as early as the fourteen century, boundaries were being marked in Europe, and people were crossing them with citizenship papers and special permissions. In most places at that time, however, borders were vague areas of dubious political control, where exact borderlines were few and far between. Today, however, most international boundaries have been clearly defined and well marked on the landscape, and they are even more of an attraction than they have been in the past.

Political boundaries as tourist attractions may be seen from two main perspectives (Timothy 1995; 2001; 2002). First, the borderline itself, including the demarcation indicators, fences, walls, and guard towers, exude considerable appeal for curiosity seekers. This is especially the case with famous borders (e.g. the former Berlin Wall and the North-South Korean DMZ) or where the methods of demarcation provide an interesting contrast in otherwise ordinary landscapes.

For example, until the 1990s, lookout platforms in West Berlin provided opportunities for spectators to look across the 'Iron Curtain' into the communist east – an experience that highlighted ideological differences in ways of living, economics, and political landscapes (Elkins and Hofmeister 1988; Koenig 1981). The 'Golden Triangle', the point where Thailand, Myanmar and Laos meet, has become a rather important tourist destination from the Thai side of the border. Thousands of tourists visit the location each year to have themselves photographed at the Golden Triangle monument on the bank of the Mekong River. Similarly, visitors to Basle, Switzerland, commonly find themselves peering at the monument that marks the place where Germany, France, and Switzerland meet at a single point. Many examples of these types of border curiosities exist in North America, Europe, Asia, Latin America, and Africa. Research and commentary demonstrate that wherever a borderline is clearly marked, visitors will have an interest in standing astride it, hopping over it, or leaning against it for photo opportunities.

Perhaps the simplest manifestation of the 'border as attraction' phenomenon is people's propensity to want to straddle borderlines, so that they can claim to have been in two places at once or at least having been abroad, even if only by a few meters (Timothy 1995). It is not uncommon to find travelers stopped at 'Welcome to…' signs and border markers photographing and standing on them. Even relict boundaries, those that are discontinued, remain important attractions, such as the East-West Germany divide that has remained a popular site in Berlin and all along the border as a series of 'borderland museums' that are being promoted as important tourist attractions (Borneman, 1998; Light, 2000). In some cases, the existence of a borderline and its historical significance becomes a tourist icon for the border community's marketing and promotional efforts.

The second way in which international boundaries attract attention among tourists and recreationists is not the line itself, but the activities, attractions, and special features of communities in the immediate vicinity of the boundary. While the line itself in these cases is not necessarily the main feature, the area's appeal is rooted in its location adjacent to the border, which creates some kind of competitive advantage from what lies on the other side. This second perspective could more accurately be described as the border as tourist destination, while the first type might best be termed the border as an attraction. Places where the border is a destination tend to have several activities and attractions in common: shopping, prostitution, gambling/casinos, restaurants, bars and nightclubs, and liquor stores.

People who live in countries where gambling is not permitted often travel across a border to neighboring countries where it is allowed (Lintner 1991). In this situation, casinos tend to dot the landscape near border crossing points or further inland, and the majority of their clientele is from abroad. In Canada, casinos have been built near ports of entry from the United States as a way of drawing American gamblers north of the border (Nieves 1996; Smith and Hinch 1996). Another prominent example was in South Africa prior to 1994, when two of the four independent homelands, Bophuthatswana and Transkei, developed successful casinos that drew thousands of gamblers from South Africa and other nearby countries, where gaming was not permitted (Stern 1987). With the re-integration of the homelands into South Africa in 1994, the casinos were permitted to continue functioning and have therefore ceased being a border phenomenon. A similar situation exists in Swaziland, an independent kingdom also surrounded by South Africa.

Alcohol consumption also tends to be associated with border communities. This is especially the case where drinking ages are lower on one side, when purchasing limits are not enforced or are legally lower, or where the cost of alcohol is less expensive. This is one of the main reasons the Mexican border towns grew to such notoriety as exotic playgrounds for Americans. During the prohibition period (1918-1933), when alcohol was not permitted to be manufactured or sold in the United States, hundreds of thousands of Americans poured over the border to quench their thirst for liquor (Arreola and Curtis 1993). Today, the bars and liquor stores are still an important part of the tourist districts there, as prices are much lower than in the United States, and IDs are typically not checked at bars, so that even American high school students are able to drink south of the border (Arreola and Madsen 1999). The same phenomenon gave rise the 'booze cruises' between Great Britain and France and between Finland and Sweden during the 1970s, 80s, and 90s wherein people crossed the English Channel and the Baltic Sea in search of alcohol at duty-free prices (Essex and Gibb, 1989; Hidalgo, 1993; Peisley, 1987).

Prostitution tends to be associated with bars and dance clubs in border areas. This activity grew quickly along with alcohol consumption in the Mexican border towns, especially those that are situated near major US military installations to fill the lonely nights of military men. Some countries have strict anti-prostitution and anti-pornography laws, which drive many visitors to neighboring countries in search of sexual gratification (Arreola and Curtis 1993; Curtis and Arreola 1991; Timothy 2001).

Shopping is among the most popular activities undertaken in border communities, usually spurred by the existence of cheaper products, lower taxes, wider arrays of goods, and different hours of operation in neighboring countries. This phenomenon can be found at international borders everywhere, even in fairly remote locations. Rarely do people cross to purchase souvenirs – rather, most products in demand are everyday, household items, such as clothing, food, shoes, cleaning supplies, electronics, jewelry, and gasoline (Michalkó and Timothy 2001; Timothy and Butler

1995). A couple of unique patterns are associated with cross-border shopping. First, there appears to be a notable spatial pattern, wherein the closer a person lives to the border, the more frequently he/she will cross for smaller items (e.g. petrol, groceries, and cigarettes). The further one lives from the border, the less frequently he/she will cross, but the items purchased will be bigger (e.g. clothing, electronics, and appliances) (Timothy and Butler 1995). Second, there is a notable seasonal pattern in most forms of cross-border consumption. Popular shopping holidays, such as Christmas and Easter, tend to see higher levels of consumption. Long-haul out-shopping trips abroad (e.g. from Europe and Japan to North America) are also on the rise. Also, levels of cross-border shopping tend to be higher during long breaks, such as summer vacation and major holidays, suggesting that there is a significant pleasure element involved rather than it being simply a rational, economic activity.

In North America, a unique form of cross-border shopping has developed in recent years, namely buying pharmaceuticals and health care. This exists in other parts of the world, too, where people travel to various nearby countries for surgery and other health care needs (e.g. Middle Easterners to Europe and North Americans to Latin America). Owing to the high cost of medications and health care in the United States, Americans have begun traveling across the border into Canada to purchase medications. Many also cross into Mexico for the same purpose, but also for medical and dental care. Among the most ubiquitous commercial establishments in the Mexican border town tourist zones are pharmacies, dentist offices and medical facilities.

Borders as obstacles to tourism

While borders may in many instances be significant tourist attractions and destinations, they also function as barriers to travel. In this sense, borders can be seen as either real or perceived impediments to travel. Real barriers are created when heavy fortifications are erected by a country to defend itself against threatening forces. Barbed wire fences, concrete walls, minefields, and armed guards contribute to the development of landscapes of conflict that are generally uninviting to cross (Timothy 2001). Strict immigration and customs policies may also function as real barriers to travel when citizens of certain countries are refused entry or are made to go through rigorous visa application processes or physical scrutiny when entering a country. It is not uncommon for people to choose countries where a visa is not needed over destinations that require one.

Travel to the countries of Eastern Europe during the communist period was affected by this phenomenon quite notably. Strict visa requirements to enter East Germany and Poland, for instance, deterred many people from visiting. Likewise, countries such as North Korea and Albania before the 1990s, were the strictest and most closed of all communist nations. US citizens, for instance, were not permitted entry into either country, and it was difficult at best for other nationalities to acquire visas. US citizens are still not permitted to enter North Korea, with only very few exceptions. Likewise, travel between Israel and Lebanon today is an uninviting prospect owing to war conditions and the heavy fortifications between the two countries, and virtually impossible for most nationalities because of austere military restrictions. If they were allowed, border crossings between Lebanon and Israel would be important in the realm of tourism owing to the heritage tourism linkages with a common past that exist on both sides and many cultural similarities.

Psychological, or perceived, barriers are the second type of border impediment to tourism. This is certainly the situation with borders that separate hostile neighbors or at borders that are heavily fortified and defended. However, even at friendly borders, people may feel a sense of nervousness or apprehension about crossing. Language and cultural differences, different currencies, and opposing political ideologies may contribute to some travelers' reluctance to cross. In addition, border formalities can be an intimidating process that might keep some people from traveling abroad. Even borders as innocuous as Italy-Switzerland or USA-Canada may erect psychological barriers when it comes to customs and immigration policies and procedures and perceived differences on opposite sides of the border (Timothy and Tosun 2003). The US-Mexico border is an excellent example of a line that separates two very different entities – the developed world from the less-developed world, language and culture, history, and political systems. For many Americans, crossing into Mexico is not easy owing to different driving laws, language differences, and fears of food and quality of hygiene.

Borders as lines of transit

The third relationship between tourism and international boundaries, and perhaps the least understood, is that of borders as lines of transit. In the majority of cases throughout the world, borderlines are simply places to go beyond to get to more important destinations. Many people pass through entry procedures and then continue on to their final destinations. Typically, border crossings that have a high level of traffic flowing through them are dotted with landscape features such as petrol stations, banks and currency exchange booths, restaurants, and insurance brokers.

Another interesting transit feature of borders is that they are similar to airports in that they are typically viewed as 'non-places' or 'placeless spaces' (Travlou 2003). For instance, it is unlikely that someone flying between New York City and Los Angeles, stopping to change airplanes at the Minneapolis airport for two hours would ever consider that he/she has been to Minneapolis. Such is often the case with international boundaries. For example, many Americans visiting Mexican border towns would be unlikely to claim that they have been to Mexico. For these people, Mexico is something that lies further inland – a place that takes longer to get to and that requires more time in the destination. Thus, borders and their adjacent communities, at least in the North American psyche, become non-places, unrecognizable as Mexico, and certainly apart from the United States.

The changing role of borders today

Finally, a look at borders and tourism would not be complete without at least a cursory discussion of the changes that have taken place during the past two decades and the ongoing geopolitical transformations that are changing the relationships between tourism and political boundaries. Broadly speaking two dichotomous patterns of change exist: a decrease in the barrier effects of borders and an increase in their role as barriers.

In the realm of decreasing barriers, one of the most prominent types of change is popularly known as supranationalism (Jessop 1995; Teye 2000; Timothy and Teye 2004). As early as the mid-twentieth century, countries began to realize the value in working together to further one another's economic development. Since that time, many supranational alliances have been formed, although some of the most prominent include the European Union (EU), the North American Free Trade Agreement (NAFTA), the Association of Southeast Asian Nations (ASEAN), the Economic Community of West African States (ECOWAS), the Caribbean Community (CARICOM), and the South Asian Association for Regional Co-operation (SAARC). Many more exist in all regions of the world, and most countries belong to more than one. These associations are sometimes better known as trading blocs, customs unions, or economic communities, but what they all have in common is a desire to collaborate in an effort to reduce trade barriers, tariffs, and import and export quotas.

While relatively few of these alliances have tourism as a major focus, almost all of them deal with issues that directly affect tourism (Timothy 2003; 2004). For example, NAFTA deals with many issues related to environmental protection, which obviously has major bearings on tourism. The EU is heavily involved directly in tourism, especially through its regional structural funds that are given out to communities and cross-border regions to develop tourism-related infrastructure, conserve cultural and ecological resources, and reduce cross-border disparities in standards of living. ASEAN has its own tourism section, which is heavily involved in promoting the entire region of Southeast Asia as a large-scale tourist destination. It also acts as a liaison in negotiations between national governments, airlines, and other forms of transportation. CARICOM is concerned with tourism and its effects in the Caribbean region, and one of the main goals of ECOWAS is to simplify and encourage cross-border travel by citizens of its member countries.

Cross-border co-operation is also taking place on bilateral and multilateral scales in the realm of environmental conservation. During the past 20 years, many international parks have been designated for the purpose of conserving natural areas that lie across or adjacent to international boundaries (Ferreira 2004; Thorsell and Harrison 1990; Timothy 2000). Likewise, many international sport and tourism events have been planned and involved people on both sides of the border, such as races, fairs, shows, and various competitions.

Another form of positive change is the opening up of previously closed societies and the opening of new tourist destinations. For example, it has only been since the

1990s that China began to allow tourists to visit Tibet and some other western parts of the country. Likewise, with the collapse of communism in many parts of the world, countries have opened up to tourism and embraced the industry as an economic development tool. Countries of Eastern Europe that were traditionally difficult to access for many westerners, including Poland, Bulgaria, Romania, and the Czech Republic, are now trying very hard to lure visitors to participate in many forms of tourism (Williams and Baláz 2000). Albania, which was communist Europe's strictest country, opened up for tourists in the early and mid-1990s. The industry is still having many difficulties there owing to insufficient infrastructure, under-skilled labor, and some political remnants from the former regime. Nonetheless, its borders are now open, even for people who were previously unwelcome.

Even North Korea is showing signs of softening its isolationist stance toward international tourism. A few family meetings have been arranged, wherein a limited number of South Koreans have been allowed to visit the North on brief trips to visit relatives they have not seen since the 1950s. In addition, as a way of increasing foreign exchange earnings, North Korea has allowed the development of the Mt. Gumgang tourist zone in the sacred mountain area in the southeast corner of the country (Kim and Prideaux 2003; Timothy *et al*. 2004). Since the inception of this project, some 862,262 South Koreans and 4,511 non-Koreans have visited the area – the only way citizens of the South can visit the North. Even the heavily fortified Demilitarized Zone (DMZ) has been opened just a crack. The Mt. Gumgang tours initially used a boat to sail around the DMZ, but since 2004, the tours have been crossing the border by bus.

A unique situation also exists in Cyprus, which has recently undergone change. Observation platforms in Nicosia were important tourist attractions on the south side of the Green Line, where people could observe life across the border in the North. Non-Cypriot tourists were permitted, reluctantly, by the Greek Cypriots to cross into the North during the day, but they had to return before sunset. Also, they were not permitted to bring any items back that they had purchased in the North. Cross-border traffic in both directions for Turkish and Greek Cypriots was virtually non-existent. Today, however, since 2003, the political climate has started to change, and Greek Cypriots now can cross into the North and vice versa. The Turkish northern portion of the island has begun establishing casinos to attract Greek Cypriots across the Green Line, and many Greek residents have started crossing to seek out the villages and homes where they grew up and the churches they used to attend before the 1974 incident (Dikomitis 2004). In the opposite direction, Turkish Cypriots now cross on a daily basis into the South to work and shop. Shopping has become so popular among the Turkish population that the government of the North recently (December 2004) decided to begin limiting the amount of items that can be purchased in the South.

Despite these widespread changes toward higher levels of openness, in some parts of the world, there is a change in the opposite direction – that is, borders are becoming stronger lines of defense and therefore greater barriers to travel. Following the terrorist attacks in the United States in 2001, that country's government immediately sealed its borders with Mexico and Canada, as well as all sea and air gateways, for several hours. This immediate reaction was the first step in a series of efforts to tighten border security on the northern and southern boundaries (Ackleson 2005). Video and laser monitoring systems have been installed in places where they did not exist previously. Additional border patrol agents have been hired, fingerprinting is now commonplace, and eye- and face-recognition technology might soon be implemented as well (Goodrich 2002). In addition, many more people are being refused visas and entry into the country. On the periphery of the European Union, stricter border security measures have been implemented as well, particularly along the EU's southern and eastern boundaries. While these efforts are meant to deter terrorists, smugglers, drug dealers, and illegal immigrants, they inevitably also affect tourists' perceptions of the border and their willingness to travel.

Conclusions

Tourism and international boundaries share a number of relationships. Probably the most common linkage is that of borders as tourist attractions or destinations. When the borderline itself creates the tourist appeal, it can be said that the border is an attraction. When the border creates contrasting conditions on two sides in terms of rules of law, taxes, prices, etc, the border area with its adjacent communities become tourist destinations. In most cases, however, the people who visit border towns are not officially tourists, because they typically do not stay over night. Instead, they may be classified as international day-trippers, or excursionists, but regardless of what they are labeled for statistical purposes, they all contribute a great deal to border communities' tourism industries.

The second obvious relationship is that of borders as real and perceived barriers to travel. Clearly, heavy fortifications and strict entrance regulations keep many people from going across the border or even approaching it. The most common obstacle, however, is the perceptual barrier created by even friendly international boundaries. Crossing procedures and differences in culture, food, language, money, and political regimes deter many less adventurous people from visiting other countries, including the borderlands. For these less-adventuresome people borders create 'functional distance', or add perceived distance to places on the other side of an international frontier. Smith (1984: 37) suggested that the added psychological distance created by the US-Canada border results in conditions where 'the volume of travel between adjacent US states and Canadian provinces more closely resembles travel patterns between two distant regions'. The concept was noted many years earlier by Reynolds and McNulty (1968) and Mackay (1958) also in the context of North America.

The most common role of borders in tourism, however, is that of transit space. In this case, people typically only cross over borders and pass through their adjacent territories on their way to other places, and rarely are borders given the status as places in this regard.

Despite the three-part relationship outlined here, the role of borders has changed dramatically in recent years. Places that were once forbidden to foreigners are now opening up and new destinations are becoming more commonplace. On the other side of the equation, however, many more borders that have traditionally hosted large-scale tourism are becoming more difficult to cross, owing primarily to safety concerns and to thwart the flow of illegal aliens.

Tourism is a complex industry and political boundaries are complex lines of contrasts, similarities, struggles, and economic opportunities. The two together make a rich area of potential research, and as long as borders exist, there will be many rich opportunities for additional inquiry.

References

Ackleson, J. (2005), 'Constructing security on the US-Mexico border', *Political Geography*, 24, 165–184.

Arreola, D.D. and Curtis, J.R. (1993), *The Mexican Border Cities: Landscape Anatomy and Place Personality*, (Tucson: University of Arizona Press).

Arreola, D.D. and Madsen, K. (1999), 'Variability of tourist attractiveness along an international boundary: Sonora, Mexico border towns', *Visions in Leisure and Business*, 17:4, 19–31.

Borneman, J. (1998), 'Grenzregime (border regime): The Wall and its aftermath', in T.M. Wilson and H. Donnan (eds) *Border Identities: Nation and State at International Frontiers*, (Cambridge: Cambridge University Press, 162–190).

Curtis, J.R. and Arreola, D.D. (1991), 'Zonas de tolerancia on the northern Mexican border', *Geographical Review*, 81:3, 333–346.

Dikomitis, L. (2004), 'A moving field: Greek Cypriot refugees returning 'home'', *Anthropology Journal*, 12:1, 7–20 (Durham).

Elkins, T.H. and Hofmeister, B. (1988), *Berlin: The Spatial Structure of a Divided City*, (London: Methuen).

Essex, S.J. and Gibb, R.A. (1989), 'Tourism in the Anglo-French frontier zone', *Geography*, 74:3, 222–231.

Ferreira, S. (2004), 'Problems associated with tourism development in Southern Africa: The case of transfrontier conservation areas', *GeoJournal*, 60:3, 301–310.

Goodrich, J.N. (2002), 'September 11, 2001 attack on America: A record of the immediate impacts and reactions in the USA travel and tourism industry', *Tourism Management*, 23, 573–580.

Hidalgo, L. (1993), 'British shops suffer as "booze cruise" bargain hunters flock to France', *The Times*, (published on 22 November 1993), page 5.

Jessop, B. (1995), 'Regional economic blocs, cross-border co-operation, and local economic strategies in postcolonialism', *American Behavioral Scientist*, 38:5, 674–715.

Kim, S.S. and Prideaux, B. (2003), 'Tourism, peace and ideology: Impacts of the Mt. Gumgang tour project in the Korean Peninsula', *Tourism Management*, 24, 675–685.

Koenig, H. (1981), 'The two Berlins', *Travel Holiday*, 156:4, 58-63, 79–80.

Light, D. (2000), 'Gazing on communism: Heritage tourism and post-communist identities in Germany, Hungary and Romania', *Tourism Geographies*, 2:2, 157–176.

Lintner, B. (1991), 'Upstaging Macau: Casino at centre of border development plan?', *Far Eastern Economic Review*, (published on 16 May 1991), page 24.

Mackay, J.R. (1958), 'The interactance hypothesis and boundaries in Canada: A preliminary study', *Canadian Geographer*, 11, 1–8.

Michalkó, G. and Timothy, D.J. (2001), 'Cross-border shopping in Hungary: Causes and effects', *Visions in Leisure and Business*, 20:1, 4–22.

Nieves, E. (1996), 'Casino envy gnaws at falls on US side', *New York Times*, (published on 15 December 1996), page 49.

Peisley, T. (1987), 'Sea ferry travel and short cruises', *Travel and Tourism Analyst*, 1, 19–29.

Reynolds, D.R. and McNulty, M.L. (1968), 'On the analysis of political boundaries as barriers: A perceptual approach', *East Lakes Geographer*, 4, 21–38.

Smith, G.J. and Hinch, T.D. (1996), 'Canadian casinos as tourist attractions: Chasing the pot of gold', *Journal of Travel Research*, 34:3, 37–45.

Smith, S.L.J. (1984), 'A method for estimating the distance equivalence of international boundaries', *Journal of Travel Research*, 22:3, 37–39.

Stern, E. (1987), 'Competition and location in the gaming industry: The 'casino states' of Southern Africa', *Geography*, 72:2, 140–150.

Teye, V.B. (2000), 'Regional co-operation and tourism development in Africa', in P.U.C. Dieke (ed.), *The Political Economy of Tourism Development in Africa*, (New York: Cognizant, 217–227).

Thorsell, J. and Harrison, J. (1990), 'Parks that promote peace: a global inventory of transfrontier nature reserves', in J. Thorsell (ed.), *Parks on the Borderline: Experiences in Transfrontier Conservation*, (Gland: IUCN, 3–21).

Timothy, D.J. (2004), 'Tourism and supranationalism in the Caribbean', in D.T. Duval (ed.), *Tourism in the Caribbean: Trends, Development, Prospects*, (London: Routledge, 119–135).

Timothy, D.J. (2003), 'Supranationalist alliances and tourism: Insights from ASEAN and SAARC', *Current Issues in Tourism*, 6:3, 250–266.

Timothy, D.J. (2002), 'Tourism in borderlands: Competition, complementarity, and cross-frontier co-operation', in S. Krakover and Y. Gradus (eds), *Tourism in Frontier Areas*, (Lanham, MD: Lexington Books, 233–258).

Timothy, D.J. (2001), *Tourism and Political Boundaries*, (London: Routledge).

Timothy, D.J. (2000), 'Tourism and international parks', in R.W. Butler and S.W. Boyd (eds), *Tourism and National Parks: Issues and Implications*, (Chichester: Wiley, 263–282).

Timothy, D.J. (1995), 'Political boundaries and tourism: Borders as tourist attractions', *Tourism Management*, 16:7, 525–532.

Timothy, D.J. and Butler, R.W. (1995), 'Cross-border shopping: A North American Perspective', *Annals of Tourism Research*, 22:1, 16–34.

Timothy, D.J. and Teye, V.B. (2004), 'Political boundaries and regional co-operation in tourism', in A.A. Lew and C.M. Hall and A.M. Williams (eds), *A Companion to Tourism*, (London: Blackwell, 584–595).

Timothy, D.J. and Tosun, C. (2003), '"Tourists" perception of the Canada-USA border as a barrier to tourism at the International Peace Garden Tourism Management', 24:4, 411–421.

Timothy, D.J. and Prideaux, B. and Kim, S.S. (2004), 'Tourism at borders of conflict and (de)militarized zones', in T.V. Singh (ed.), *New Horizons in Tourism: Strange Experiences and Stranger Practices*, (Wallingford, UK: CAB International, 83–94).

Travlou, P.S. (2003), 'Airport terminals and hotel lobbies: Gazing Athens from in-transit spaces', (New Orleans: Paper presented at the annual meeting of the Association of American Geographers 5–8 March).

Williams, A.A. and Baláz, V. (2000), *Tourism in Transition: Economic Change in Central Europe*, (London: I.B. Tauris).

Chapter 2

Tourism Policy Tools Applied by the European Union to Support Cross-bordered Tourism

Holger Faby
European Institute of Tourism, Germany

Introduction

Europe continues to grow together – on 14 June 1985, five European countries[1] signed the 'agreement of Schengen' (Luxembourg). The Schengen treaty is a part of EU law, which allows for common EU immigration policies and border system. Altogether 26 countries (all EU states except Ireland and the UK, but including Iceland, Norway and Switzerland) have signed the agreement and 15 have implemented it so far. Border posts and controls have been removed between Schengen countries and a common 'Schengen visa' allows access to the area. However, the treaty does not cover residency or work permits for non-EU nationals. The first goal was to eliminate border checkpoints and controls within the Schengen area and harmonise external border controls.

Europe's bordering areas have been neglected with a historical background: In the past, national policy often neglected border areas, especially those that had been considered as peripheral within national boundaries. The presence of borders impedes border communities from each other and hinders coherent territorial management of economic, social, and cultural issues (European Communities 2002). But the single market and the economic and monetary union (EMU) are created with the target of enlargement and to turn cross-border co-operation in Europe and European integration to a future capable and lasting way.

European community[2] initiatives

The initiatives of the European Community are aid and action programmes set up to complement structural fund operations in explicit problem areas. Community

---

[1]   Belgium, France, Germany, Luxembourg, and The Netherlands.

[2]   As official publications of the European Union are referenced to the earlier name 'European Community', both names will be applied for the European Union.

Initiatives are set up by the European Commission and implemented and coordinated under national control. Each initiative is financed by one fund only – in the sum they make up about 5.4 per cent of the grand total budget of the European Structural Funds (Table 2.1).

Within the period 2000–2006, four initiatives can be differentiated (European Communities 2001a, 2001b, 2001c, 2003, 2004): The first one, INTERREG III, promotes cross-border, transnational, and interregional co-operation. The second one, EQUAL, aims to eliminate the factors leading to inequalities and discrimination in the labour market. LEADER +, the third initiative, concentrates its support to bring together those active in rural societies and economies to bring on new local strategies for sustainable development. The fourth initiative, URBAN II, supports on innovative strategies, i.e. to regenerate cities and declining urban areas.

National borders, which obstruct a balanced development and integration of Europe, are to be overcome with the help of this initiative. For regions in extreme peripheral location along the borders to the new member states and to the neighbour states thereby special attention applies.

Table 2.1   Budget of the community initiatives for the period 2000–2006

Initiative	billion Euro
INTERREG III	4.9
EQUAL	2.8
LEADER +	2.0
URBAN	0.7
Total	**10.4**

Source: European Communities 2001b

Because INTERREG III has been the most relevant initiative for the tourism development in European border areas, the following chapters concentrate on its structure, principles, and co-operation areas.

The INTERREG Initiative

Actually, the INTERREG initiative (covering 2000–2006) is in its third phase and follows on from the success on INTERREG I (1989–1993) and INTERREG II (1994–1999).

INTERREG III is a European Union (EU) initiative, which aims to stimulate interregional co-operation in the EU between 2000–2006 – and it is part of the policy of the economic and social co-operation and coherence politics of the community to foster the balanced development of the continent through cross-border, transnational, and interregional co-operation. It is financed by the European Regional Development Fund (ERDF) in addition with national, regional, and private means. Special emphasis has been placed on integrating remote regions and those, which share external borders with the candidate countries.

In detail, INTERREG III is divided in three strands: (European Communities 2005 a, 2005b): Strand A does mean cross-border co-operation by promoting integrated regional development between neighboring border regions, including external borders and certain maritime borders. Strand B contains transnational co-operation by contributing to harmonious territorial integration across the Community. The third strand, Strand C, promotes inter-regional co-operation by improving regional development and cohesion policies and techniques through transnational/interregional co-operation.

INTERREG III principles

The co-operation under INTERREG III is based on the following principles: The Programming territories and regions, which like to cooperate, must present the Commission with a Community Initiative Programme (CIP). They have to define their joint development strategy and demonstrating the cross-border value added by the planned operations. The programming has to take the general guidelines on the Community policies and Structural Funds[3] into account. The local, regional, and national authorities, the social and economic partners encourage a bottom-up approach to development. Transparent publicity allows the widest possible degree of participation by public and private stakeholders. The implementations of INTERREG III have to be consistent and must be synchronised with the other financial instruments concerned. The ERDF is responsible for all operations undertaken on the Community's territory, while other of the Community's external-policy financial instruments.

The challenge for INTERREG III is to take cross-border co-operation even further, in particular by financing the running costs of specific joint structures set up for that purpose, such as the European economic interest groupings (EEIGs). These structures are necessary for drawing up, managing and monitoring the programme.

INTERREG III: Different challenges and chances for the tourism in the European Union

On the one hand side, most of the tourists in the European Community do cross borders while they are travelling (including day trips). Because of the 'agreement of Schengen', of the freedom of crossing European borders, of establishment and

---

[3]   These are: job creation, improving competitiveness, sustainable development, environment, equal opportunities, compliance with Community competition rules.

of provision of services the EU already represents a favourable area and favourable destinations for international tourism.

On the other hand, side there are the official places/professionals in tourism in EU's border regions (i.e. national, regional and local tourist boards), which do not always work together in the European Community's context (Faby 2004). This can be justified with language barriers, cultural differences, and various types of state (e.g. centralism in France versus decentralized organization of the administration in Germany). The reasons are various, however, it is indisputable that the tourism sector is of primary importance for the Community: The European Commission, recognising the important role of tourism in the European economy, has been increasingly involved in tourism since the early 1980's, in co-operation with the Council, the European Parliament, the Economic and Social Committee and the Committee of the Regions (Commission of the European Communities 1987). Thus also all INTERREG strands (A, B, C) offers outstanding chances for European co-operation in tourism; the strands offer possibilities with their programs of improving on different levels cross-bordered co-operation in tourism lastingly. In the following, the essential features of the three strands are picked out and short examples of co-operation projects in tourism will be given.

INTERREG IIIA: Cross-border co-operation in tourism

The cross-border co-operation of neighbouring areas is intensified with strand A. This strand is the most important part of the INTERREG III initiative because of its essential integrating role for the European Union and the future Member States. It implements joint strategies for the development of cross-border regions or economic and social peripheral locations. The areas concerned are areas lying along the Union are certain coastal areas and internal and external land borders (Map 2.1). Measures can also be funded in certain non-border areas adjacent to those already mentioned. Only one programme is established per border between two States.

Projects must correspond to the target of the INTERREG IIIA program. It is the essential aim to drive the common development of the program areas forward in economic, socio-cultural, and ecological view. To this it requires a cross-border co-operation intensified and systematized further, which pursues the mining of still existing bordering conditional hindrances and disadvantages and an efficient use of the cross-border potentials.

Beyond this the projects must be in the position to be put in order into one of the main emphasises. These are in detail: Municipal and rural development, economy support, tourism and culture, labour market and education, nature and landscape, network formation and communication. Not only the main emphasis 'tourism and culture' is suitable for the support of tourism projects. In addition, the others main emphasises make the support of tourist projects possible.

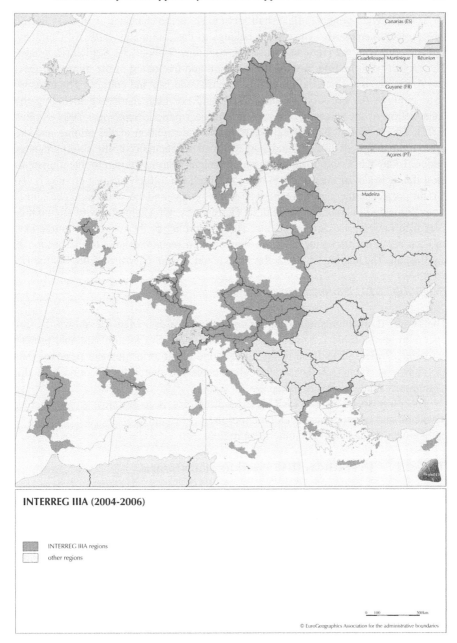

INTERREG IIIA (2004-2006)

INTERREG IIIA regions
other regions

0   100   500km

© EuroGeographics Association for the administrative boundaries

Map 2.1   INTERREG IIIA: Co-operation regions

Source: European Communities 2005c
© EuroGraphics Association for the administrative boundaries
© Europäische Gemeinschaften, 1995-2005

The following project outline shall serve to show the character of an INTERREG IIIA project exemplarily. The project belongs to the community initiative Germany, Luxembourg with the German-speaking community Wallonische Region (Belgium). The project with the title 'Tourist up valuation and marketing of the mining in Eifel and Ardennes' is assigned to the main emphasis 'tourism and culture'. The measure is to improve the cross-border networking and common marketing including the construction of the material infrastructure (project period: September 2004–August 2007). The economy and the culture have the old tradition of the mining and the opening up of goods like ore and slate stamped substantially in the Eifel-Ardennes region. Today the 'Geotourism' represents a weight-bearing column for the tourism in the Eifel. It is aim of the project to bring this tradition to the visitors more nearly. To this the project partners improve the slate tunnel 'Recht' and the 'Mühlenberger' tunnel from Rhineland-Palatinate and from the German-speaking community of Belgium in Bleialf to new tourist cardinal points. The networking of the two tourist attractions and the common marketing invite to the common reconnaissance of the history of civilization. The leading partner is the slate tunnel 'Recht' (Recht/St. Vith, Belgium).

INTERREG IIIB: Transnational co-operation in tourism

The strand B distinguishes at all 13 co-operation areas. Map 2.2 shows the co-operation areas, Table 2.2 picks up the five key priorities for transnational project activities. Project ideas have to meet the scope of the priorities and peasures that are shown in Table 2.2. These strategic themes are shared by all regions of the Community and are closely linked.

The purpose of all INTERREG IIIB-principles is to promote a sustainable, balanced, and harmonious development of the European Union. Special attention is being given to the island and outermost regions.

Table 2.2   INTERREG IIIB: Priorities and measures

	Measure 1	Measure 2
Priority 1	More attractive metropolitan areas in the global and European	Coherent and poly-centric pattern of complementary cities, towns, rural areas, coastal and peripheral regions
Priority 2	Sustainable mobility management	Improved access to the Information Society
Priority 3	Land use and water systems	The prevention of flood damage
Priority 4	Stronger ecological infrastructure, reduced ecological footprint	Protection and creative enhancement of the cultural heritage
Priority 5	Promoting cooperation between sea and inland ports	Facilitating cooperation across and between maritime and inland regions

Source: European Communities 2005e

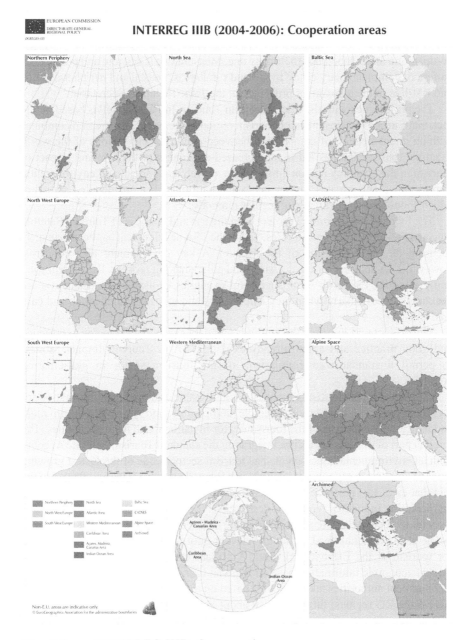

Map 2.2   INTERREG IIIB: Co-operation areas

Source: European Communities 2005d
© EuroGraphics Association for the administrative boundaries
© Europäische Gemeinschaften, 1995–2005

Some of the listed Priorities and Measures support projects in the area of transnational co-operation in tourism (Table 2.2), i.e. Priority 1, measure one and two. An example for a planned tourism project is placed in the North West Europe-co-operation area[4]. It belongs to Priority one, Measure two (Coherent and polycentric pattern of complementary cities, towns and rural areas, coastal and peripheral areas) and it still is in the assessment procedure at the Joint Technical Secretariat.

The project idea with the title 'European Tourism Impact Model' is based on the fact, that Europe remains the main tourism region in the world. This region combines with an intensification of competition between countries and regions of the world to attract tourists and has led to an increased awareness of the role and impact of tourism in the economy and on employment as well as its social and environmental implications. This creates further needs for harmonised, more detailed statistics, which should also be available at regular intervals.

The focus of the submitted project is on the use and analysis of Tourism Satellite Account (TSA) in different countries. So far, the TSA development has been primarily focusing on the production of the account. A key development area concerns, however, TSA analysis for the decision- and policymaking process. Papers reporting and evaluating TSA with some successful examples of usability and analysis are encouraged, particularly when they employ realistic industrial application and case analysis.

The project will improve the methods and data available, exchange experience and good practice and adapt already existing TSA or Tourism Impact Models to permit a standardized comparison of the tourism sector in participating countries.

This will provide tourism enterprises with useable and practical information for their decision making to provide credible, comparable, and systematic information, and indicators on the role of the tourism sector in national and local economies. The integral part of this project will be integrating the standard Tourism Satellite Account with an environmental assessment. The tourism sector is one of the fastest growing areas of the service economy in many North West EU regions. This is reflected in the fact that extensive resources are targeted on marketing the offer of EU regions, and on marketing individual regions as destinations for domestic tourists. However, it is difficult to accurately assess either the scale or the rate of growth of the sector as a whole at EU regional level, although relevant pilot work has started on accounting for tourism at the EU national level. A critical component of the problem is a limited economic information base on regional tourism. Consequently, EU regional development authorities take decisions based on a very limited suite of information. Across the EU regions, little is known about how tourism actually generates value added; the extent to which it supports foreign earnings; or how far growth in visitor

---

[4]   Project team in the case of the granting: Partner 1 (Lead Partner): Wales Tourist Board, UK; Partner 2: North West Development Agency, UK; Partner 3: Wirtschaftliche Strukturforschung (GWS mbh), Germany; Partner 4: Europäisches Tourismus Institut GmbH an der Universität Trier, Germany.

spending supports investment or the creation of employment, both directly and indirectly. Moreover, little is known about the environmental consequences of tourism activity.

Expected concrete results are:

- Establishing a network of organisations and research practitioners across INTERREG partner regions through which to develop sub-national tourism satellite accounts;
- Pilot sub-national tourism satellite accounts for regions within the network;
- Development of a best practice guide to account development and uses;
- Integration of developed network with more general activity to develop national tourism satellite accounts across the EU.

INTERREG IIIC: inter-regional co-operation in tourism

Strand C supports co-operation between players all over Europe and not necessarily just those in neighbouring regions. INTERREG IIIC helps to make regional development policies and tools more efficient by enabling a vast exchange of information, the sharing of experiences and the creation of structures of co-operation between regions. The Programme management differences four programme areas (Map 2.3).

One of the regular INTERREG IIIC South projects is COESIMA (Coopération Européenne de Sites Majeurs d'Accueil). It is a network which surrounds the following seven places of pilgrimage: Lourdes (F), Loreto (I), Czestochowa (PL), Patmos (GR), Fatima (P), Altötting (D) und Santiago de Compostela (E). All seven places of pilgrimage have a tradition lasting for centuries as important pilgrim places and count about 20 million visitors per year. However in the nearer past the numbers of visitors have decreased. Reasons for this are the changes of the expectations of the visitors, difficulties at the steering of the visitors as well as at the customization of the services and reception structures. To counteract the sinking numbers of visitors, the needs of the guests shall be investigated

The network COESIMA has set itself the aim therefore to conduct these activities better in favour of the areas concerned. This shall get possible by a narrow co-operation of the different pilgrim places and a better, professional training of the partners of the network.

The work program of the network has four priorities: Identifying the visitor expectations, adapting the equipment to the needs of the visitors, revaluation/upgrade of the cultural possession and to set up common and international marketing of the seven places of pilgrimage.

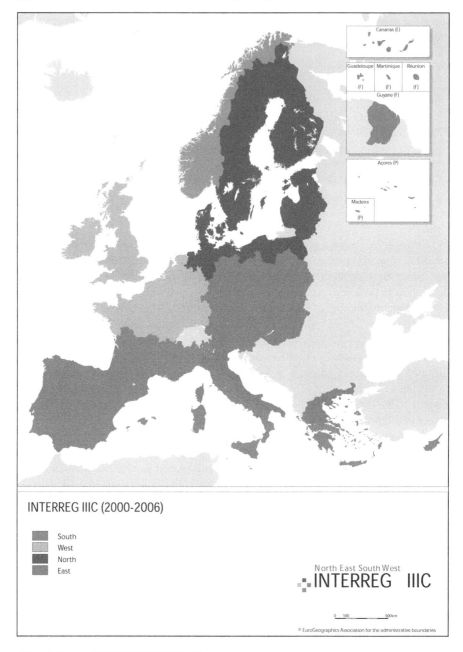

Map 2.3   INTERREG IIIC: Programme areas
Source: European Communities 2005f
© *EuroGraphics Association for the administrative boundaries*
© *Europäische Gemeinschaften, 1995–2005*

Conclusions

This chapter discusses the European Community Initiative INTERREG and analyses the suitability and efficiency of tourism co-operation projects. To this the essential characteristics of INTERREG were picked up and the strands A, B, and C were covered with typical tourism project examples. Tourism projects can excellently be integrated to the European Community Initiative. At this strand, B and strand C are very suitably for strategic tourism co-operation projects on the one hand side. On the other hand side strand A and strand C are very suitable for operative projects whose results are visible and measurable for tourists.

The promoting period for INTERREG III has started on November 2000 and will end on December 2008. Projects can be applied for up to December 2007, the projects must be completed at the latest on December 2008. The present program period of the structure funds and -instruments runs out at the end of December 2006. The commission, the member states and the regions are engaged in the preparations for the next program period 2007–2013, called 'INTERREG IV'. The structure funds and her tools shall improve the conditions for economic growth in the union and have furthermore three aims: convergence, competitiveness, and co-operation[5].

Co-operation in tourism without frontiers is a difficult and rarely spontaneous process. To ensure that border regions are competitive and to help them benefit from the opportunities of cross-border co-operation, the Union offers with the INTERREG initiative a whole and most essential series of tools for tourism co-operation projects.

References

Commission of the European Communities (ed.) (1987), *Conference on Tourism Horizon 1992*, (Brussels).

European Communities (ed.) (2001a), *A guide to bringing INTERREG and Tacis funding together*, (Luxembourg).

European Communities (ed.) (2001b), 'INTERREG III 2000–2006', <http://europa.eu.int/comm/regional_policy/sources/slides/interregintro.ppt>, accessed on n.d.

European Communities (ed.) (2001c), 'INTERREG IIIC 2000–2006', <http://europa.eu.int/comm/regional_policy/sources/slides/interregiiic.ppt>, accessed on n.d.

European Communities (ed.) (2002), *Structural policies and European territory – Co-operation without frontiers*, (Luxembourg).

European Communities (ed.) (2003), 'The INTERREG III Community Initiative – How to prepare programmes. A practical guide for preparing new, and amending existing, INTERREG III Community Initiative Programmes as a result of Enlargement', <http://europa.eu.int/comm/regional_policy/interreg3/documents/practical_guide_for_interreg_14_3_03_en.pdf>, accessed on n.d.

---

[5]   Current information about this is available in the Internet: http://www.europa.eu.int/comm/regional_policy/funds/2007/index_en.htm.

European Communities (ed.) (2004), 'COMMUNICATION FROM THE COMMISSION TO THE MEMBER STATES of 2 September 2004 laying down guidelines for a Community initiative concerning trans-European co-operation intended to encourage harmonious and balanced development of the European territory INTERREG IIII', *Official Journal of the European Union* C 226/2, 10 September 2004.

European Communities (ed.) (2005a), 'INTERREG III frequently asked questions', <europa. eu.int/comm/regional_policy/interreg3/foire/faq_en.htm>, accessed on 10 September 2004.

European Communities (ed.) (2005b), 'INTERREG III' <http://europa.eu.int/comm/regional_ policy/interreg3/abc/abc_en.htm>, accessed on 10 September 2004.

European Communities (ed.) (2005c), 'INTERREG IIIA co-operation areas' <http://europa. eu.int/comm/regional_policy/interreg3/down/pdf/europe.pdf>, accessed on 10 September 2004.

European Communities (ed.) (2005d), 'INTERREG IIIB co-operation areas' <http://europa. eu.int/comm/regional_policy/interreg3/images/pdf/int3b_uk_a4p.pdf>, accessed on 10 September 2004.

European Communities (ed.) (2005e), 'Towards stronger transnational co-operation in North-West Europe for a balanced and sustainable territorial development', <http://www. nweurope.org>, accessed on 10 September 2004.

European Communities (ed.) (2005f), 'INTERREG IIIC areas', <http://www.interreg3c.net/ sixcms/media.php/5/int3c25_a4p_print_040122_1400.pdf>, accessed on 10 September 2004.

Faby, H. (2004), 'Untersuchung von kartographischen Medien und Nutzerbedürfnissen als Basis für zielgruppenorientierte touristische Internet-Anwendungen', *Kartographische Bausteine* BD KB 27, (Dresden).

Chapter 3

Interregional Co-operation between Europe's Island Regions

A Case Study of the GEDERI Project

Dimitri Ioannides

Missouri State University, USA

Cross-border and multinational alliances have gone from near non-existence 70 years ago to become the norm in the twenty-first century, with important implication for tourism
(Timothy and Teye 2004: 584)

Introduction

An obvious outcome of growing integration within the European Union (EU) in recent years has been the implementation of numerous cross-border cooperative and collaborative initiatives involving two or more countries. Commonly, these measures relate to community-wide regional policy, which aims to reduce the gap between the most and least economically developed regions (Feldin 2003). Through the European Regional Development Fund (ERDF), which was established thirty years ago, cross-border initiatives have addressed issues such as deindustrialization, urban regeneration, transportation, rural development, quality of life, labor force training, environmental resource management, and cultural heritage and tourism (Church and Reid 1996: 1298).

A number of cross-border measures targeting balanced development appear under the guise of INTERREG, a community-wide program funded by the ERDF, which specifically aims to encourage co-operation within the EU. INTERREG, the largest of four community initiatives in terms of funding, divides into three strands namely INTERREG IIIA – cross-border co-operation, INTERREG IIIB – transnational co-operation, and INTERREG IIIC – inter-regional co-operation. INTERREG IIIC 'aims to improve the effectiveness of regional development policies through large scale information exchange and co-operation networks. It is open to co-operation between regions from the whole of the EU independently of their location and participation of third countries is also possible ...' (Feldin 2003: 68).

One project recently funded by this program is the so-called GEDERI (Gestion et Developpment des Regions Insulaires), which focuses on the development and management of Europe's island regions. Specifically, GEDERI recognizes that European islands share problems and concerns including economic marginality, a limited resource base, poor accessibility, chronic depopulation and a brain drain, plus environmental and/or socio-cultural threats brought on, among others, by excessive tourist development.

This chapter explains why the EU has decided a cross-border initiative such as GEDERI, which focuses specifically on insular regions, is a worthwhile endeavor for tackling their shared problems and concerns. It proceeds to examine the main tourism-related tasks which have been planned or are currently under-way as part of this ongoing project. Suggestions are provided on how the project aims to contribute to the sustainable development of Europe's islands. The chapter also highlights some of the key impediments standing in the way of realizing the project's objectives.

Europe's islands regions

Throughout the EU there are hundreds of inhabited islands, which before the recent addition of ten new member states in 2004, made up 3.4 per cent of the community's area and were home to 13.5 million (EURISLES 2002). Thirteen million islanders lived in 21 island regional authorities while the rest inhabited the numerous coastal islands scattered throughout the EU. Only in Italy and Greece do islands account for a sizable proportion of the entire national population (more than 10 per cent). Sardinia, Sicily, and the Canaries all have populations over one million, but the majority of European islands have fewer than 250,000 inhabitants and, indeed, many of them fewer than 50,000. On most islands the population density is low, sometimes exceedingly so (under 100 inhabitants per square kilometer), and to make matters worse, many suffer from chronic depopulation. This, in turn means that Europe's islands usually do not have large enough internal markets to support various economic activities (such as manufacturing) and, because of isolation from the mainland, this problem is accentuated.

Enterprises on islands are normally 'subject to far greater costs than their mainland competitors. This engenders several consequences for island enterprises: reduced profit margins, limited competitiveness, more complex and costly inventory management, inadequacy of production plants' (*Manifesto of EU Island Chambers of Commerce* 2005), not to mention diseconomies of scale (Ioannides et al. 2001). The islands often lack natural resources and most do not have access to cheap energy supplies. If the islands are located too far from the mainland, they may have to rely on oil imports to run their power stations, which given the excessive rise in price of this commodity in recent times makes these environments uncompetitive in terms of their ability to diversify their economy. The fact that many islands do not have adequate water resources further stifles their ability to attract economic activities and new residents.

Unsurprisingly then, like many rural areas, islands often have a narrow economic base often relying on primary activities such as fisheries and farming. As discussed later, over the last few decades the tertiary sector, especially tourism, has also emerged as a major economic activity for many insular regions. While the average share of primary activities in the EU before 2004 was only 4.3 per cent, on islands like Orkney this sector accounted for 27 per cent and on Crete for 33 per cent. This explains why so many insular regions are significantly poorer in terms of GDP per capita when compared to the EU 15 average. For instance, Vorio Aigaio (North Aegean) had a per capita GDP that was less than two thirds the EU average (EURISLES 2002).

Often, unemployment is also a major problem because of lack of economic opportunities and the seasonal nature of many of the available jobs. In 1999 Sicily's unemployment amounted to 24.8 per cent and even on a fairly affluent island like Bornholm it was close to 10 per cent. In turn, this lack of economic opportunities explains why many islands are rapidly losing population as their working-age inhabitants search for opportunities elsewhere. Unsurprisingly, this means that many insular regions experience a 'brain drain', as their higher educated inhabitants find their opportunities for employment and economic advancement severely limited. In turn, this problem limits the capacity of insular areas to generate innovative activities (Ioannides and Petersen 2003).

A particularly thorny handicap for islanders is their physical isolation from the mainland. In many cases this isolation is exacerbated because islands are located further from their own state's mainland than that of another country. The Greek island of Lesvos, for instance, is located much closer to the Turkish mainland than Athens. The EURISLES (2002) report *Off the Coast of Europe* indicates that isolation and low level of accessibility do not arise solely because islands are surrounded by water. Rather, 'transport is first and foremost a function of choice. By definition, islanders cannot use road or rail to communicate with the outside, which means that they do not enjoy the benefit of competition between these various transport modes. They are not, therefore, in a position to benefit from the freedom of movement of persons and goods in conditions comparable to those of the inhabitants of the European mainland' (p. 25).

Because islanders have little choice in terms of transportation they often suffer the consequences of saturation during peak visitation seasons. In some cases, if they have not reserved a seat on a ferry, they can find themselves trapped on their islands. The EURISLES report mentions that transport problems are exacerbated because of time and price. Consider that if people depend on a ferry service to travel from their island to the mainland they must endure not only a fairly slow and arduous journey but have to put up with delays in departure not to mention time required for check-in, loading and unloading. Because in many cases there is lack of competition between air or sea carriers the price for traveling to and from an island is disproportionately high compared to traveling between two mainland cities. For instance, air transport from Copenhagen to Bornholm remains an expensive option as only a single airline, CIMBER Air, provides a service.

Although the negative effects arising from the islands' peripheral location attract considerable attention in the academic literature, it is vital to point out that many of these insular environments are endowed with a number of positive attributes. Because they are far from metropolitan areas their inhabitants are more likely to maintain their traditional norms than their mainland counterparts (Kousis 1989). They are also plenty of islands, which because of their isolation have witnessed limited urbanization and industrialization. Consequently, because of the slow pace of development many of them have managed to preserve their natural resources such as their fauna and flora as well as their architectural heritage. It is these attributes, among others, that put these islands on the map, capturing the attention of many potential visitors. The following section focuses attention on the characteristics of tourism development on the islands of the EU.

Tourism in island environments

As described, European islands often have a narrow economic base and depend heavily on primary sectors. However, one economic activity that has become increasingly common in insular environments throughout the EU over the last 4-5 decades has been the tertiary sector, especially tourism. This industry's growth, which especially in some Mediterranean destinations has been phenomenal, hardly comes as a surprise considering, as mentioned earlier, that many islands are well placed to attract visitors because of their ample natural ingredients including clean sand and sea, and plenty of sunshine (Apostolopoulos and Gayle 2002; Ioannides et al. 2001). Most of the growth on these destinations has occurred because of the advent of mass tourism. However, in recent years a number of these destinations have begun diversifying their products by promoting other attributes (such as their built heritage or their fauna and flora), which target individual travelers, many of them during the off-peak seasons. Additionally, cold water destinations like the islands of the North Sea or the Baltic are able to draw special interest visitors because they too either possess interesting natural habitats and/or historical resources.

Because many observers regard tourism as one of only a few realistic options for promoting economic diversification in insular environments it is unremarkable that it has emerged as a key economic sector, responsible for a sizeable portion of the gross regional product and the labor force. In 1999, the tertiary sector, including tourism, accounted for 72.5 of all workers in the Balearics (EURISLES 2002). Many Mediterranean islands have seen a number of economic benefits from tourism, which in some cases like in the southern Aegean, has reversed chronic depopulation and led to considerable economic growth.

The Balearics, and especially Mallorca, are one island destination that has come to depend heavily on tourism over the last four decades or so. According to Meaurio and Murray (2001) during the 1990s, unemployment declined and wages increased significantly. Because of the improved quality of the tourism product in the Balearics, following a series of innovative policy-making during the 1990s, the number of arrivals has continued to grow. Generally, visitors to these islands are satisfied with the product as reflected by the high number of repeat visits. Additionally, the

Balearics display a high degree of local ownership in the tourist accommodation sector and significantly, some of the major Spanish hotel chains are headquartered on these islands (Bardolet 2001).

Yet another positive aspect of tourism development in many insular areas of the EU is that it is credited with boosting entrepreneurial activity. On most destinations the accommodation establishments are small and medium tourism enterprises (SMTEs), most of which are family owned and operated (Ioannides and Petersen 2003). According to Buhalis and Diamantis (2001: p. 146) 'nearly half the persons employed in the tertiary sector [on the Greek islands] are employed in tourism, especially in the vast array of SMTEs'.

While there are obviously a number of positive aspects associated with insular tourism development, the industry also regularly causes adverse problems for the islands' economies, cultures, and environments. These problems, which have been discussed at length by many observers (Apostolopoulos and Gayle 2002; Britton 1978; Ioannides et al 2001; Kousis 1989) include: the accentuation of spatial imbalances between tourist resorts normally located in coastal areas and the islands' hinterlands; an excessive rise in the cost of living, including the price of housing; labor shortages; excessive seasonality; economic leakages; a loss of traditional ways of life; increased crime; and a host of environmental concerns such as deforestation, soil erosion, threats to marine and already meager water resources, loss of architectural heritage, and ribbon development, which leads to aesthetic pollution. Then there is the problem of waste management, which EURISLES (2002) maintains is far more serious than in mainland areas. It is extremely costly to locate facilities to collect and treat waste on the islands and in many instances their geology and soils does not make it realistic to handle waste internally. The option of collecting waste from the islands is a possibility but poses problems logistically given the scattered 'archipelago-type nature of certain regions' and the excessive cost of transportation.

The need for a coordinated strategy for Europe's islands

The report *What Status for Europe's Islands* (EURISLES 2000) indicates that 'from being the appendix of a Nation State', since the inception and expansion of the EU, 'islands have become the fringes of a continent, and they now have to fight for their case in a totally different political and administrative environment'. Most of the EU's institutions and legal frameworks demonstrate marked hesitance toward acknowledging the islands' difference to mainland regions. Especially those islands, which lack autonomy, often find themselves bypassed by the EU's mainstream programs and policies. Illustrating this problem, the Vice President of INSULEUR (the Network of the Insular Chambers of Commerce and Industry of the EU) Mr. Juan Gual de Torella recently pointed out that 'island markets have different characteristics from the continental ones and competition rules cannot be strictly the same. [Thus] European policies and regulations, and especially the rules about state aid, should take into consideration this context and help to counterbalance the islands' disadvantage' (INSULEUR 2005).

Given this situation, not to mention the islands' handicaps described earlier, the growing push by many observers to assert the unique identity of insular regions has been unsurprising. In turn, this led to Article 158 of the Treaty of Amsterdam, which states that the handicaps of EU island regions must be addressed through specially tailored policies and legal frameworks. The *Manifesto of EU Island Chambers of Commerce* seeks to implement Article 158 by, among others, fostering an environment allowing for the development of small and medium establishments (SMEs) on islands and promoting entrepreneurial activities, one of the necessary ingredients for economic diversification. Further the manifesto proposes that the EU must provide the steps for the islands to overcome the traits that 'engender negative differentials' and argues that 'member states should be empowered to grant fiscal and social incentives, tailored to meet each individual island context'.

Bodies responsible for islands in Europe

Within the EU various bodies deal, either directly or indirectly, with island affairs. One of the most important is the Islands' Commission, which was established in 1981 and is one of seven geographical commissions of the Conference of Peripheral Maritime Regions of Europe (CPMR) (CPMR 2005). The CPMR is the umbrella organization responsible for balancing development within the EU through a polycentric approach involving as many sub-state players as possible. The CPMR's membership amounts to 149 regions from a total of 27 states (including non-EU countries), all of which are 'located on one of Europe's main sea basins' (CPMR 2005: 1).

In all, 25 regional island authorities, including every EU island authority plus the Isle of Man, belong to the Islands' Commission. Many of the members of the Island Commission are also associated with various regional networks linking neighboring islands. Examples are the B-7, an association of the seven Baltic islands, and MEDOC, which includes Sardinia, Sicily, the Balearics, and Corsica. There are also associations like the EU's Network of Insular Chambers of Commerce and Industry and those representing small islands like the ones off the coast of Finland. The Islands' Commission maintains close links with all of these.

Among the objectives of the Islands' Commission is to ensure the various countries and the institutions of the EU acknowledge the islands' problems and, thus, implement appropriate policies to solve these. Secondly, the commission aims to 'foster interregional co-operation between islands, specially on issues in direct relation with their insularity' (CPMR 2005). Thus far, the Islands' Commission has been credited with ensuring that the Amsterdam Treaty of 1997 recognized the islands through Article 158 (see discussion in previous section). It also has successfully created networks addressing technical co-operation such as ISLENET and EURISLES. The former, which is managed by the Western Isles Council of Scotland, addresses energy-related matters.

Corsica-based EURISLES (European Islands System of Links and Exchanges) deals with statistical and documentary databases and undertakes various research projects. Established in 1992, this organization was originally funded by the European Commission, while nowadays it is entirely subsidized by its member regions. In all, EURISLES includes 12 island regions, namely the Balearics, Canaries, Madeira, Corsica, Sardinia, Sicily, Vorio Aigaio, Notio Aigaio, plus the non-European islands of Guadeloupe, Martinique, the Azores, and Réunion. It is hoped all of the EU's island regions will eventually become members of this network.

The EURISLES website reports that one of the major problems when trying to document the unique handicaps of island regions (such as their isolation, remoteness, and limited resources) is that data are often incomplete and incompatible for comparative purposes. 'The first task of EURISLES, therefore, was to create a network linking selected local information centres such as regional statistics offices, local university centres, or regional development agencies' (EURISLES 2005). The network is coordinated by staff based in Ajaccio, Corsica.

One of the most ambitious measures put forth in recent years to encourage a cooperative effort between various Europe's island regions is the GEDERI project. This is managed by EURISLES and encompasses a number of themes relating to insular regions. The next section provides an overview of the objectives of the GEDERI Project, especially as they relate to tourism.

The GEDERI project

The GEDERI project is partly funded through the EU's INTERREG IIIC program. It focuses on the development and management of Europe's island regions. Specifically, this project provides the impetus for various islands to exchange experiences and information concerning a variety of topics. In all, eight themes are included in this project:

1. the significance of accessibility for islands (both to and from but also within the islands);
2. the issue of promoting sustainable tourism and identifying carrying capacities;
3. finding strategies for curbing depopulation;
4. identifying measures to balance the needs of the workforce with training opportunities;
5. recognizing ways to use the image and identity of each island to market local products;
6. identifying measures to enable islands to respond to crises posed by natural disasters;
7. developing appropriate R and D and higher education policies to inspire the islands' economic development;
8. and establishing an integrated strategy for developing and managing the islands.

In all, GEDERI includes 12 project participants. In addition to the CPMR – Islands' Commission, the others are Corsica, Sicily, the Balearics, Crete, Bornholm, Gozo, Gotland, the Western Isles, the Ionian Islands, Åland, and Sardinia. In terms of their development level, degree of tourist activity, population, geographic area, and location these islands vary significantly, but the advantage is that researchers and policymakers can draw from their diverse experiences as they seek to identify viable responses to the constraints and potentials of insularity. Over a three-year period (2004–2007), meetings have been scheduled on various islands to address each of the eight themes. Ultimately, it is hoped the findings arising from these thematic meetings will lead to effective actions for addressing the handicaps of insularity.

As indicated above, one of the project's underlying themes has to do with tourism, a task that has become the responsibility of the Center for Regional and Tourism Research (CRT), located on the Danish island of Bornholm. Specifically the CRT aims to develop and test novel conceptual and methodological approaches in a comparative context as these relate to the development of tourism within a sustainable context. The expected output from the CRT's endeavors is ultimately to produce at least one comprehensive report plus policy recommendations to the Islands' Commission concerning tourism's future development trajectory within the context of sustainability.

Progress thus far

For approximately 18 months, researchers based at the CRT have addressed a number of research questions, the most important of which are:

1. What constitutes sustainable tourism development within the EU's insular regions?
2. What steps can be undertaken to ensure that tourism is developed and managed sustainably?
3. How can the carrying capacity of each island be identified?
4. Are tools like Limits to Acceptable Change (LAC) and Visitor Impact Management (VIM) appropriate for managing tourism development in the future? and
5. What measures can be undertaken to ensure that carrying capacities are not exceeded?

Though answers to these questions are beyond the scope of this chapter, it is important to mention that work on this topic has begun, already resulting in two preliminary reports, on sustainable development and on carrying capacity respectively (Ioannides and Billing 2005a, 2005b). These background papers highlight a number of issues, including the reasons why the achievement of existing measures to promote sustainable development have met with limited success. They indicate the necessity within the study's context to go beyond definitional matters, which seek to identify the meaning of sustainability and capacity constraints and, rather, focus on developing explanations as to how sustainability practices can be implemented in practical terms and how tourism's growth can be managed in a balanced manner. Further, they highlight the need to develop a standardized methodology for estimating the carrying capacities of each island destination.

The background papers bring to the surface a number of recommendations including the need to develop a comprehensive database of standardized indicators for all island regions relating to both fixed and variable carrying capacity parameters (Coccossis and Mexa 2004). A reliable database is imperative for developing a typology of islands, which will reflect, based on a continuum, the stage that each island has reached in terms of its tourist life cycle (Parpairis 2004). For instance, at one end of the spectrum, islands like the Western Isles that have experienced very low levels of tourist development can be placed in one category while those that have gone through decline and are now experiencing rejuvenation, like Mallorca, can be placed in another. Generating classes of islands will undoubtedly prove helpful for estimating carrying capacities that are context-specific and for developing broad policies for implementing sustainable development options.

Until now, the CRT has recommended that each of the islands participating in the project should collect as much information as possible including plans, policies, and statutory instruments that, either directly or indirectly, address tourism's development within the context of sustainability. These include reports and policies specifically relating to tourism, such as tourism marketing plans or policies emphasizing quality-oriented tourism, land use plans, documents relating to the protection of natural areas or the built environment, transportation plans, and environmental statutes. Further, since it is recognized that not all islands have detailed policy documents in place, there is a need to conduct interviews with various stakeholders who have a direct or indirect bearing on the sustainable development of islands. These stakeholders will represent the public and private sectors as well as not-for-profit groups and communities. It is vital to collect information from representatives of all these groups since they undoubtedly have conflicting agendas regarding their interpretation of sustainable tourism. Knowing the attitudes of the different groups is necessary to devise and implement policy recommendations, which seek to reconcile these conflicting agendas.

Necessary data and obstacles to completing the study

One way to identify the level of tourism development on each of the GEDERI islands and, consequently to be able to develop a typology of insular areas, is by examining various economic, social, and environmental characteristics for each destination. Such data can be combined to develop a composite measure, which McElroy and de Albuquerque (1998) term the tourism penetration index (TPI). The advantage of the TPI is that it combines into a single variable, measures of economic, sociocultural, and environmental penetration. For example, one can combine data like tourist expenditure per capita (an economic measure), density of tourists per 1000 population (a social measure), and the number of bed spaces per square kilometer (an environmental measure) to create the TPI (Figure 3.1).

To derive such a composite TPI for all three variables, McElroy and de Albuquerque suggest calculating the average of the three TPIs as they relate to each respective measure. For instance, if the TPI for visitor spending per visitor (economic measure) is .745 and the average daily visitors per 1000 population (social measure) is .683, and the hotel rooms per square kilometer (environmental) is 1.000 then the composite TPI for that particular destination would amount to the average of all three measures (i.e., .809).

The equation for calculating the TPI for each of the three variables (economic, social, and environmental) based on a database relating to a number of destinations involves the following:

$$TPI_{ij} = (X_{ij} - minX_i)/(maxX_i - minX_i)$$

Where:

TPI_{ij} is the degree of tourism penetration for the jth island with respect to the ith variable, X_{ij} is the value of the ith variable for island j and $maxX_i$ and $minX_i$ stand for the maximum and minimum values of the ith variable of all islands in the sample. The value of TPI for each variable (economic, social, and environmental) can vary from 1 (maximum penetration) to 0 (minimum penetration).

Figure 3.1   Calculating the TPI

Source: based on McElroy and de Albuquerque (1998)

For each of the islands in the GEDERI study the TPI will range from a minimum of 0 (least developed in terms of tourism) to a maximum of 1.0 (most developed). Thus, islands like Corfu (Kerkyra) in the Ionian will likely have one of the highest TPIs while the Western Isles will have one of the lowest. In turn, each of the islands can be ranked from least to most penetrated, in terms of tourism development.

A simple typology based on TPIs is a first step for identifying targeted policy instruments addressing sustainable solutions that more closely apply to the level of tourist development on each island (or island grouping). The typology would help policymakers understand with a higher degree of clarity whether or not there is a point (temporally and spatially speaking) at which the islands experience impacts on their natural and sociocultural environments that are so catastrophic as to reduce their attractiveness to potential visitors. Through the recognition that there are indeed varying types of insular destinations in the EU, one can better pinpoint strategies relating to tourism planning, which can be used for ensuring a future development path that more closely follows the tenets of sustainability.

While the development of TPIs appears a straightforward exercise, the problem is that even the very basic data required for developing simple indices are not readily available for every single island. This is despite EURISLES' concrete efforts to create a database of common variables for all of the EUs islands. Unfortunately, not all islands collect data on a consistent basis and even when they do, the measures are not always useful for comparative purposes since methodologies for reporting the variables may differ. It is particularly hard to obtain measures relating to spending per visitor since most European islands do not report these on a regular basis, if at all; this means that it is particularly hard to estimate penetration indices relating to economic data. To further complicate matters, in certain cases the variables collected for islands relate to a whole insular grouping (and not a particular island) or, occasionally, when the island is part of a region that includes mainland areas, it is impossible to obtain information about the insular area alone.

Undoubtedly, over time the efforts of EURISLES to establish a comprehensive database relating to island tourism will pay off. However, in the short run, the incomplete amount of information not to mention the unwillingness in the case of some island authorities to respond to surveys relating to the project have stalled the efforts of GEDERI to come up with comprehensive policy recommendations for the future development of tourism on Europe's islands.

Conclusions

Within the EU, cross-border collaborative and cooperative efforts aimed at reducing development gaps between central and marginal areas have undoubtedly begun benefiting the community's insular regions. However, there is concern that these efforts fall well short of dealing with many of the islands' problems and potentials that are linked to their ultra-peripheral situation within the community. Especially in recent years, various observers have recognized that islands share traits including a number of handicaps setting them aside as a significant subgroup worthy of context-specific responses, including policies and legal frameworks. This chapter has focused on the effort of the CPMR through its affiliate Islands' Commission to implement the GEDERI project, which specifically concentrates on the management and development of insular regions.

The GEDERI project focuses on various topics, one of which is the promotion of tourism within the context of sustainable development. As part of this theme, researchers have been assigned the task of identifying the carrying capacity of each island participating in the project. It is worth mentioning at this point that members of the CRT who are charged with this exercise, realize all too well that it is impossible to pinpoint how many visitors are too many for these destinations and, thus, the concept of carrying capacity is used more as a guiding principle than a precise scientific process, fixated on identifying upper limits of visitation (McCool and Lime 2001). Ultimately, the aim is to provide, based on best-practice, policy recommendations for balancing the economic objectives deriving from tourism with environmental and socio-cultural concerns.

It has been argued that, unfortunately, one of the stumbling blocks to realizing the aims of the GEDERI project has been and continues to be the inadequacy of information from the various islands participating in the project. This is despite the ongoing efforts of EURISLES to generate a comprehensive database for all of Europe's islands. To worsen matters, the CRT was recently informed that the whole project has been frozen for the time being because some of the member islands have not paid their dues and because not everyone involved shares the same degree of enthusiasm about the potential outcomes. It is not clear at this moment if the project will be revived under a different guise any time soon.

Regardless of whether the project is revived or not it reflects that for the first time the EU has begun taking note of the islands' unique circumstances, choosing to treat them collectively as a subgroup worthy of special treatment. The GEDERI Project, a true trans-boundary effort, reflects that Europe's islands have moved closer to the center stage of the EU's agenda. Hopefully, in the next few years the project or some similar endeavor will be revived and the original objectives will be realized through tangible policies and the creation of statutory and fiscal instruments, which will effectively target the islands' unique circumstances.

References

Interreg IIIC (2005), 'About Interreg IIIC', <http://www.interreg3c.net/sixcms/detail. php?id=310>, accessed 2 November 2005.

Apostolopoulos, Y. and Gayle, D.J. (eds) (2002), *Island tourism and sustainable development: Caribbean, Pacific, and Mediterranean Experiences*, (Westport, Connecticut: Praeger).

Britton, R.A. (1978), 'International Tourism and Indigenous Development Objectives: A Study with Special Reference to the West Indies', (Unpublished PhD thesis at the University of Minnesota).

Church, A. and Reid, P. (1996), 'Urban power, international networks and competition: The example of crossborder co-operation', *Urban Studies*, 33:8, 1297–1318.

Coccossis, H. and Mexa, A. (eds) (2004), *The Challenge of Tourism Carrying Capacity Assessment: Theory and Practice*, (Aldershot, UK: Ashgate).

Conference of Peripheral Maritime Regions (2005), (website), <http://www.crpm.org/>, accessed 10 January 2006.

Buhalis, D. and Diamantis, D. (2001), 'Tourism development and sustainability in the Greek archipelagos', in D. Ioannides and Y. Apostolopoulos and S. Sonmez (eds), *Mediterranean*

Islands and Sustainable Tourism Development: Practices, Management and Policies, (London: Continuum. 141–170).

Eurisles, European Islands System of Links and Exchanges (2005), (website), <http://www.eurisles.org/> , accessed 1 November 2005.

Eurisles (2002), *Off the Coast of Europe: European Construction and the Problem of the Islands*, (Corsica: Eurisles for CPMR).

Eurisles (2000), 'What Status for Europe's Islands', <http://www.eurisles.com/statut_iles/EN/cadre.htm> , accessed 11 January 2006 (Corsica: Eurisles for CPMR).

Feldin, L. (2003), 'Analysis of the transnational INTERREG Strand's Contribution to EU Spatial and Regional Development', (MSc Thesis in Human and Economic Geography, Gotteborg University).

Insuleur, 'Network of the Insular Chambers of Commerce and Industry of the European Union 2005 Insuleur News', <http://www.insuleur.net/eng/news.html>, accessed 10 January 2006.

Ioannides, D. and Apostolopoulos, Y. and Sonmez, S. (2001), *Mediterranean Islands and Sustainable Tourism Development: Practices, Management and Policies*, (London: Continuum).

Ioannides, D. and Billing, P. 'Seminar Paper on Carrying Capacity: For INTERREG IIIC 'GEDERI' 2005a', <http://www.gederi.org/doc/2/Bonholm/EN%20GEDERI%20Carrying%20Capacity%20-%20Seminar%20Paper.pdf>, accessed 10 January 2006, (Bornholm, Denmark: Center for Regional and Tourism Research).

Ioannides, D. and Billing, P. 'Theme Paper on Sustainable Tourism for INTERREG IIIC 'GEDERI' 2005b', <http://www.gederi.org/doc/2/Bonholm/EN%20GEDERI%20Sustainable%20Tourism%20-%20Theme%20Paper.pdf>, accessed 10 January 2006, (Bornholm, Denmark: Center for Regional and Tourism Research).

Ioannides, D. and Petersen T. (2003), 'Tourism and "non-entrepreneurship" in peripheral destinations: A case study of small and medium tourism enterprises on Bornholm', *Tourism Geographies*, 5:4, 408–435, (Denmark).

Kousis, M. (1989), 'Tourism and the family in a rural Cretan Community', *Annals of Tourism Research*, 16, 318–332.

McCool, S. and Lime, D. (2001), 'Tourism carrying capacity: Tempting fantasy or useful reality', *Journal of Sustainable Tourism*, 9:5, 372–388.

EU Island Chambers of Commerce (2005), 'Manifesto of EU Island Chambers of Commerce 2005', <http://www.insuleur.net/doc/manifesto-ENG.pdf>, accessed 9 January 2006.

Meaurio, A. and Murray, I. (2001), 'Indicators of Sustainable Development in Tourism: The Case of the Balearic Islands', <http://www.world-tourism.org/sustainable/IYE/Regional_Activites/Seychelles/Balears-Indicators.htm>, accessed 9 January 2006 (Paper presented at the Conference on Sustainable Development and Management of Ecotourism in Small Island Developing States (SIDS) and Other Small Islands. Mahé, Seychelles: December 8–10).

McElroy, J.L. and de Albuquerque, K. (1998), 'Tourism penetration index in small Caribbean islands', *Annals of Tourism Research*, 25:1, 145–168.

Parpairis, A. (2004), 'Tourism carrying capacity assessment in islands', in H. Coccossis and A. Mexa (eds) (2004), *The Challenge of Tourism Carrying Capacity Assessment: Theory and Practice*, (Aldershot, UK: Ashgate. 201–213).

Part II
Destination Marketing and Management

Chapter 4

Inbound Tourism: Systematic Approaches of International Tourism Behaviour

The Case of Germany

Christian Rast
ift Leisure and Tourism Consultancy, Germany

Stephan Kroll
California State University, USA

Introduction

Inbound tourism is tourism by non-resident visitors for the purpose of leisure, visiting friends and business. In this chapter, which is divided into seven sections, the specific problems associated with the marketing of inbound tourism will be explained and discussed, that approaches cross-border tourism from the point of view of the hosting destination.

Following this introduction, section 2 defines and characterizes inbound tourism (very often called also incoming tourism; especially in the German language area this term is still more used than inbound tourism). In Section 3, an overview of the current research on inbound tourism in Germany is presented, and it is explained why that research has been unsatisfactory. The low level of research compared to research on domestic tourism is disconcerting given that knowledge about differences of tourism behaviour of different target groups is essential for successful international tourism marketing.

Section 4 proposes an integral and systematic approach of analyzing inbound tourism and its international components to fill the research gap. Some of the components that influence inbound tourism in a given country are: needs of tourists in different international target markets, other competing destinations, the supply of the tourist industry in the tourist generating countries, transit routes and destination regions (such as tour operators, bus coaches, airlines, destination management organizations, hoteliers etc.), and the broader international environment. A focus purely on supply factors within the destination region without taking demand factors and other supply factors into account is too narrow of an approach.

The next two sections summarize existing data on the inbound tourism in Germany in general (section 5) and on the inbound tourism from two important generating countries into Germany in particular (section 6). While inbound tourism into Germany is still small in numbers compared to internal and outbound tourism, it is characterized by a growing demand for overnight stays over the last years. Moreover, the great differences in the tourist behaviour between tourists from different countries as the Netherlands and the United States will be pointed out. Section 7 concludes.

Inbound Tourism: Definition and characteristics

The World Tourism Organization (WTO) defines inbound tourism as 'non-resident visitors within a given country for the purpose of leisure, visiting friends and business' (WTO, n.d.). Inbound tourism is, in addition to domestic tourism and outbound tourism, one of three basic types of tourism of a country. For a given country X, the WTO categorizes tourism in the following way (WTO, n.d.):

- Internal tourism: tourist destinations are in country X (inbound and domestic tourism);
- National tourism: tourists are from country X (domestic and outbound tourism);
- International tourism: tourists cross borders either from or into country X (outbound and inbound tourism).

The focus of this chapter, inbound tourism, is part of internal and international tourism of country X (see Figure 4.1).

The international component is important for research on marketing for inbound tourism. A country has to market its national tourism in an international environment, which is not only influenced by the needs of the tourists in different international target markets, but also by competing destinations, parts of the tourism supply in the target markets (such as tour operators, airlines etc.), political, technical and economic circumstances and more. Nowadays this means that the ongoing economic and political globalization has to be considered, especially for tourism. This globalization includes a changing technological, social, economic, and political environment. Note, for example, the increasing importance of e-business, or the development of transnational spaces with a common law and economic system such as the European Union. However, the international component also leads to a reinforcement of the problems associated with the marketing of tourism.

Marketing tourism services, a good that is intangible, heterogeneous, inseparable and perishable, is already complicated, but in addition, the international character of inbound tourism intensifies the specific problems associated with the marketing of tourism services. Greenley and Matcham describe these problems (1983, p. 59–60):

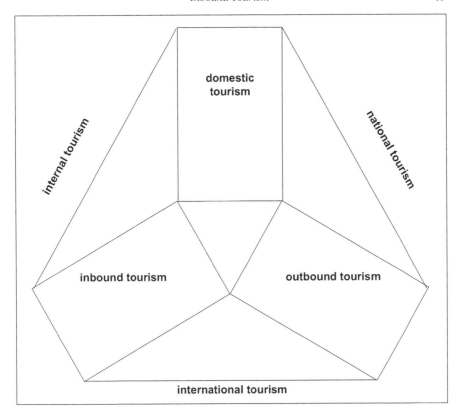

Figure 4.1   Forms of tourism

Source: WTO, n.d., p.2

> The *nature of incoming tourism,* that the core product of tourism (countryside, waterways, stately homes etc.) cannot be modified, is contrary to established marketing principles, where the product is modified to fulfil the needs of the target markets. The only opportunity to carry out modifications is via the augmented product, in form of enhancing facilities and modes of transport.

Indeed, this is a problem for marketing tourism in general, but especially for the case of inbound tourism, because of its international character with a broader range of target markets and different target groups with different cultural backgrounds, while the core product can just be modified. Arlt (2002) shows that the travel motives and behaviour of East-Asian tourists in Germany and Europe differ completely from European or US tourists, because of their more collectivistic rather than individualistic attitudes.

Two other characteristics of the market for inbound tourism are that its demand and supply sides are highly heterogeneous and globalized. Globalization results in the development of new and growing markets in Eastern Europe or Asia, in addition to traditional centres of international tourism demand in North America and Western Europe. The problem is to fit the core product to a wide range of tourism needs in a wide range of countries. In order to accommodate the various tourists needs, the marketing strategy needs to be varied, based upon the special requirements of different countries.

In addition, the tourism industry itself is highly fragmented and diversified and consists of many different organizations such as tourist boards on different levels (international, national, regional and local), tour operators, hotel companies, independent hoteliers or airline, coach and railway companies. Tourism marketing is carried out at one level by the different tourist boards, at another level by the tour operators, and yet another level by in tourism involved companies that have different marketing strategies and goals. Consequently, the heterogeneous nature of the tourism industry provides potential marketing problems for the companies marketing inbound tourism. Again, this is also true for tourism in general, but it is more complicated in the case of inbound tourism, because there is more international co-operation between suppliers from different countries and cultural backgrounds.

These specific problems of marketing inbound tourism and the changing international environment through the continuing globalization explain the need for a systematic research approach of inbound tourism, especially if a nation wants to be successful as an international tourist destination in the future. The next section examines the existing level of such research for Germany.

Level of research on inbound tourism in Germany

Until recently, the level of research on German inbound tourism has not been very advanced. The focus of tourism research in Germany has mainly been on domestic and outbound tourism, e.g., travel motives of Germans, effects of tourism in German destinations and favoured destinations of German tourists abroad.

One reason for the lack of research on inbound tourism in Germany, which started only in the early 1980s (Becker, 1984), is that inbound tourism in Germany is not as significant as the strong domestic and outbound tourism. Although the numbers of inbound tourism demand have risen over the last years, just 16.6 per cent of arrivals and 12.2 per cent of the overnight stays were by international tourists in Germany in 2003 (see section 5 of this chapter).

However, the most important reason for the poor standard of research on inbound tourism in Germany is the lack of relevant data. The Statistisches Bundesamt (German National Statistical Office) is the only source that provides free data about the tourism demand of international tourists in Germany. The data, however, does not provide relevant information for the international tourism marketing of Germany and/or German destinations since only monthly arrivals and overnights stays of international tourists in accommodations with nine or more beds and on campgrounds are counted.

Consequentially, the quantitative measurement of German inbound tourism is inaccurate and underestimated (Roth, 1984 and Horn/ Luckhaup 1999). This is especially true for German tourism regions in rural areas, which have traditionally a high tourism demand. Here, the structure of accommodations is typically characterized by a high share of small businesses up to eight beds, which are led as secondary job sector businesses, e.g. holiday on farms. Moreover, the data of the Statistisches Bundesamt offers no qualitative information on behaviour of international tourists in Germany, e.g., motives, spending, ways of booking a trip, types of tourism (business travel, vacation, cultural trips etc.). While it remains of limited use for the international tourism marketing of German destinations (Scholz, 1994), it is, however, the only available source that provides regional und chronological comparable data of the German inbound tourism (Becker, 1984).

Therefore, most of the few studies on inbound tourism in Germany are based on the data sets from the Statistisches Bundesamt with the limitations described above. Roth and Wenzel (1983) provide one of the most comprehensive analysis of German inbound tourism. Kurz (1981) and Becker (1984) summarize developments of inbound tourism in Germany during the 1960s, 1970s and early 1980s. Horn and Luckhaup (1999) examine the development of German inbound tourism in the 1990s.

In 1988, the commercial market research company IPK International established international tourism information systems, the World and European Travel Monitor (WTM and ETM). These systems offer, for a fee, chronological and nationally comparable information about international tourism and the inbound tourism in several countries. In addition to the more detailed information about the volume of international tourism, e.g., with the inclusion of visits of friends and relatives (VFR), the WTM and the ETM provide qualitative data of international tourist behaviour.

Almost all international market research studies of national tourism organizations, such as the German National Tourism Bureau (DZT), are based on data from WTM and ETM, which are multi-client studies. Since even the clients of the purchased market research studies are not allowed to publish the results without permission by IPK International, there is hardly any qualitative information about inbound tourism in most countries, such as in Germany, available, something that has been criticized in the literature, especially by Ungefug (1996). He concludes that this is the main reason for the poor international marketing orientation of the German destination management organizations on the regional and local level. Some publications on German inbound tourism that use qualitative data (for example, Roth (1984), Feldmann (1987), Roth (1988) or Scholz (1994)) show country-based differences in the travel behaviour of international tourists in Germany. In 2003, the DZT published for the first time an information brochure about inbound tourism Germany (DZT 2004a and 2005a), and in 2004 and 2005 it revised its 'DZT Marktinformationen' ('Market Information') of more than 20 target markets, in which data of the DZT market research studies, such as the ETM, are included (DZT 2004 b–d and 2005b–u).

Social scientists in the fields of tourism have made little effort to collect qualitative data about inbound tourism in Germany, and existing research methods of analyzing inbound tourism and international tourist behaviour in other, primarily Anglo-American countries have not been considered. For example, Pizam and Sussmann (1995) examine the behavioural characteristics of Japanese, French, Italian, and American tourists on guided tours by a questionnaire of British tour guides. In another example, Greenley and Matcham (1983, 1986) analyze the special problems of marketing inbound tourism and the missing marketing orientation in the service of inbound tourism by a questionnaire of British tour operators.

In summary, the studies about inbound tourism in Germany are mostly antiquated and based on secondary data that are of little value for international tourism marketing. The few available studies with qualitative information about the travel behaviour of international tourists in Germany show important differences in the travel behaviour between the tourists from different countries. Existing international research methods have not been considered yet. Section 4 presents the basics of an integrated and systematic approach, which is applied to German inbound tourism in sections 5 and 6.

Integral and systematic approach of analysing international tourism behaviour for inbound tourism

The main characteristic of inbound tourism is that the tourism supply of a destination has to be marketed in an international environment. This means that the tourist-generating regions are abroad and that the tourists have to cross international boundaries on their transit routes to arrive at the tourist destination region they want to visit. Therefore, inbound tourism of a given country is influenced by different needs and cultural backgrounds of tourists in various international target markets, other competing destinations, the supply of the tourist industry in the tourist generating regions, transit routes and tourist destination regions (such as tour operators, bus coaches, airlines, destination management organizations, hoteliers etc.) and the broader international environment, such as political, technological and economic circumstances and so on (see Figure 4.2).

There are many factors affecting the tourism demand. Freyer (2001) refers to a bundle of different factors affecting the tourism demand and tourist behaviour. He classifies the factors into:

- Social factors (e.g., values and norms, social order, ability for mobility, leisure patterns);
- Environmental factors (e.g., climate, urbanization);
- Political factors (e.g., legislation, foreign exchange, passport und customs regulations);
- Economic factors (e.g., unemployment rates, incomes, overall economic situation);

- Individual factors (e.g., curiosity, communication, relaxation, entertainment);
- Factors of the tourism supply (e.g., products, prices, marketing efforts).

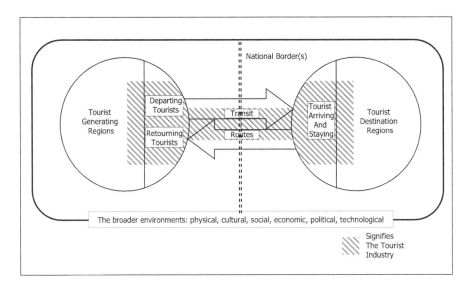

Figure 4.2   The system of tourism

Source: after Leiper (1979, S. 404)

Since many of these factors are distinct across different countries, tourism demand and tourist behaviour have to be examined individually for each country. Moreover, a comparison of the destination country with its main international competitors has to be conducted in order to determine comparative advantages.

Faulkner (2001) proposes a strategic approach to marketing inbound tourism. He divides the factors that affect demand of inbound tourism and international tourist behaviour into two categories (see also Figure 4.3).

Factors that affect the propensity of populations to travel:

- Socio-cultural factors such as leisure patterns, migratory history or holiday entitlement;
- Demographic factors such as population, age profile and household structure;
- Economic factors such as economic growth, disposable income and unemployment levels.

Table 4.1 shows the number of total outbound trips of the 20 most important international markets of German inbound tourism demand in 2003. Moreover, the outbound tourism propensity, the number of outbound trips per inhabitant, is presented for each country. As expected, the biggest markets for the international tourism in absolute numbers are highly populated western industrialized countries. However, except for the UK, outbound travel intensity is relatively low in these countries compared to smaller Western and Northern European countries, in which on average inhabitant travels at least once a year abroad. Switzerland has the highest outbound travel intensity with 179 per cent. Eastern European countries such as Russia, Poland, Czech Republic, or Hungary, while having still relatively low outbound travel intensities, have become increasingly important markets for the international tourism demand. The different levels of outbound travel propensity even in economic comparable countries like France and UK show that there must be other factors, which are responsible for the level of outbound tourism.

Table 4.1   Number of outbound trips and outbound travel propensity in selected countries in 2003

Countries	Total outbound trips in millions	Outbound travel propensity
UK	54,2	110%
US	48,3	17%
France	24,0	41%
Netherlands	19,9	122%
Italy	19,1	33%
Canada	16,0	51%
Japan	13,3	10%
Switzerland	12,9	179%
China	12,9	1%
Belgium	12,5	122%
Russia*	10,9	8%
Spain	9,9	25%
Austria	9,6	118%
Sweden	9,1	108%
Poland	7,9	20%
Denmark	6,7	125%
Finland	5,7	110%
Czech Republic	5,5	54%
Norway	4,9	108%
Hungary	3,8	37%

*Source: DZT Marktinformationen 2004c and 2005b-u, * for Russia data of 2002*

Factors that affect the comparative advantage of a destination vis-à-vis other destinations (Faulkner 2001):

- Geographical proximity (costs and level of inconvenience of getting to the destination in terms of both time and money). Besides the physical or real distance between tourist origin and tourist destination regions there are also other important influences on the geographical proximity, e.g. the degree of boundary permeability in terms of real barriers (political conflicts, demarcation methods, fortifications, restrictions by host and home countries) and perceived barriers (border formalities and restrictions, travel costs, cultural differences). The combination of these barriers effects results in what is known as functional distance, affecting human interaction across political boundaries and hereby inbound tourism (Timothy 2001).
- Comparative prices of various aspects of the product, which can be influenced by differences in wages and price inflation and by exchange rates.
- Political stability of a country.
- The effectiveness of destination marketing in stimulating the markets' awareness of, and interest in, travelling to the destination. The special role of events for short-term effects on visitor numbers and longer-term promotional effects for the destination is stressed.
- Intrinsic characteristics or man-made and natural attributes of the destination that affect its appeal.

While Faulkner's approach does not provide a comprehensive coverage of all factors affecting the inbound tourism demand, it still has the potential to provide a framework for the research on inbound tourism. The factors affecting the comparative advantage of the examined destination vis-à-vis other destinations are especially useful in explaining international tourism demand and travel behaviour.

In the next two sections, inbound tourism in Germany is used as a case study. Section 5 summarizes the relevant general data for Germany, while section 6 examines the behaviour of tourists from the two most important markets of German inbound tourism, the Netherlands and the United States.

This section presents key figures for the inbound tourism demand in Germany. The analysis uses data mainly from three sources: Statistisches Bundesamt, beginning in 1992, EUROSTAT (2005), and DZT (2005a–u).

In 2003, nearly 41.6 million overnight stays of international tourists in accommodations with nine or more beds and on camping grounds (excluding overnight stays in accommodations of the social health tourism (clinics etc.) were reported in Germany. This means, that 14 per cent of total overnight stays in Germany in 2003 were by international tourists, a relatively small number in comparison to other countries in the European Union (EU). Table 4.2 lists inbound tourism, measured in overnight stays, as the percentage of the entire internal tourism for the EU-countries.

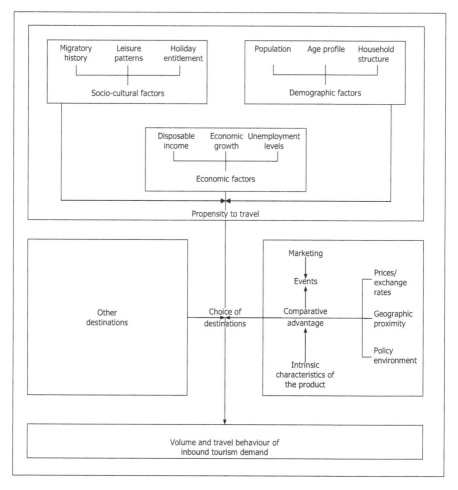

Figure 4.3   Factors affecting inbound tourism at a destination

Source: modified after Faulkner 2001, p.302

Inbound tourism in Germany: Overview

Overall, the share of non-resident overnight stays in the EU was 41 per cent, with a minimum of 14 per cent in Germany and 17 per cent in Poland and a maximum of above 90 per cent in small countries like Cyprus and Luxemburg. Geographically, the share of inbound tourism is relatively low in the Nordic and Middle-European countries, such as Germany, Poland, Sweden, Finland, United Kingdom or the Netherlands, and relatively high in small countries and/or in countries that are traditional tourism destinations, like Cyprus, Luxemburg, Greece, Austria, Spain or Portugal.

Table 4.2   Inbound tourism as share of internal tourism in countries of the
EU in 2003

Country	Overnight stays of residents in millions	Overnight stays of non-residents in millions	Share of non-resident overnight stays
EU-25*	1.232,7	856,2	41%
Austria	27,0	68,2	72%
Belgium	13,1	15,9	55%
Cyprus	1,0	13,5	93%
Czech Republic	22,8	16,5	42%
Denmark	16,5	9,9	37%
Estonia**	0,8	2,3	74%
Finland	11,8	4,3	27%
France	179,4	103,7	37%
Germany	251,4	41,6	14%
Greece	13,1	39,1	75%
Hungary	8,6	10,0	54%
Ireland**	11,7	20,0	63%
Italy	204,4	138,9	40%
Latvia	0,8	1,0	54%
Lithuania	0,8	0,8	50%
Luxemburg	0,2	2,5	92%
Netherlands**	56,0	26,4	32%
Poland	37,5	7,8	17%
Portugal	16,7	24,9	60%
Slovakia	7,1	4,9	41%
Slovenia	3,2	4,0	56%
Spain	124,7	217,9	64%
Sweden	34,3	9,7	22%
United Kingdom	189,9	72,4	28%
Malta	no data available		

*Source: Eurostat (2005), *without Malta, ** data for 2002*

Figure 4.4 illustrates that from 1992 to 2003 the number of international arrivals
and overnight stays of international tourists in Germany have slightly risen, with
a maximum in 2000 when over 18.0 million international arrivals, and nearly 39.7
million international overnight stays in Germany were reported. That demand in 2000
was driven by two mega-events, the World Exhibition 2000 in Hannover (Lower-
Saxony) and the Passionsfestspiele in Oberammergau (Bavaria), which attracted
more international tourists to Germany than usual. However, even in the years 2001
and 2002, when German inbound tourism, especially from North America and Asia,

was affected by the security risks of international terrorism and the war in Iraq, the level of inbound tourism demand was higher than in 1999.

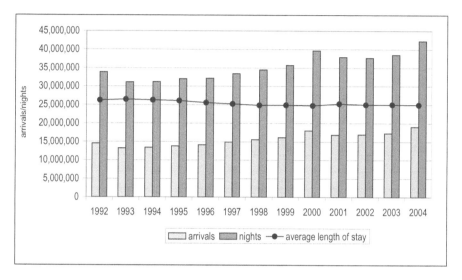

Figure 4.4   International arrivals, nights and average length of stay of international visitors in Germany in 1992–2003

Source: Statistisches Bundesamt, different years

The data of the Statistisches Bundesamt also show that the average length of stay has slightly fallen over the same period. More precisely, the average length of stay in a specific form of accommodation is falling. As mentioned above, the German tourism statistics only consider arrivals and overnight stays in accommodations with nine beds or more. Consequently, an American couple, doing a round trip through Germany with 10 overnight stays in five different hotels in five different places with a duration of two nights each is counted according to the German tourism statistics with 10 arrivals and 20 overnight stays in total. The average length of stay of the American couple according to the German tourism statistics is two nights, even though the American couple spent ten days in Germany.

Table 4.3 presents the regional structure of internal and inbound tourism in Germany in the year 2003, expressed by the number of arrivals and overnight stays in the sixteen German states (*Bundesländer*). It highlights the importance of inbound tourism within the several states, expressed by the market share of inbound tourism as a percentage of internal tourism within each state. Additionally, it describes the market share of the states on the total inbound tourism demand in Germany. A few observations are noteworthy:

Table 4.3   International arrivals and overnight stays in Germany by states in 2003

	Arrivals				Overnight stays			
	All	International	Market share of inbound tourism in state	Market share of inbound tourism in Germany	All	International	Market share of inbound tourism in state	Market share of inbound tourism in Germany
Baden-Württemberg	13.096.338	2.357.766	18,0%	13,6%	37.069.432	5.207.132	14,0%	13,5%
Bavaria	21.890.555	4.240.774	19,4%	24,5%	69.325.984	8.873.677	12,8%	23,0%
Berlin	4.952.798	1.277.365	25,8%	7,4%	11.329.459	3.356.891	29,6%	8,7%
Brandenburg	2.956.091	213.032	7,2%	1,2%	8.452.025	472.802	5,6%	1,2%
Bremen	713.964	127.157	17,8%	0,7%	1.282.077	259.823	20,3%	0,7%
Hamburg	2.956.012	560.197	19,0%	3,2%	5.406.542	1.092.868	20,2%	2,8%
Hesse	9.378.116	2.316.371	24,7%	13,4%	23.822.564	4.471.508	18,8%	11,6%
Mecklenburg-West Pomerania	5.145.948	218.022	4,2%	1,3%	22.140.077	500.640	2,3%	1,3%
Lower Saxony	9.529.774	852.584	8,9%	4,9%	32.305.507	1.900.694	5,9%	4,9%
North Rhine-Westphalia	14.258.705	2.629.639	18,4%	15,2%	35.498.763	5.962.992	16,8%	15,5%
Rhineland-Palatinate	6.180.416	1.260.644	20,4%	7,3%	17.939.006	3.656.899	20,4%	9,5%
Saarland	662.809	82.257	12,4%	0,5%	2.066.346	227.327	11,0%	0,6%
Saxony	5.125.603	423.932	8,3%	2,5%	14.239.746	922.534	6,5%	2,4%
Saxony-Anhalt	2.142.285	128.925	6,0%	0,7%	5.407.056	293.773	5,4%	0,8%
Schleswig-Holstein	4.512.002	444.732	9,9%	2,6%	20.668.082	899.874	4,4%	2,3%
Thuringia	2.793.329	165.839	5,9%	1,0%	8.174.834	424.836	5,2%	1,1%
Germany	106.294.745	17.299.236	16,3%	100,0%	315.127.500	38.524.270	12,2%	100,0%

Source: Statistisches Bundesamt (2004)

- Inbound tourism in Germany is concentrated in Southern Germany, especially in Bavaria and Baden-Württemberg, and Western Germany, namely in Hesse and North Rhine-Westphalia.
- Besides the city states of Berlin, Hamburg and Bremen, the share of inbound tourism within the states is particularly high in Rhineland-Palatinate with a share of over 20 per cent on the total tourism demand.
- In absolute and relative numbers, inbound tourism in Eastern Germany (Brandenburg, Mecklenburg-West Pomerania, Saxony, Saxony-Anhalt and Thuringia) has a disproportionately small representation. Only 2.6 million of all international overnights stays and just over 1.1 million of all international arrivals in Germany occurred in Eastern Germany, which contributes to 6.7 per cent of all international arrivals and 6.8 per cent of all overnight stays in Germany (as a comparison, Eastern Germany, excluding Berlin, has 19.1 per cent of the German domestic arrivals and 20.2 per cent of the domestic overnight stays, 16.5 per cent of the German population and 30.2 per cent of the German land area). Inbound tourism in Eastern Germany is concentrated in Saxony, because of the international trade fair city of Leipzig and the capital and scenic city of Dresden.

Map 4.1 illustrates the spatial concentration of German inbound tourism in Berlin, Southern Germany and the industrialized and metropolitan centres of West Germany, like North Rhine-Westphalia and Hesse, regarding the absolute numbers of international overnight stays. In addition, the destination inbound tourism intensity in each state is shown, expressed by the number of international overnight stays per 1,000 inhabitants in 2003. The average destination inbound tourism intensity in Germany was 536 international overnight stays per 1,000 inhabitants.

The maximum numbers of the destination inbound tourism intensity in Germany are in Berlin (990 international overnight stays per 1,000 inhabitants) and Rhineland-Palatinate (901 international overnight stays per 1,000 inhabitants). In addition, in Hesse (734 international overnight stays per 1,000 inhabitants), Bavaria (716 international overnight stays per 1,000 inhabitants) und Hamburg (632 international overnight stays per 1,000 inhabitants) the inbound intensity was above the German average. However, the East German states Saxony-Anhalt (115 international overnight stays per 1,000 inhabitants), Thuringia (178 international overnight stays per 1,000 inhabitants) and Brandenburg (183 international overnight stays per 1,000 inhabitants) had the lowest inbound intensity.

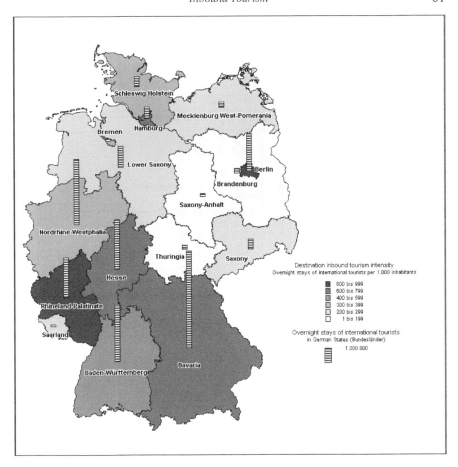

Map 4.1   **International arrivals, overnight stays, and destination inbound tourism intensity by state in Germany in 2003**

Source: Statistisches Bundesamt 2004

In Figure 4.5, developments of overnight stays by national and international tourists are shown, indexed to 100 in 1992. Both indices have grown about the same between 1992 and 2001, but since 2001, they have diverged, with the index for international tourists going up and for national tourists going down. The number of overnight stays by international tourists was somewhat sluggish in 1993 but recovered nicely, except for 2001, the year of the terror attacks in the United States. Note again, in 2000 the international tourism in 2000 was driven by two mega-events in German, the World Exhibition EXPO 2000 in Hanover and the *Passionsfestspiele* in Oberammergau (Bavaria).

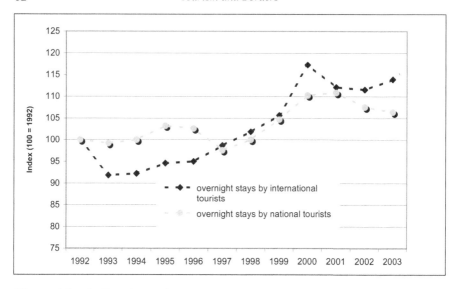

Figure 4.5   Indices for national and international overnight stays in Germany 1992–2003

Source: Statistisches Bundesamt, several years

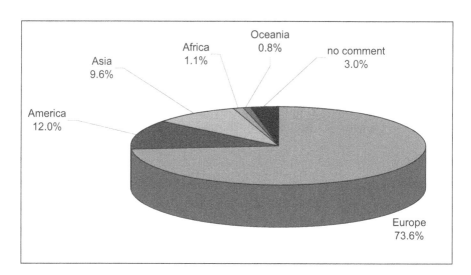

Figure 4.6   Share of overnight stays of selected continents in 2003

Source: Statistisches Bundesamt 2004

German inbound tourism is dominated by tourists from other European countries (see figure 4.6), comprising nearly three quarters of all international overnight stays. America (including North-, Central- and South America) had a share of 12 per cent (share of the US is nearly 10 per cent or 81 per cent from all American overnight stays) and Asia had a share of nearly 10 per cent.

The most important single international market for the German inbound tourism is the Netherlands with nearly 5.8 million overnight stays in Germany in 2003. They are followed by the US with over 3.7 million, United Kingdom (including Northern Ireland) with over 3.3 million, Switzerland with almost 2.5 million and Italy with nearly 2.2 million overnight stays in 2003. Furthermore, Austria, France, and Belgium generated more than 1.7 million overnight stays. As a result, the share of these eight most important markets is 59 per cent of the German inbound tourism demand. In addition, Germany received more than 1 million overnight stays from Denmark, Sweden, and Japan in 2003 (see figure 4.7).

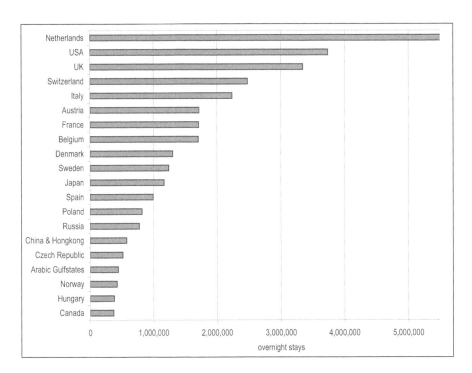

Figure 4.7   Overnight stays of the top 20 international tourism markets of Germany

Source: Statistisches Bundesamt 2004

As Figure 4.8 indicates, overnight stays from the majority of the most important markets of German inbound tourism grew between 1998 and 2003. The exceptions were the US, Japan (both probably mainly due to fear of terrorism and the Iraq War), and Poland.

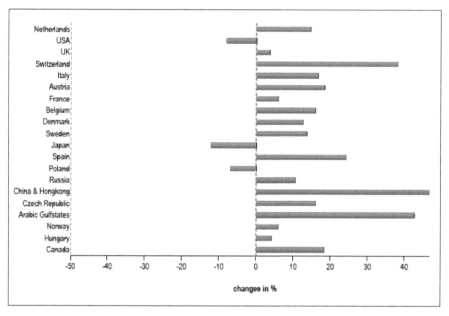

Figure 4.8   Development of overnight stays from the most important 20 markets in the period 1998 to 2003 (changes in per cent)

Source: Statistisches Bundesamt (1999, 2004)

Note again that the quantitative volume of the inbound tourism market in Germany is statistically inaccurate since only arrivals and overnight stays in accommodations with nine or more beds are counted. In contrast, market research studies like ETM or WTM report all trips, even of accommodations up to eight beds, and visits to friends and relatives, where no professional tourist accommodation is used. Therefore, they give a more detailed overview about the quantitative amount inbound tourism in a country.

ETM 2004 reported 27.2 million trips of European tourists to Germany in 2003. 14.3 million trips were vacation trips: 5.2 million short vacations (up to 3 nights) and 9.1 million long vacations (4 and more nights). These numbers illustrate that the German inbound tourism demand is not mainly characterized by business trips as it is often suspected (Scholz 1994). While the proportion of business trips with 28 per cent (7.7 million) is high compared to the European average, vacation trips still make up 53 per cent. The proportion of short holidays is exceptionally high.

Germany is a preferred short holiday destination for Europeans. An additional 5.3 million trips were VFR-trips or other trips (see Table 4.4).

Table 4.4   Purpose of trips of the European travelling abroad and to Germany in 2003

Purpose of trip	Europe		Germany	
	in millions	in per cent	in millions	in per cent
Holidays:	228.6	68%	14.3	53%
Short holidays	35.9	11%	5.2	19%
Long holidays	*192.7*	*57%*	*9.1*	*34%*
Business trips	47.6	14%	7.7	28%
Visits of friends and relatives (VFR) and other visits	61.4	18%	5.3	19%
All trips	*337.7*	*100%*	*27.2*	*100%*

Source: ETM 2003/2004 in DZT 2004a, p. 15

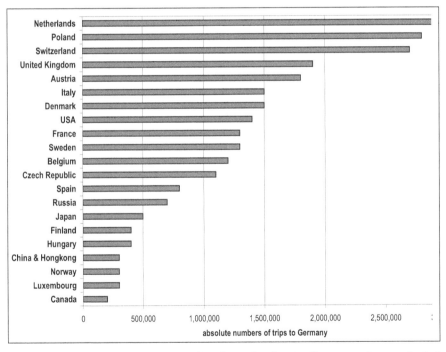

Figure 4.9   Absolute numbers of outbound trips to Germany from selected countries (2003)

Source: DZT 2004c and 2005b-u, Data for the Russia 2002

An overall comparison of the data recorded in the DZT market research studies and the numbers of overnight stays counted by the Statistisches Bundesamt confirm the importance of the same markets for the German inbound tourism market, although the numbers differ somewhat (see Figure 4.9). Of particular interest are the high numbers of trips from Poland to Germany. The main reason for this is the demand for private accommodations, mainly for recreational purposes (54 per cent of all trips from Poland to Germany are vacation trips).

In summary, inbound tourism in Germany:

- is growing in arrivals and overnight stays, but still less important relatively to domestic tourism demand and to other countries in the EU;
- is concentrated on the regions in Southern Germany, the (post-) industrialized regions in West Germany and the metropolitan areas of Hamburg and Berlin;
- is driven by the demand from the neighbouring European countries. China and the Arabic Gulf states have the highest growth rates (but starting at a low level of absolute numbers of overnight stays);
- is mainly conducted for recreational purposes. The majority of trips are vacations, although business trips are more important than in most other European countries.

International Tourist Behaviour: The case of Dutch and US-American tourists in Germany

This section discusses the travel behaviour of the two most important inbound markets of Germany, the tourists from Netherlands and the US The analysis is based on secondary data from the Statistisches Bundesamt and the DZT. Using insights from the analysis, further research recommendations for an integral and systematic approach are presented at the end of chapter 7.

Figures 4.10 and 4.11 show the development of arrivals and overnight stays of Dutch and American tourists in Germany in accommodations with nine or more beds during the period of 1998 to 2003.

The reported arrivals and overnight stays of the Dutch tourists grew steadily over the period 1998 to 2003. Overall, the arrivals from the Netherlands to Germany increased by about 19 per cent (+ 343,000 arrivals), the volume of overnights increased by about 15 per cent (+ 739,000 nights).

The number of the arrivals from the US to Germany increased by about 23 per cent (+ 453,000 arrivals) between 1998 and 2001, while the volume of overnights increased by about 24 per cent (+ 983,000 overnight stays). Beginning in the year 2001 these numbers declined, mainly due to fear of international terrorism and the second war in Iraq. From 2000 to 2003, the arrivals from US tourists reported in Germany declined more than 30 per cent (– 735,000 arrivals), while the overnights shrunk by about 26 per cent (– 1,309,000 overnight stays). Overall, the arrivals from the US to Germany dropped down by about nearly 15 per cent (– 283,000 arrivals) during the period of 1998 to 2003, and overnights declined by about 8 per cent (– 326,000 overnight stays).

The development of the US tourism demand in Germany is an example of the impact of the political environment on the volume of inbound tourism demand, as mentioned by Faulkner (2001), Freyer (2001) and Leiper (1979). The negative spill-over effects of the war on tourism across political boundaries, even to countries that are not directly involved in the conflict, are described and explained by Timothy (2001), who researched the changing international travel patterns during the Gulf War in 1991 and 1992.

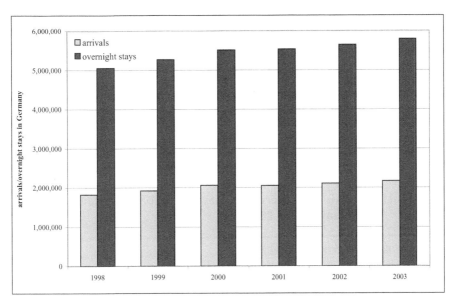

Figure 4.10 Arrivals and overnights stays from Dutch tourists in Germany 1998 to 2003

Source: Statistisches Bundesamt, several years

Table 4.5 summarizes selected aspects of the travel behaviour of Dutch and US tourists on their trips abroad and in Germany, prepared by the DZT in the 'Marktinformation Niederlande 2005' (DZT 2005k), 'Marktinformation USA 2004' (DZT 2004d) and 'Marktinformation USA 2005' (DZT 2004u). The data for the Netherlands is based on the ETM 2003 and for the US on the WTM/Performance-Monitor 2003, both conducted by the market research company IPK International (for the US in co-operation with D.K. Shifflet).

Not surprisingly, the outbound travel propensity of the Netherlands is higher than of the US The 15.9 million Dutch people generated 19.9 million trips abroad in 2003, while the 287.6 million US-Americans generated only 48.3 million trips abroad in 2003.

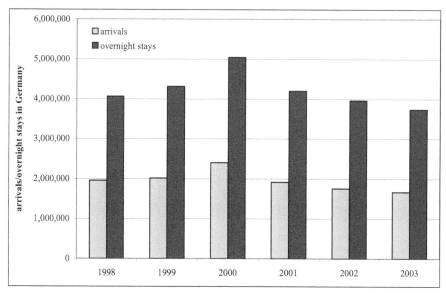

Figure 4.11   Arrivals and overnights stays from US-American tourists in Germany 1998 to 2003

Source: Statistisches Bundesamt, several years

For tourists of both countries, Germany is one of the preferred European destinations. For the US, however, note that only 44 per cent visited just Germany during a trip to Europe, while more than half visited other European countries as well (DZT 2004d).

The inbound tourism demand from the Netherlands to Germany is characterized mainly by vacation trips, whereas VFR- and other trips as well as business trips do not play an important role. In fact, Dutch inbound tourism demand in Germany consists of tourists that prefer individual tours (79 per cent). 51 per cent of the trips by Dutch to Germany are short. Besides hotels, Dutch tourists have a special affinity for vacation homes and flats as well as to camping. The typical Dutch tourist spends 47 € per night, which is a comparatively low value, similar to those of tourists from Russia (51 €) (DZT 2004c).

Only a minority of trips in 2003 from the US to Germany were also vacations. Other purposes for travelling such as business or visiting friends and relatives are more important for US citizens than they were for Dutch. Hotels, however, are clearly the most preferred accommodation for the US tourists, who stay on average 15,6 nights in Germany. The share of packaged holiday tours is comparatively high with 30 per cent. The typical US tourist spends 262 € per night, whereby the costs for transport are included.

Table 4.5   Aspects of travel behaviour of Dutch and US-American tourists on their trips abroad and in Germany in 2003

Aspect of travel behaviour	Netherlands	USA
Total number of outbound trips in millions	19,9	48,3 (11,4 to Europe)
Outbound travel propensity	122%	17%
Most preferred destinations for outbound trips	22% France, 14% Germany, 12% Spain, 8% Italy	28% Canada, 24% Europe and 16% Mexico Most preferred European destinations: 23% UK, 17% France, 12% Germany, 10% Italy, 7% Benelux
Total number of trips to Germany in millions (in %)	2.9 (100%)	1.4 (100%)
Vacation trips to Germany in millions (in %)	2.3 (79%)	0,4 (29%)
VFR- and other trips to Germany in millions (in %)	0.3 (11%)	0.4 (30%)
Business trips to Germany in millions (in %)	0.3 (10%)	0.6 (41%)
Share of packaged holiday tours to Germany	7%	30% (in 2001)
Preferred accommodations on holidays in Germany	37% Hotel, 25% Vacation home 24% Camping , 8% Vacation flat, 1% Friends and relatives	43% Hotel, 34% Guest Houses, Bed and Breakfast and others, 23% Friends and relatives and other private accommodations
Average length of stay in nights in Germany	5,1	15,6
Share of short vacation trips (up to 3 nights) to Germany	51%	No data available
Share of long vacation trips (4 and more nights) to Germany	49%	No data available
Total spent per vacation trip per person to Germany	241,– €	2.675,– €
Spent per day and person on vacation in Germany	47,– €	262,– €

Source: DZT 2005k, 2004d and 2005u

The analysis of selected aspects of the travel behaviour has clearly shown differences in the travel behaviour of Dutch and US tourists on their trips to Germany, which can be explained by the presented factors affecting the inbound tourism demand and travel behaviour.

Steinmetz (1999) explains the special affinity of Dutch tourists for using vacation homes and flats because of their individualistic way of life. Hence, it is very important for Dutch tourists to be free in their arrangement of their daily routine even on vacation. Moreover, she mentioned short distances and the high mobility of Dutch people for their preference of doing several short vacation trips annually.

In the case of US tourists in Germany, the high share of VFR-trips shows the importance of the migratory history for the inbound tourism demand. Sixteen per cent of the US population is of German origin (DZT 2004d, p. 3). Additionally it is important that many US citizens served their military duty in Germany.

The high number of business trips can be explained by the strong economic connections between Germany and the US The high share of packaged holiday tours from the US to Germany is typical for overseas tourists in Germany. The share of packaged holiday tours from Japan to Germany, the other strong overseas market, is about 43 per cent (DZT 2004b, p.15). The reasons might be the great distance and language as well as cultural barriers, which makes it more convenient for these tourists to book packaged holiday tours.

The differences in the tourism behaviour of Dutch and US tourists translate into important consequences for the international marketing of German destinations:

- Because of the high share of US tourists that visit also other destinations besides Germany, a more international and cooperative approach between German and international destinations (e.g., in France, Austria, Switzerland) is essential for successful tourism marketing on the US market.
- For successful tourism marketing, it is important to know if the tourists prefer individual or organized packaged tours for their trips to Germany. Consequently, different target groups have to be addressed with different products. For example, on the US market, tour operators are a more important target group for tourism marketing activities than in the Netherlands.
- The special affinity of Dutch tourists for vacation flats, vacation apartments and camping as preferred accommodations as well as their preference to make short holidays in Germany need to be taken into consideration during the marketing process, e.g. with special camping guides in Dutch language, short-trip offers for weekends in vacation flats, and communication of the easy and fast accessibility of many German destinations.

Conclusions

The current level of research and the quality of available data about inbound tourism in Germany needs development. A systematic and integral approach for research on inbound tourism is still missing, and is sorely needed for successful international tourism marketing.

The detailed analysis of inbound tourism demand of Dutch and US tourists in Germany based on secondary data has shown differences in the travel behaviour that have important consequences for the international marketing of Germany and its regional destinations.

The research in this area is still in its infancy, and important shortcomings of data sets have to be addressed in the near future. For example, data used in the case studies in section 6 is based on Germany as a single national destination and not as a conglomerate of many regional destinations. The behaviour of international tourists in Germany, however, most likely differs across different destinations, which is important for a more international marketing orientation of tourism organizations on a regional and local scale. Hitherto, there are no regional data on travel behaviour of international tourists in Germany (for example from metropolitan areas or individual German states). Further research in that direction would also complement the current market research activities of the DZT.

In addition, no qualitative data about international tourists' satisfaction from their trips to Germany or to German destinations are available yet. Consequently, a monitoring system for the evaluation of the satisfaction of international tourists in Germany does not exist.

Furthermore, inbound tourism demand is strongly influenced by the actions of tourism suppliers, as can be seen in the case the high proportion of US or Japanese tourists on vacation packages. Therefore, it would be useful to investigate the market behaviour of international tour operators and travel offices that offer trips to Germany, including tourism suppliers within the country, like German destination management organizations, hoteliers, or tourism consulters.

An integral and systematic approach of inbound tourism should not only examine inbound tourism demand directly, but it would be also useful to do research on other important groups of the inbound tourism demand and/or supply, such as tourism experts and suppliers within the country or tour operators. This should include:

- A permanent collection of qualitative data of the inbound tourism demand in Germany below the national level; for example, on a regional or local scale, which could complete the market research studies of the DZT.
- Research on strategies and opinions of other important groups involved in German inbound tourism, groups that influence the inbound tourism demand directly with their supply or services, e.g., tour operators, tour guides, travel writers and travel book authors. This should include also investigations on strategies and opinions of important groups in German inbound tourism within the country, e.g. the opinions of German tourism suppliers, destination managers, tourism scientists, and consulters.

It also would be useful if the DZT published more of the results from their market research studies, which give important and current information about the travel behaviour of international tourists in Germany.

It is important to note that a systematic and integral approach for research on inbound tourism should also include the function of borders as real or perceived barriers for inbound tourism. This is especially true in times, where global transformations leads to international economic and political alliances with a high degree of permeability of borders on the one hand, such as the 'borderless' European Union (EU) within its territory, and to declining international relations with increasing political conflicts (e.g. wars, terrorism, illegal immigration) connected with the construction of 'new' barriers across political borders on the other hand, e.g. increasing border regulations in the U.S or the external borders of the EU.

References

Arlt, (2002), *Die Eingeborenen sind wir. Ostasiaten als Inbound-Touristen*, in *Reisen & Essen = Voyage – Jahrbuch für Reise- und Tourismusforschung 5*, (Köln, 144–153).

Becker, Chr. (1984), 'Der Ausländertourismus und seine räumliche Verteilung in der Bundesrepublik Deutschland', *Zeitschrift für Wirtschaftsgeographie*, 28, 1–10.

Deutsche Zentrale für Tourismus e.V. (DZT) (2004a), *Incoming-Tourismus Deutschland – Zahlen – Fakten – Daten 2003* (edn 2004), (Frankfurt, 23).

Deutsche Zentrale für Tourismus e.V. (DZT) (2004b), *Marktinformation Japan 2004*, (Tokio, 32).

Deutsche Zentrale für Tourismus e.V. (DZT) (2004c), *Marktinformation Russland 2004*, (Moskau, 19).

Deutsche Zentrale für Tourismus e.V. (DZT) (2004d), *Marktinformation USA 2004*, (New York, 28).

Deutsche Zentrale für Tourismus e.V. (DZT) (2005a), *Incoming-Tourismus Deutschland – Zahlen – Fakten – Daten 2004* (edn 2005), (Frankfurt, 23).

Deutsche Zentrale für Tourismus e.V. (DZT) (2005b), *Marktinformation Belgien, Luxemburg 2004*, (Brüssel, 35).

Deutsche Zentrale für Tourismus e.V. (DZT) (2005c), *Marktinformation China, Hongkong 2004*, (Peking, 48).

Deutsche Zentrale für Tourismus e.V. (DZT) (2005d), *Marktinformation Dänemark 2004*, (Kopenhagen, 18).

Deutsche Zentrale für Tourismus e.V. (DZT) (2005e), *Marktinformation Finnland 2004*, (Helsinki, 19).

Deutsche Zentrale für Tourismus e.V. (DZT) (2005f), *Marktinformation Frankreich 2004*, (Paris, 41).

Deutsche Zentrale für Tourismus e.V. (DZT) (2005g), *Marktinformation Großbritannien, Ireland 2004*, (London, 43).

Deutsche Zentrale für Tourismus e.V. (DZT) (2005h), *Marktinformation Italien 2004*, (Mailand, 27).

Deutsche Zentrale für Tourismus e.V. (DZT) (2005i), *Marktinformation Japan 2004*, (Tokio, 32).

Deutsche Zentrale für Tourismus e.V. (DZT) (2005j), *Marktinformation Kanada 2004*, (Toronto, 18).

Deutsche Zentrale für Tourismus e.V. (DZT) (2005k), *Marktinformation Niederlande 2004*, (Amsterdam, 21).

Deutsche Zentrale für Tourismus e.V. (DZT) (2005l), *Marktinformation Norwegen 2004*, (Oslo, 21).

Deutsche Zentrale für Tourismus e.V. (DZT) (2005m), *Marktinformation Österreich 2004*, (Wien, 29).

Deutsche Zentrale für Tourismus e.V. (DZT) (2005n), *Marktinformation Polen 2004*, (Warschau, 18).

Deutsche Zentrale für Tourismus e.V. (DZT) (2005p), *Marktinformation Schweden 2004*, (Stockholm, 21).

Deutsche Zentrale für Tourismus e.V. (DZT) (2005q), *Marktinformation Schweiz 2004*, (Zürich, 30).

Deutsche Zentrale für Tourismus e.V. (DZT) (2005r), *Marktinformation Spanien 2004*, (Madrid, 31).

Deutsche Zentrale für Tourismus e.V. (DZT) (2005s), *Marktinformation Tschechische Republik 2004*, (Prag, 21).

Deutsche Zentrale für Tourismus e.V. (DZT) (2005t), *Marktinformation Ungarn 2004*, (Budapest, 17).

Deutsche Zentrale für Tourismus e.V. (DZT) (2005u), *Marktinformation USA 2004*, (New York, 28).

EUROSTAT (ed.) (2005), *Tourismus in der erweiterten Europäischen Union*, (Statistik kurz gefasst, 13/2005, 7).

Faulkner, B. H. W. (2001), 'Developing Strategic Approaches to Tourism Destination Marketing: the Australian Perspective', in F. William (ed.) *THEOBALD, Global Tourism* (2nd edn), (Oxford: Butterworth-Heinemann, 297–316).

Feldmann, G. (1987), 'Der Ausländertourismus in der Bundesrepublik Deutschland – eine Analyse des Reiseverhaltens ausgewählter Gästegruppen', (Germany: University of Trier, Thesis).

Freyer, W. (2001), *Tourismus – Einführung in die Fremdenverkehrsökonomie*, (7th edn) (Oldenbourg, Munich, Vienna, 470).

Greenley, G. E. and Matcham, A. S. (1983), 'Problems in Marketing Services: The Case of Incoming Tourism', *European Journal of Marketing*, 17, 57–64.

Greenley, G. E. and Matcham, A. S. (1986), 'Marketing Orientation in the Service of Incoming Tourism', *European Journal of Marketing*, 20, 64–73.

Horn, M. and Lukhaup, R. (2000), 'Herkunft und Ziele ausländischer Gäste', in co-ed. Chr. Becker und H. Job, Institut für Länderkunde (ed.), *Nationalatlas Bundesrepublik Deutschland*, (Bd. 10. Freizeit und Tourismus) (Heidelberg/ Berlin, 104–107).

Kurz, G. (1981), 'Die Entwicklung des grenzüberschreitenden Reiseverkehrs der BR Deutschland von 1965 bis 1979', *Schriftenreihe des Fremdenverkehrs*, (Heilbronn: Institut für angewandte Verkehrs- und Tourismusforschung, 4, 97).

Leiper, N. (1979), 'The Framework of Tourism – Towards a Definition of Tourism, Tourist, and the Tourist Industry', *Annals of Tourism Research*, 6 :4, 390–407).

Pizam, A. and Sussmann, S. (1995), 'Does nationality affect tourist behavior?', *Annals of Tourism Research*, 22, 901–917.

Roth, P. and Wenzel, G. (1983), 'Der Ausländertourismus in der Bundesrepublik Deutschlan: Die Entwicklung von 1950–1981', (Starnberg: Studienkreis für Tourismus e.V.) page 93.

Roth, P. (1984), 'Der Ausländerreiseverkehr in der Bundesrepublik Deutschland. Eine Darstellung der Struktur ausländischer Gäste in der Bundesrepublik, ihre Nachfragepräferenzen,

Reiseziele und Motive, dargestellt am Beispiel der ausländischen Gäste aus den USA, Großbritannien und der Schweiz', *Zeitschrift für Wirtschaftsgeographie*, 28, 157–163.

Roth, P. (1988), 'Aktuelle Aufgaben und Ergebnisse der Marktforschung für den Deutschen Fremdenverkehr aus der Sicht der Deutschen Zentrale für Tourismus (ed.), Beiträge zur Fremdenverkehrspraxis', *Schriftenreihe des Fremdenverkehrsverbandes Nordsee – Niedersachsen-Bremen – e.V.* 73, (Oldenburg: Fremdenverkehrsverband Nordsee – Niedersachsen – Bremen e.V., 9–26).

Scholz, J. (1994), 'Ausländerreiseverkehr in die Bundesrepublik Deutschland: Eine analytische Betrachtung des internationalen Reiseverkehrs nach Deutschland unter besonderer Berücksichtigung der Bundesrepublik als europäisches Reiseziel', *Zeitschrift für Wirtschaftsgeographie*, 38, 193–209.

Steinmetz, M. (1999), *Die Niederländer und die eigenen vier Wände* (ed.), *Auslandstourismus in Nordrhein-Westfalen. Die Niederländer: Bedeutung, Chancen, Perspektiven*, (Köln: Tourismusverband Nordrhein-Westfalen e.V., 27).

Statistisches Bundesamt (n.d.), 'Binnenhandel, Gastgewerbe, Tourismus', *Fachserie*, 6: 7.1 (Wiesbaden).

Timothy, D. J. (2001), *Tourism and Political Boundaries*, (Routledge, London, New York, 219).

Theobald, W. F. (2001), *Global Tourism* (2nd ed.), (Oxford: Butterworth-Heinemann, 503).

Tourismusverband Nordrhein-Westfalen e.V. (1999), *Auslandstourismus in Nordrhein-Westfalen. Die Niederländer: Bedeutung, Chancen, Perspektiven*, (Köln, 27).

Ungefug, H.-G. (1996), 'Das veränderte Reiseverhalten der Ausländer in Deutschland und der Deutschen im Ausland. Reiseverhalten im Wandel der Zeit: Deutsche auf dem Trip zur Selbstverwirklichung – Ausländer auf dem Weg zur Selbständigkeit', (Berlin: Internationale Tourismus-Börse ITB Berlin 1996 vom 9. bis 13. März, Presse-Information ITB/53/d).

World Tourism Organization (WTO) (n.d.), *Empfehlungen zur Tourismusstatistik* (Madrid).

World Tourism Organization (WTO)(2004), *World Tourism Market Trends 2003* (Madrid).

World Tourism Organization (WTO) (1993), *Empfehlungen zur Tourismusstatistik* (Madrid).

Further recommended readings and information sources:

- Roth, P. and Wenzel G. (1983), provide a good overview of the development of inbound tourism in West Germany from 1951 to 1980.
- Scholz (1994), analyzes inbound tourism with Germany as a European destination.
- The website of the German National Tourist Office (DZT) www.deutschland-tourismus.de (English version www.germany-tourism.de) offers the download of the brochure 'Incoming Tourismus: Zahlen, Daten, Fakten' (since 2004 also available in English). For more information, the department of market research of the German National Tourist Office provides current market studies for more the most important inbound markets of German inbound tourism, which are renewed annually.
- The website of the European Travel Commission (ETC) www.etc-europe-travel.org, a co-operation of more than 20 European National Tourist Offices, provides current market research studies of European oversea markets in Asia, North America, South America and the Middle East from 1998–2004. Furthermore, the ETC offers direct connections to several national statistic offices in Europe, America, Asia and Oceania and transnational organizations, which are engaged in tourism, like the World Tourism

Organization (WTO), World Tourism and Travel Council (WTTC), European Union and EUSTAT. The ETC also provides the website www.etcnewmedia.com, where regularly new data of the use of new media (Internet) on international generating markets are published.

Chapter 5

Organizing Destination Management: France and Germany Compared[1]

Heinz-Dieter Quack
European Institute of Tourism, Germany

Introduction

France is the most frequently visited tourist destination worldwide. Nearly 75 million foreign guests visited France in 2003 compared to around 18.4 million visitors in Germany. The number of domestic tourists amounted to about 146.5 million in France, compared to 94.2 million in Germany. There were 1.5 billion overnight stays in France, compared to 315.1 million in Germany (cp. *Statistisches Bundesamt Deutschland 2005; Deutscher Tourismusverband* 2005a, 2005b; *Französisches Tourismusministerium* 2005).

However, the much higher number of arrivals in France does not necessarily mean tourism plays a greater economic role in France than it does in Germany. After all, employment figures in tourism are quoted to total 2.8 million in Germany, around 600,000 more than in France (cp. *Deutscher Tourismusverband* 2005a, 2005b).

In Germany, as well as in France, tourism represents a major sector of the economy; which means it is an important factor in politics, too. In 2003, tourism accounted for 6.6 per cent of the national GDP in France, a figure that totaled 7.9 per cent in Germany (cp. *Deutscher Tourismusverband 2005a, 2005b, Französisches Tourismusministerium 2005*).

Along their common border, both countries share the same geographical landscapes (Upper Rhine Graben with the Vosges Massive to the west and the Black Forest to the east; valleys of the Saar and the Moselle). Additionally, the border regions have always been the scenes of military conflict. Archeological relics dating back to the Celtic settlement, and Roman history up until the Second World War show the historical significance of the area. The whole area boasts numerous ancient and more recent fortresses. At the same time, the area is characterized by common but also diversified cultural aspects, influencing cross-border co-operation until today.

---

[1]   The paper is based on an empirical research project running from 2002 to 2004, supported by AGIP, Hannover, Germany (ref. AGIP2002_486), located at Braunschweig/ Wolfenbuettel University of Applied Sciences. The author would like to express his gratitude to the project team, especially to Dipl.-Geogr. Birgit Muskat for her contribution.

Table 5.1   Comparison of German and French tourism data

Tourism data	GERMANY	FRANCE
Foreign visitors	18.4 million	75 million
Domestic visitors	94.2 million	146.5 million
Total Arrivals	112.6 million	221.5 million
Overnight stays	315.1 million	1.5 billion
Employment	2.8 million	2.2 million
GDP	7.9 %	6.6 %

Source: Own representation, with data form Statistisches Bundesamt Deutschland 2005;
Deutscher Tourismusverband 2005a, 2005b; Französisches Tourismusministerium 2005

Several contributors to this book already pointed out the development of cross-border alliances in tourism. Regarding their common heritage, it should make sense for France and Germany to co-operate closely in tourism matters.

Consequently, this paper aims to give a brief report on how destination management is currently organized in Germany and France (spatial and judicial structures). After a short debate on the conditions and backgrounds of destination management in both countries, this paper will unveil own empirical data on organizing destination management in both countries and reflect on opportunities and threats of cross-border alliances.

Outer structures of tourism organization in France and Germany

In Germany and France, the development of tourism is politically classified as a cross section task (cp. Flasshoff 1998, p. 17; PY 2002, p. 97). The different constitution of the two political systems results in different structures of tourism. Therefore, responsibilities and the implementation of destination management affairs differ.

In *Germany*, the federal political system has lead to the principle of bottom-up legislation. Federal institutions are engaged in tasks that cannot be completed on a local or regional level. While most cities and towns declare tourism as one of their own tasks, federal engagement in tourism management and marketing concentrates widely on financial support for the incoming activities of the DZT (= Deutsche Zentrale für Tourismus, German Tourism Board). Politically speaking, tourism is clearly a business dealt with by the federal states, with central government giving some support. Therefore, not only does each federal state draw up its own tourism plans, but so do most regions and communities.

The data on the number of German tourism organizations varies highly. Freyer names 6,000 places, appealing for a (local) tourism promotion. Among them are 2,400 tourist centers, 150 thematic routes, and 160 regional associations (cp. Freyer 1998, p. 184 f.). According to Bleile (2002, p. 3), there were '280 tourist associations on the regional and federal state level' in 2002. The German Tourism Association (Deutscher Tourismusverband/ DTV) in a recent statement refers to 4,000 organized tourist centers, partly combined in co-operations with no data to local or regional organizations available.

At the top of the organizational structure are the German tourism lobby of *Deutsche Zentrale für Tourismus/DZT* and *Deutscher Tourismusverband/ DTV*. Below that, on a federal state level, tourist interests are represented through *Landesfremdenverkehrsverbände* (federal state tourism associations) or *Landesmarketinggesellschaften* (federal state organizations for marketing), which are above the third and regional level. Regional organizations are separated into district communities as well as private and communal capital providers. On the following municipal or borough level diverse tourism associations and organizations with different legal forms can be found. On the sixth and lowermost level there are the local service providers representing their specific interests (cp. Bleile 2000, p. 3 f. and Freyer 1998, p. 185 ff.).

The centralistic and strictly hierarchical political system in *France* has lead to a clear top-down organization of tourism destination management. There is a strong development strategy, which is defined and predominantly financed by the central state, according to a range of strictly defined tasks for every level (central, regional, departmental, local). This range contains market research, definition, and recruitment of new target groups, and competitive intelligence investigations on the central level. The definition of infrastructural developments takes place on a regional level, while service related face-to-face customer care is realized on a local level.

In France, the organizational structure of tourism is divided into four basic levels. In the first place, there is the organization *Maison de la France*. It is comparable to the German National Tourist Board (DZT). Basically, this economic association of interest focuses upon the commercialization of France. Tourism institutions, e.g. local tourist offices or tourist information centers, private tourism stakeholders, hotels, tour operators and experts of the tourism development belong to its members. The *Maison de la France* supports all members with consulting, information, advertising campaigns, economical promotion, and public relations. Subsidies and membership fees are their financial basis (e.g. Maison de la France 2005). The second organization unit is formed by regional tourism committees (*Comites regionaux de tourisme*). In principle, they are equal to the tourism associations of the federal states in Germany. Each French region is represented by such a tourism association (incorporated societies). Their main task is to coordinate all activities of their local members on a regional level. Their aims are the development of the economic factor of tourism and local marketing. With the support of a local development scheme, acquired by a committee appointed by the minister of tourism, political framework directives of the tourism department should be realized. The committee consists of members of the consulate and the tourism associations of the departments. The above scheme operates in the following sectors: sciences, planning, economy, constitutions, accommodation, and technical distribution assistance. The public and private sectors subsidize local tourism associations.

The tourism associations of the departments operate at the same level as the departments, which represent a further division of the regions. They are called *Comité departemental de tourisme*, organized in corporate societies, and assigned to one department, just like the local tourism associations. Their main task consists in the development of tourism of the appropriate departments. They are also responsible for the coordination of tourist initiatives, the realization of the political framework directives of tourism given by the state and the local tourism associations, the briefing and consulting of the funding agencies and the creation and distribution of tourist products. Members of the tourism associations of the departments are consular unions, standing committees on industry and trade, representatives of the local tourism associations and tourism offices. They are financed in the same way as the local tourism associations.

On a local level the tourism associations act as so-called tourist information offices (*Office de tourisme*) or *Syndicats d'initiative*. They are non-profit organizations self-financed out of events or supported by subsidies of the commune and membership fees. Overall, the tourism associations pursue the same tasks as the tourist information offices in Germany.

A special type of the tourism associations is the *Offices municipal de tourisme* and the *Offices de tourisme intercommunal*. An *Office municipale de tourisme* is organized under the legal form of an *Etablissement Public a Caractere Industrielle et Commerciale* (EPIC). EPIC represents a partnership with a communal character and aims at the tourist development of the destination. To achieve this goal EPICs adopt the local marketing, the activity coordination of the different funding agencies, and act as an adviser for all participants in tourism. As well as bring together several communities an *Office de tourisme intercommunal* is also responsible for development of new destinations. Managed by collaborations of different communities it is possible to overcome existing political borders. That helps to distribute and generate competences and funds more effectively. An appointed mayor represents each location. The entirety of tourism associations is brought together under the *Federation nationale des offices de tourisme et syndicates d'initiative* (cp. Micheaud, J.-L. 1995, 13ff, Cadieu 1999, p.149 ff.).

As one can see, there are some substantial differences in the organization of destination management in both countries. In order to give a statistically based review, own investigation from 2002 to 2004 will be presented.

Methodology of the survey

In the light of the inconsistent data provided by German tourism organizations, it was necessary to establish current and reliable data to provide a basis for further examination. Consequently, 5,399 German tourism organizations were identified and divided hierarchically in up to six spatial levels. The individual levels, as a rule, align to political borders such as local and federal state borders. Co-operation across local borders or federal state borders is barely evident, but currently discussed in detail.

Altogether empirical analysis took place in two waves of inquiry. The first inquiry was from September until November 2003. The target group of the first inquiry were the pre-investigated 5,399 tourism organizations in Germany. It was carried out via a facsimile questionnaire. Those institutions that did not have access to a facsimile machine were interviewed via mail. A total of 900 answered inquiries for the first analysis were sent back indicating a return quota of 17 per cent. The target group of the second inquiry was the respondent participants of the first inquiry. From December 2003 until January 2004, the questionnaire was sent via mail, with limited facsimile machine usage. 549 tourism organizations were contacted in total, resulting in 355 replies, i.e. a return quota of 65 per cent.

All 3,600 French tourist offices of France were selected as benchmark partners. However, there is no overall public register of the organizations available although the number of tourist offices is well publicized. Therefore, with the support of the institute *Maison de la France* and by Internet investigation, addresses were collected.

In contrast to the German inquiry, the French one does not represent a total survey. Because of financial reasons, the inquiry was implemented via e-mail. For this method, it was necessary to obtain all e-mail addresses of the tourist offices; but this was possible for only 1,400 offices.

With the possibility to answer via mail, fax or mail 204 interviewees replied, i.e. a return quota about 15 per cent. Contrary to the German inquiry, there was only one inquiry wave in France.

Organizations' legal forms

The legal structure of organizations can be defined according to demand or the economy of acquisition. According to corporate goals, the organization by the economy of acquisition is value-orientated and quantified by economic success. Contrarily, non-profit organizations are orientated towards the economy of demand and pursue non-profit aims. These organizations include all organizations whose target is not acquisition, so they are not established to obtain profits for the investor (cp. Bea/Göbel 2002, p. 410). Superficially, there is the appropriation of a service.

The following picture surveys the current distribution of the legal forms of German tourism organizations. The majority, about 83 per cent, is organized by the economy of demand. It is composed of the legal form of the offices (own or directed establishment), registered societies, and joint ventures. Tourism companies with the legal form of a Ltd. company (GmbH), and others such as small PLC (AG), Ltd. with a limited partnership (GmbH & Co. KG) and others are orientated by acquisition and represent a rate of 16 per cent.

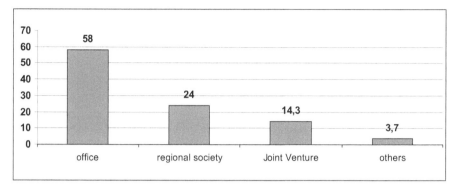

Figure 5.1   Legal forms of German tourism organizations (in per cent)

Source: Own survey

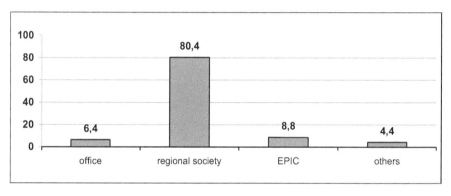

Figure 5.2   Legal forms of French tourism organizations (in per cent)

Source: Own survey

The German survey shows divergent conditions: there is a wide range of organizations responsible for only a part of a commune, leading to second best results in tourism marketing and often waste of resources. There is no clear definition of specific tasks for a certain level, leading sometimes to parallel projects and often to insufficient return. The overall number of organizations in France is smaller, with a significantly higher efficiency. Since there is a strict definition of specific tasks and a defined responsibility for the spatial units, the Return on Investment (ROI) seems to be much higher than in Germany.

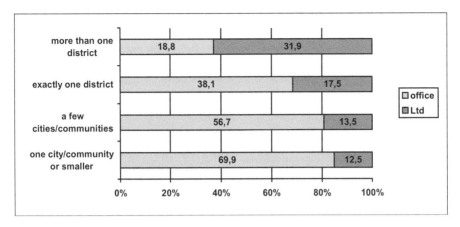

Figure 5.3   Dependency on competency and legal form in Germany

Source: Own survey

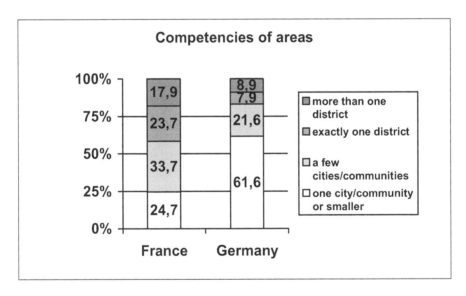

Figure 5.4   Spatial competencies of German and French tourism organizations

Source: Own survey

Generally, it seems that French tourism organizations feature a lean and in this respect more efficient structure than the organizations in Germany.

Organizations' self-understanding

The most straightforward method to define the self-understanding of tourism organizations is to ask them to list their primary tasks and authorities.

Surprisingly, a large amount of organizations define themselves as stand-alone tourist destinations (Germany: 48 per cent, France: 60 per cent). In terms of tourism science, tourist destinations are well defined as places that:

- are widely known for their attractive surroundings;
- offer all kinds of accommodation and F&B facilities;
- can communicate attractive experience settings, and
- are spatially defined by consumer behavior (i.e. excursions).

The following tables provide answers as to why tourist destinations consider themselves as a defined destination and why not.

Table 5.2   Reasons for being a defined tourist destination

GERMANY		FRANCE	
	Percent		Percent
Internal reasons, financing, organizational structure	35,9	Variety of tasks, advertising media, supply	45,7
Tradition, political independence	23,2	Name recognition, behavior of demand	28,3
Variety of tasks, advertising media, supply	19,9	Internal reasons, financing, organizational structure	13,0
Name recognition, behavior of demand	9,8	Tradition, independence	8,7

Source: Own survey

As we can see clearly in Tables 5.2 and 5.3, there is a significant gap in the understanding of what a tourist destination is composed of, not only between tourism science and tourism practice, but also between different organizations. Eventually, the understanding of what makes a place a distinguished tourist destination, varies distinctively between the German and the French survey. From a consumer-based view, the only correct answer to this question should be the term 'name recognition, behavior of demand'. As is to be seen in Table 5.2, only 28.3 per cent of those French organizations (and worryingly 9.8 per cent of German organizations) that understand themselves as stand-alone-tourist-destinations refer to that term.

Table 5.3   Reasons for *not* being a defined tourist destination

GERMANY		FRANCE	
	Percent		Percent
Component of a region	47,7	Component of a region	73,1
Internal reasons, financing, organizational structure	20,5	Internal reasons, financing, organizational structure	15,4
Too few competences of regions	15,8	Tourism demand is too weak	7,7
Tourism demand is to weak	12,9	Too few competences of regions	3,8

Source: Own survey

Cross Border Tourism

The demonstrated differentiated organizational structures in France and Germany implicate some specific disadvantages for German incoming tourism. Although tourism marketing in France seems to be much more successful than in Germany, there are still some weaknesses in the French organizational style as well.

Generally, these aspects are pointed out as follows:

Outer structure of tourism organization

In both countries, there is a hierarchical method of defining organizations. In France however, there is a clear definition of tasks for each level. In Germany, the organizational structures do not adapt to the needs of their members and additionally do not exist in harmony with tourism market requirements. The borderlines of tourism destinations are not orientated to customers' tourism mind maps but rather to political frontiers. As to be seen in the hierarchical disposition, the structure is defined by strong sectionalism. Management activities are fulfilled twice or even more often due to the multi-level composition.

Dependency on short-term political needs

Bleile criticizes the close link between German tourism organizations and politics. Thus there is, for example, a strained financial dependency of tourism responsible on the individual communities, administrative districts and federal states. German tourism organizations are running the risk of administrating tourism in a bureaucratic fashion, because of the non-profit commitment of their demand-orientated background. This attitude is strongly in conflict with microeconomic cost management, which is required for tourism organizations, as well as efficient management concepts (cp. Bleile 2000, S. 6).

Lack of orientation toward customer needs

The majority of German tourism organizations, their internal structures as well as their market behavior, is not consumer-driven. The French organizations can point out the centralistic structure as an advantage, since it allows short-term redefining of marketing strategies.

Despite this, 48 per cent of the German and surprisingly 60 per cent of the French tourism organizations consider themselves as defined tourist destinations. It is remarkable that the majority both in Germany and France refer to non-market defined reasons in their arguments. This might be seen as a strong hint to the supposition that French tourism organizations in their internal structure and self-understanding are rather close to the German organizations. Thus, the only strategic advantage of French organizations remaining is to be seen within the French states centralistic structure.

It is this structure that is widely discussed at present, since recently tourism demand in France is declining. Reasons for this can be found in wide-ranging changes in consumer behavior i.e. increasing demands on the quality of accommodation, customer satisfaction, and experience settings. It is often criticized that communication styles within the French centralistic structure are too slow and recommendations given by central authorities are too vague.

Concerning terms of organizational structures of destination management, a mix of between strict bottom-up or top-down strategies seems to be more effective. It is on a more regional level, on which tourism managers can achieve a glance of customer needs as well as claims of the supply chain.

Conclusions

Organizational structures of destination management both in France and Germany are clearly politically dominated. In this way, the structures do not refer to consumers but to politicians' needs. Cooperating in cross-border areas means to put together two widely different styles of organizing not only destination management, but also the whole state. Whereas co-operation on the community level works better now than twenty years ago, jointly acting on the regional level still requires the European Union's financial support. Decision-making in tourism matters in Germany is situated on a regional or federal state level, whereas in France the President of the French Republic holds this power.

Therefore, the above-mentioned strategic advantage of French tourism organizations all in all turns out to a clear strategic disadvantage in terms of cross-border co-operation.

References

Bea, F.X./ Göbel, E. (2002), *Organisation: Theorie und Gestaltung*, (Stuttgart).

Bleile, G. (2000), 'Management des Wandels – Plädoyer für eine neue Tourismusorganisation', *Schriftenreihe Tourismus* 4, (Freiburg i.Br.: Akademie für Touristik Freiburg).

Deutscher Tourismusverband (2005a), 'Beschäftigungs- und Umsatzvergleich 2003 mit ausgewählten Branchen' <http://www.deutschertourismusverband.de/content/files/zahlen_ daten_fakten_2003.pdf>, accessed on 11 January 2005.

Deutscher Tourismusverband (2005b): 'Gästezahlen und Übernachtungen 2003 in Beherbergungsstätten mit neun und mehr Betten und Touristik-Camping (Vergleich zu 2002)', <http://www.deutschertourismusverband.de/content/files/zahlen_daten_fakten_2003.pdf>, accessed on 11 January 2005.

Eggert, A. (2000), *Tourismuspolitik*, (Trier: Trierer Tourismus Bibliographien, 13).

Flasshoff, W. (1998), *Die Entwicklung der Fremdenverkehrsförderung im Rahmen der Gemeinschaftsaufgabe 'Verbesserung der regionalen Wirtschaftsstruktur'*, (Trier: Materialien zur Fremdenverkehrsgeographie, 47).

Kreilkamp, E./ Pechlaner, H./ Steinecke, A. (2001), *Gemachter oder gelebter Tourismus? Destinationsmanagement und Tourismuspolitik*, (Wien: Management und Unternehmenskultur, 3).

Kubiak, C. (2004), *Destinationsmanagement in Deutschland und Frankreich. Benchmarking der rechtlichen, organisatorischen und wirtschaftlichen Aspekte*, (Salzgitter: FH Braunschweig/ Wolfenbüttel, unveröffentlichte Diplomarbeit).

Luft, H. (2001), *Organisation und Vermarktung von Tourismusorten und Tourismusregionen: Destination Management*, (Meßkirch).

Maison De La France (2005), 'Tätigkeitsbericht', <http://de.franceguide.com/mieuxconnaitre/ qui.asp?z1=3Ybc1gx7>, accessed on 11 January 2005.

Micheaud, J.-L. (1995), *Les institutions de tourisme*, (Paris).

Mundt, J. (2004), *Tourismuspolitik*, (München).

Py, P. (2002), *Le tourisme: Un phénomène économique*, (Paris).

Quack, H.-D./ Baur, N./ Volle, B. (2004), 'Optimierung von Organisationsstrukturen im Destinationsmanagement', <http://cms.fh-wolfenbuettel.de/fks/home/ index?selpage=Sonstiges&s=5081123,5081439>, accessed on 11 January 2005.

Statistisches Bundesamt (2005), 'Ankünfte und Übernachtungen von Gästen in Beherbergungsstätten und auf Campingplätzen in 1000', <http://www.destatis.de/basis/d/ tour/tourtab8.php>, accessed on 11 January 2005.

Chapter 6

Destination Management in Cross-border Regions

Karin Hartmann
Mobility Consulting Team of Swissenergy, Switzerland

Introduction

All over the world, the economic importance of tourism has grown strongly over the last few years. In many regions, particularly in rural, structurally weak areas, the tourism industry is often one of the most important job providers and catalyst for economic and infrastructural development. Tourism has constantly increased during the last few years, and according to the UNWTO, this growth is predicted to continue in the years to come as well.

Nevertheless, European destinations in particular suffer from a massive tourism decline. There are various reasons for this: crowding out because of increasing worldwide competition, new trends in consumer behaviour as well as altering economic, political and social conditions. Therefore, one of the most important challenges for the future will be to guarantee the competitiveness within the global market. It will be increasingly important to concentrate and coordinate the economic, infrastructural, natural, cultural, and social strengths and potentials of a region.

To face this new challenge, a tourism management model, the so-called destination management, has been under discussion by tourism managers and tourism researchers since the mid 1990s. The following discusses the key issues of destination management as a strategy for touristic development of areas. In the face of increasing global competitive conditions, a tourism management model, the so-called destination management, has been under discussion for a couple of years. Particularly in Austria and Switzerland, Destination managers have been using this management model intensively into praxis. In the late 1990s, a cross-border area, the international Lake Constance area, ventured destination management with the foundation of the 'Internationale Bodensee Tourismus GmbH'.

The implementation of a destination management process can help to make cross-border co-operation more effective, more professional, more sustainable, binding, and closer to the market. To achieve this result, professional organisational structures need to be implemented, the cross-border core businesses must be defined, and electronic cross-border marketing networks must be established.

In the following, the example 'Lake Constance Area' will be analysed with regard to the conditions necessary for the implementation of a destination management process and to the kind of opportunities and problems related to this. The results are based on interviews with experts and tourism stakeholders of the Lake Constance area as well as on interviews with experts on cross-border co-operation in other border regions. Furthermore, an analysis of literature and statistical data of the Lake Constance area was affected.

Destination management

The concept of destination management originates in American ski and golf resorts. Many of these destinations are organised and managed by one dominating enterprise that sets their strategy or even owns their entire infrastructure (Bieger 2002, p. 2). This requires a diverging approach to planning, financing and enforcement of destination-wide measures compared to the small scale structures of European destinations, which are characterized by decentralised ownership and decision structures (Bieger 2002, p. 2). Accordingly, that concept cannot be imposed to European tourist areas but it calls for the conception of a new kind of destination management customized to European conditions. The fact that in Europe many different stand-alone enterprises offer their services shifts the focus to coordination and co-operation aspects. They are the key issues of any successful destination management.

Destination

The term 'destination' itself remains an Anglo-Saxon loanword in German literature, being widely used in its touristy. It chiefly stands for places, areas, countries or even continents which are visited by tourists because of their inherent natural attractivity and any derived offering.

In destination management, however, 'destination' has come to denote an actual tourism product. This product usually covers a whole range of touristic offerings and services, representing the actual competitive unit within incoming tourism. Bieger defines 'destination' in the sense of a:

> Geographic area (community, region, country, continent) that the respective visitor (or a visitor segment) selects as a travel destination. It encompasses all necessary amenities for a stay, including accommodation, catering, entertainment, and activities. It is therefore the actual competitive unit within incoming tourism which must be run as a strategic business unit. (Bieger 2002, p. 56).

The different layers of a destination (community, region, country, continent) may however overlap. To an American traveller for instance, Europe might be a destination, to a tourist from Northern Germany it might be the Lake Constance region, and to a visitor from Zurich it might be the town of Constance itself. Different target groups have different perceptions and different purposes of travel and thus define their destinations accordingly. It is commonly assumed that, the greater the travel distance from the traveller's country of origin, the bigger a destination must be in order to receive perception (Bieger 2002, p. 57).

Table 6.1 shows, according to Bieger's definition, substantial differences between destinations in the context of destination management and traditional destinations. The representation refers to the area covered.

Table 6.1   Destinations in the sense of destination management vs. traditional destinations

Destinations in the sense of destination-management		Traditional Destinations
• demand based	⇔	• Offer based
• Independence of political and operational boundaries	⇔	• Dependence of traditional political boundaries
• Competitive unit	⇔	• Uncoordinated provision of services
• Uniformity of the natural and the derived offering perceptible from the outside	⇔	• Uniformity for the neutral observer not perceptible

Source: Bieger 2002

The problem with Bieger's definition seems to be that it doesn't clarify the spatial delimitation of a destination. Presumably, the visitor defines more or less randomly a geographic space according to his personal needs. A generalized spatial delimitation of destinations thus seems impossible. From an organisational point of view as well as for product definition reasons the service providers, however, are required to establish a sensible form of delimitation (Scherhag 1999, p. 39). He suggests the delimitation of touristic destinations to be defined so that they are perceived by the demand side as units of maximal homogeneity. At the same time, however, destinations ideally should be operationally coherent and manageable units. Hence, delimitation seems sensible along identical or at least similar touristic structures, such as geographical regions (e.g. Black Forest, German North Sea Coast), coherent sociohistoric areas, and similar structures of offer and demand.

Furthermore, delimitation can be carried out under the aspect of interlacing relationships (e.g. with regard to the excursion behaviour of tourists or due to complementary offers) (Iwersen-Sioltsidis/Iwersen 1997, p. 113). This procedure appears appropriate whenever existing local or regional destinations that already have a strong branding provide the basis for destination development (Müller 1998). As a further delimitation method, Spieß (1998) suggests the principle of self-organisation, i.e. the ongoing process of incoming agencies defining their own destination along existing communication and co-operation relationships (Spieß 1998, p. 27).

The concept of destination management

The above definitions propose the underlying concept of destination management: By cooperating, local agents and service providers intend the unified and comprehensive development, organisation, and marketing of their destination (Grasshof 2000, p. 39). Although provided by different, independent operators, the entire range of services on offer in one destination should appear to the visitor as one single homogeneous touristic product (Bieger 1998, p. 8).

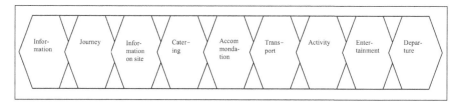

Figure 6.1   Service concatenation in tourism

Source: Bieger 1998, 2

The highest aim of any destination management concept is to ensure its sustainable and enduring competitiveness. This may require adaptations of existing organisational structures and a renunciation of the traditional isolationistic mentality. In order to defy the overly saturated market, a destination must have ample offerings, focussing on its strong points. This can only be realised through efficient organisational structures and professional management and marketing (Kühne 2000, p. 131).

As opposed to North-American destinations, the economically small to average European destinations cannot be centrally organized and managed as holding companies. In Europe, a common business strategy must be elaborated in collaboration with many different lobbies. It is important to bundle all available resources, use synergies, and thereby develop the destination as a whole (Grashoff 2000, p. 40). 'The decisive point in the development of destinations is to turn the proximity of places, companies and infrastructural operators in traditional tourism areas into a self-propelled, strategically manageable competition unit for the international market' (Bratl/Schmidt 1998, p. C1). Its successful promotion requires a common destination name as a corporate identity platform for a unified market presence of all tourism service providers, while a USP (Unique Selling Proposition) or an UAP (Unique Advertising Proposition) should ensure a better distinction from other tourism destinations (Kühne 2000, p. 128–129).

Table 6.2 Development stages of the co-operation and organisation in tourism

	1950s and 1960s: Development stage	1970s and 1980s: Boom stage	1990s and millennium turn: Concentration stage
Demand	‣ Vacation is regeneration ‣ Inland and neighbouring foreign countries as the main travel destination ‣ Passenger cars as an economical mean of transport ‣ Exceeding demand with capacity restrictions ‣ Provider market ‣ Growing demand	‣ Vacation is variation and adventure ‣ Aeroplanes become an economical mean of transport for the whole world ‣ Visitor wants its specific offer ‣ Excess supplies grow ‣ Growing demand	‣ Vacation- and consumption style shapes leisure time based life philosophy ‣ Consistent experiences and convenient service concatenation are wanted ‣ Service and atmosphere gain importance ‣ Globally orientated visitor ‣ Multi-optional consumption behaviour ‣ Partially stagnating, but globally growing demand
Competitors	‣ Provider market with weakly marked competition ‣ European states and neighbouring countries as the main competitors ‣ Growing capacities	‣ Demand market with strong competition prevails ‣ International destinations push their way into the principal markets of traditional European vacation countries ‣ Capacity extension ‣ Growing marketing standards	‣ Global crowding out ‣ Visitors must be attracted continuously ‣ New destinations push onto the market ‣ Global capacity extension ‣ Continuously growing marketing standards ‣ Starting of concentration process
Co-operation and organisation	Advertising-community ‣ House- and community orientated ‣ Common local and area brochures ‣ Events ‣ Fair participation ‣ 'Verschönerungsgemeinschaften' (free translation: Associations to upgrade municipalities)	Regional tourism organisation ‣ Local based strategies ‣ Area brochures ‣ Neutrality principle opposite members ‣ Common advertising and sales promotion ‣ Regional offer development in ski tourism ‣ Dominance of the local organisations ‣ Growing external effects (environment, local) ‣ Broad integration of different interests	Destination management ‣ Concentration on development and organisation of core businesses ‣ Active marketing and direct sale ‣ High-quality and brand orientation ‣ Giving up neutrality principles ‣ Regional computer reservation systems ‣ Regionally coordinated infrastructure development ‣ Differentiation between tourism and politics ‣ Leadership and coordination on region level

Sources: Bratl and Schmidt, 1998, p. B2

To Bratl and Schmidt (1998), destination management represents the third generation of the development of co-operation in tourism. It corresponds to the highly competitive contemporary market conditions as well as to the consumer's increased requirements (Bratl/Schmidt 1998, p. B2). Table 6.2 shows the development stages of co-operation and organisation in tourism since the fifties.

Since there are only very few places which can withstand contemporary competition as independent destinations (e.g. Zermatt, St. Moritz), a reorganisation of touristic structures is necessary. This includes both a spatial and organisational redesign and a reallocation of the tasks on all operational levels.

Tourism organisations

The term 'tourism organisation' applies on one hand to touristic organisational structures, denoting the overall organisation of a destination, i.e. 'the particular organisational structures, the coordinative significance within a destination'. On the other hand, it also applies to an institution, e.g. a tourism organisation that fulfils comprehensive tasks within the touristy of the destination and coordinates activities in destination management (Bieger 2002, p. 71). In the following, the term 'tourism organisation' is used for the institution providing all cooperative offerings (Bieger 2002, p. 72). The term 'touristic organisational structures' is used for the overall organisation of a destination. This includes the regulation of tasks and competences among the different stakeholders within the destination.

Tourism organisations are necessary on different levels of regional authorities (communities, regions, countries). Their tasks and priorities vary according to their level. Traditionally, these different levels are hierarchically arranged and the task distribution is alike, differing only in activities and weighting (Bieger 1997, 73). This has caused considerable redundancies that complicated the efficient use of inherently scarce resources. Therefore, a sensible destination management requires the redesign of existing organisational structures and a reallocation of the tasks and responsibilities as well as the financial resources within the destination, extending to a co-operation with other hierarchical levels, i.e. the provincial authority. In the course of this reorganisation, mergers of existing local or regional tourism organisations often are necessary and useful. There is not, however, any panacea on how the redesign of touristic organisational structures need to be shaped in detail. It is conceivable to transfer the tasks within a destination management process to an already existing organisation or to merge existing regional or local tourism organisations. It is crucial that a tourism organisation accepted by all participants is appointed to establish communication and co-operation structures, to regulate responsibilities and task allotment and to survey the realization of all joint measures (Grashoff 2000, p. 41).

Bratl and Schmidt plead for the assignment of a so-called *destination management company* (DMC). It assumes the market-oriented leadership and the management of the destination, carried by the strongest service providers of the destination, relieving it from too diverse integration and representation efforts interfering with its neutrality policy (Bratl/Schmidt 1998). The DMC should in the first place concentrate on the development and the promotion of core businesses and the supervision and coordination of marketing activities (cf. Bratl/Schmidt 1998).

Bleile's (2000) proposition bears some resemblance, suggesting the assignment of regional, district and interstate *tourist service centres* (TCS) for Germany. These are meant to operate as privately runned enterprises, for example as an 'AG' or a 'GmbH' (similar to 'Ltd's'), acting as a marketing centre of competence. The funding should mainly come through own revenues, for example from fees for services provided to public or private customers (Bleile 2000, p. 106).

Pechlaner in turn speaks of networks of tourism organisations and destinations. In this context he considers the management of this compound a 'head office' of a (regional) network in the sense of a market-orientated grouping of competences and potentials on local and regional levels' (Pechlaner 2000, p. 38). Networks are supposed to represent 'the key to a competitive capacity guaranteeing the advantages of small organizational units yet using the advantages of a common market presence at the same time' (Pechlaner 2000, p. 36).

Requirements for a destination

In their manual 'Destination management' (1998), Bratl and Schmidt name different criteria that are supposed to give the practitioners a first orientation whether their area, based on the ratio of offer and demand, has the possibility to implement a destination management strategy (Bratl/Schmidt 1998, p. D2–D4). Based on Müller's theoretical requirements, Kühne also provides an overview of the practical requirements of a destination (Kühne 2000, pp. 139). Müller lists six theoretical requirements of a destination (Müller 1998) which are a comprehensive, synchronized and well-coordinated concatenation of offerings and services, at least one marketable brand, a maximal independence from traditional political boundaries, a competent and qualified management extensive quality development and control system, and a sufficient funding for the realization of tasks, in particular for branding in selected markets.

Bratl and Schmidt set different criteria by which the achievements at management level of a region can be evaluated. 'Strong' areas show a continuous development of the core businesses, where at least one core business exists (i.e. all regional subsets are organized in a package). They have a functioning direct reservation system and at least 20 per cent of the overnight stays can be directly contributed to the association's marketing efforts. Furthermore, a regional quality control programme exists for all service providers and not only within an individual provider group i.e. lodging. An internationally established brand carrying sympathy and affinity of experience exists or a continuing development of the market appearance is at least being worked on (Bratl/Schmidt 1998, p. D3).

Whether or not the above-mentioned requirements can be met, largely depends on the presence of a common intent, the willingness to cooperate, and mutual trust within the destination. The joint development of a tourism concept may contribute to the acceptance of plans and decisions. By creating provider networks, synergies can be used while safeguarding the distinctness and self-sufficiency of individual companies.

The necessity to increasingly cooperate in the future also arises, however, if tourism organisations strive for a better position on the market through alliances with neighbouring tourism organisations and regions.

Furthermore, the professionalisation of all activities must be aimed at. Particularly in the scope of offering design, quality management, and marketing (brand strategy), it is crucial to act more professionally and closer to the market. In the context one also speaks of a 'depolarization' of the tourism activities design in order to achieve a greater efficiency and effectiveness. In this case the total withdrawal of politics from the touristy will not be helpful; much rather the influence of politics should be reduced and a more market-orientated entrepreneurial organisation of tourism activities should be encouraged (Fontanari 2000, p. 84). This can be reached through an increased outsourcing of specific tasks (e.g. marketing) from public administration (e.g. by private-public partnerships). This is to be seen in particular also under the aspect of the tight financial situation of public authorities, which complicates a comprehensive fulfilment of all tasks or even make it impossible.

Destination management in border regions: The example of Lake Constance area

In Europe, cross-border-co-operation in tourism has only been intensified and professionalized since the mid 1980s. This development is strongly related to European regional politics. Particularly the promotion of projects by different collaborative initiatives in order to support the political activities of the European Union in the regions had a positive influence on the cross-border co-operation in general and on tourism in particular.

Due to historical development, cross-border regions are at a disadvantage. A considerable number of problems of border regions are caused by their location at the periphery, away from economic centres, and the consequential neglect in comparison to more central parts of the country. In many border regions, the infrastructure is insufficiently developed, the economical efficiency is on a low level, and cross-border contacts are hindered or even not existent because of different systems of law and administration.

In the late 1950s, a couple of border regions decided to organise their affairs across borders in order to face these problems more effectively. However, an important push was given to the organised cross-border co-operation, particularly in connection with the launch of the European Union's collaborative initiatives. During this period, numerous co-operations and partnerships in different fields had been established.

Cross-border co-operation in tourism did also profit from this development. Particularly since the beginning of the collaborative initiative INTERREG (program of the European Union to promote the cross-border co-operation) in the early 1990s, a number of projects have been initiated and furthered. The main aim of these co-operations was to create and commercialise cross-border activities in addition to national offers and thereby to reach an esteem of the border as a unique selling proposition.

In the Lake Constance area, a couple of years ago, it was decided to follow a new course of cross-border co-operation and to arrange the co-operation more effectively, more sustainable and closer to the market as well as under the aspect of a better coordination of the bearers with the foundation of a destination management company. This chapter illustrates the Lake Constance example and gives a closer view of the implementation process of a destination management strategy in a border region.

For over 200 years, the Lake Constance area has been ranking amongst the most important tourism areas in Europe. Meanwhile, the so-called 'Schwäbische Meer' has become an internationally known destination that attracts millions of tourists every year. Common history, culture, and language make the Lake Constance area a homogeneous area. This accounts for the abundance of cross-border contacts that have been established long before the politically motivated transborder co-operation was sought. At present, there is an indefinite number of cross-border contacts, on both the administrative and the non-administrative level. Significant co-operations covering the entire area include the 'Internationale Bodenseekonferenz' (IBK) the 'Arbeitsgemeinschaft Alpenländer' (ARGE Alp) and the 'Bodenseerat' (which constitutes an Euregio and operates on the non-administrative level).

The cross-border co-operation in tourism has a very long tradition. The 'Verband der Gasthofbesitzer am Bodensee und Rhein' ('Association of Lake Constance and Rhine Innkeepers') was established in 1893, followed by the foundation of the 'Internationaler Bodenseeverkehrsverein' (IBV) ('Lake Constance Tourism Association') in 1902. Both associations have transcended national borders from their outset. Therefore, the augmenting cross-border co-operation on the political level has affected the tourism development only indirectly. The IBK for instance has made efforts to improve the 'Bodenseerundwanderweg' ('Lake Constance Hiking Circuit') and introduced the 'Euregio Bodensee day pass'. However, no commission within the IBK is assigned to its tourism development. Nevertheless, the improved co-operation in numerous fields (e.g. in culture or environmental protection), the increasing political commitment to cross-border co-operation as well as the prospects of European Union subsidies (INTERREG) are the result of the political co-operation and have a positive effect on tourism.

Map 6.1   The international destination Lake Constance

Source: Own illustration, based on: INTERREG III Alpenrhein/Bodensee/Hochrhein

The Lake Constance area as a destination

In a sense, the lake Constance area could be considered a model area for cross-border co-operation in tourism for several reasons: natural attractiveness, comprehensive derived offering, the lake Constance as its common geographical feature, homogenous language and culture, and well-established cross-border co-operation in tourism. Instead of impeding tourism development, the area's internationality has obviously become a unique selling proposition and is one of its most prominent assets. However, even though cross-border co-operation in tourism works well on a local level and on the level of tourism organisations, the lake still is perceived as an obstacle for the co-operation between enterprises. This applies to its economies in general and to its touristy in particular. It seemed to be crucial that an action plan should be elaborated since many years, which also should take into account the prerequisites for a successful destination management.

Moreover, tourism around the lake has not developed uniformly. The German side, the so called 'German Riviera', stands for a traditional recreation and spa destination, while the majority of the Swiss and Austrian destinations offer cater for business, event and short holiday tourism. To combine such diverse business fields within one brand will be one of the major challenges in the near future. After all, the Lake Constance area has not been spared by the harsh economical climate and its effects on travel behaviour. No less than the touristic future of the region is at stake. The fact that the German side was represented by two tourism associations – the 'Tourimusverband Bodensee Oberschwaben (TBO)' (an organisation of Baden-Württemberg state) and the 'Internationale Bodenseeverkehrsverein (IBV)' for the entire cross-border area (D/CH/AT/LI) – called for more efficient structures. After two years of discussions and hearings, the TBO and the IBV decided to merge their operational businesses while keeping their political independence. The foundation of a private organisation brought about a useful degree of depoliticalisation and professionalisation, particularly for marketing tasks. To this purpose, the marketing association 'Internationale Bodensee Tourismus GmbH' was founded in 1997.

The 'Internationale Bodensee Tourismus GmbH (IBT)'

Founded 1997 in Constance, the IBT's business framework was funded by INTERREG II. By 1999, it had fully assumed its operational business. Its only associates are the 'Internationale Bodenseeverkehrsverein' and the 'Tourismusverband Bodensee-Oberschwaben'. Both associations have a relatively lean executive board that is communicating with the IBT. The municipalities remain members of the association, which still receives their membership fees. These contributions are then passed on to the IBT as subsidies for its operational business. The state contribution amounts to approximately 35 per cent of the total IBT budget. The decision-making process within IBT is restricted to their annual meeting; hence, a direct influence on IBT operations is not possible.

The organisational structure of the international Lake Constance destination

The Lake Constance destination is divided into different sub-destinations that are joint to so-called 'Zielgebietagenturen' (IBT 2001, p. 5). These sub-destinations are, on the Swiss side, the destination 'Ostschweiz' (joint to 'Tourismusverband Ostschweiz'), on the Austrian side the destination 'Bodensee-Alpenrhein' (joint to 'Bodensee-

Alpenrhein-Tourismus') and Liechtenstein (joint to 'Liechtenstein Tourismus'). These tourism organisations already existed before the foundation of the IBT. On the German side, three sub-destinations are planned: Allgäu und Oberschwaben, Untersee und Hegau (cross-border sub-destination Germany/Switzerland) and the German shore of Lake Constance (Lindau to Konstanz, including Überlinger See).

The IBT focuses on product management and on the marketing for core businesses. These core businesses are defined by the IBT, and it is only there that the IBT creates products (like the 'BodenseeErlebniskarte' for example, see chapter 3.4) in co-operation with their partners (the enterprises). Moreover, the destination level is a 'hub for marketing and sale' and 'coordinates all marketing activities of the destination' (IBT 2001, p. 29).

The task of the sub-destinations, on the other hand, is to establish their core businesses and competences. The sub-destination Allgäu and Oberschwaben with its special offers in health tourism serves as an example. This sector is supposed to be established and attended by the sub-destination in the first place.

On the local level, the basic offers for tourists have to be established and developed in the first place. Tourism organisations on a local level are supposed to concentrate on visitor care and commitment as well as on product development (e.g. events) (IBT 2001, p. 30).

With the exception of larger places like Constance or Lindau, marketing tasks should only be undertaken by the IBT. This is because communities do not usually have a sufficient marketing budget to establish an effective marketing concept. Therefore, it would be necessary to combine funds (operational, local, regional). Moreover, the IBT is planning to raise more funds by means of campaigns (for example campaigns to extend the season in spring/fall).

At present, this is more of a 'theoretically desirable' distribution of tasks as intended in the marketing master plan. It will depend on the finances whether more tasks can be taken over by the IBT. Especially in the marketing field, a lot of convincing still has to be done.

Terms of reference within IBT

The IBT sees itself as a destination management company even though the offered services cannot be managed centrally. Therefore, the IBT is focussing on achieving 'a binding co-operation between the different tourism organisations and between the most important players' (IBT 2001, p. 21). Its management intends to act as 'a moderator and initiator of processes within the destination that lead to a binding collaboration between the most important players complying with market conformity' (IBT 2001, p. 21). Strategic networks are being established in order to focus on core competences and to operate more closely to the market by pooling resources. Furthermore, especially the marketing activities and brand building are considered IBT's main tasks. In the following, the terms of reference are examined more closely:

Planning: The main planning instrument besides the 'Bodenseeleitbild' is the 'Marketing Masterplan' for the international destination of Lake Constance. Essentially, the IBT provides this 'Marketing Masterplan'. Codetermination is restricted to the marketing committee, in which the sub-destinations is represented. Being a functional planning instrument, the 'Marketing Masterplan' defines marketing objectives, identifies target markets, and determines the instruments for market development (product policy, communication, distribution, market analysis).

Product Policy: Provider networks constitute the basis of IBT's product policy. These provider networks concentrate on the common organisation, development, and marketing of their core businesses. In collaboration with the so-called 'stars' (i.e. the strongest service providers of a destination), the IBT develops different exploitable products in different categories. At the same time, these products are subject to product branding. To the present day, one provider network within excursion tourism was successfully implemented, namely the 'BodenseeErlebniskarte'. There are a great number of excursion opportunities in the Lake Constance area. Thus, it made sense to build a provider network for this field and to develop a product in co-operation with this provider network, which at the same time represents an ideal marketing instrument. For this purpose, the strongest service carriers (measured by the number of visitors) and the so-called 'image carriers' were invited to a round-table discussion. The coordinating process was done exclusively with these parties. However, this platform actually is accessible to all service providers. The objective of the 'BodenseeErlebniskarte' is to pool all excursion sites of the destination in an all-inclusive offering, sharing an electronic chip card system. The visitor purchases an all-in-price card in size of a credit card that allows him to visit any of about 200 sites during three, seven or fourteen days. Besides the 'BodenseeErlebniskarte', there are additional provider networks in the field of accommodation under way.

Marketing: The above mentioned provider networks and their products are one of the most important marketing tools for product policy. Within a few years time, it is planned, to implement more provider networks in the fields of active, health-oriented, cultural, pleasure and business tourism. The IBT's uses different communication channels. An important advertising media is the so-called 'Bodensee Erlebnisplaner'. This travel brochure is a cross-border digest of sights, trips, hotels, etc. Brochures are also applied for advertising provider networks like the 'BodenseeErlebniskarte' and the 'Bodenseehotels'. Additionally, there is a promotion pool used to finance sales promotions such as fairs and roadshows. Public relations is essential both in internal and external communications. Besides press journeys, press information and press conferences, it is essential to inform service providers and to improve the communication within the destination.

Conclusive remarks on the destination management process in the Lake Constance area

The foundation of the 'Internationale Bodensee Tourismus GmbH' in 1997 marked the beginning of the 'destination management age' at the Lake Constance area. The start-up of such an endeavour is a crucial stage as many instances have shown, since there is a need to increase awareness for what destination management is supposed to be and what the differences to traditional instruments are. At Lake Constance, discussions have been intense and they still are. Especially in Germany, the concept of destination management has not been developed as far as in Switzerland and Austria.

Therefore, the foundation of the IBT and their decision to implement a destination management strategy was an important step on the way to their destination's success. Even in an area with over a hundred years of experience in cross-border co-operation, the implementation of a destination management process remains a very complex task considering its coordination and the strategic unification of the touristy key players in four nations. Lake Constance area is neither an own economic area nor a homogenous travel destination. Hence, especially on the level of tourism enterprises adequate structures in cross-border co-operation also need to be established.

However, the success of the 'Bodensee-Erlebniskarte' showed that common cross-border platforms meet a demand since this kind of cross-border co-operation is sustainable because it offers reasonable economic security and it is more than just a shortsighted marketing activity. The area's networks should be able to exploit their key asset of internationality to an increased degree. The IBT also contributed to a better quality and professionalism in marketing than before. However, many projects could not be realized so far. For example, problems arose from the different and incompatible national hospitality booking systems. Another problem is the funding of marketing tasks. The IBT wants to achieve the pooling of all funds available. At least 40 per cent of the touristy budgets (on local, regional and sub-destination s level) are supposed to be used for common communication activities on the superimposed level of 'International Lake Constance Area' (cf. IBT 2001). However, for the time being it is still far from accomplishing this objective. Additional funding is required especially for a successful branding strategy. In the future, this issue will constitute a major challenge.

The brand 'Bodensee' is well known in Germany, Switzerland, and Austria, but the conundrum will be what propositions should be linked to it? Therefore it seems vital to develop the brand 'Bodensee' in a way that all destinations around the lake get the opportunity to benefit from the brand in spite of the diversity of their business fields.

Destination management as a strategy for touristic development in border regions

Particularly in the fields of product policy and marketing, destination management offers solutions to improve the cross-border co-operation in tourism. The highest aim of a destination management strategy is to ensure the enduring competitiveness of the destination.

Thus, cross-border co-operation should act on the following principles:

- Long-term planning and developing of the destination;
- Commitment to cross-border co-operation in tourism on all levels;
- Creation of marketable cross-border products;
- Professional management and marketing;
- Efficient distribution of financial funds, particularly for marketing tasks within the entire destination.

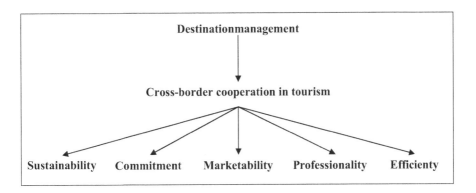

Figure 6.2   Principles of cross-border co-operation

Source: Own illustration

The initial situation varies in each border region. Therefore, no tailor-made strategy for the implementation of a destination management process exists. A shared interest in development within the cross-border destination and a desire for new, more professional solutions to improve the cross-border co-operation are essential.

At Lake Constance, the basis for successful development in tourism is the natural attractivity, which already made it popular 200 years ago. The lake provides a central, unifying geographical feature and the German language a common basis of understanding. These are important reasons for the early forming of communities of interests 100 years ago, acting across borders from the start. This has been the basis for a continuous and stable co-operation on a local and regional level and on the level of tourism organisations until today.

Therefore, the beginning of cross-border co-operation on a political level did not have a major influence on the development of tourism at Lake Constance. The development was ensued bottom-up (bottom-up-strategy) from the beginning, i.e. it was planned and realized in co-operation with the local authorities and the local trade under the aspect of marketability and did not depend on contracts and negotiation efforts between the national partners. The touristic development does therefore not depend on the arbitrariness of political decisions, which is an important condition for long-term touristic development in border regions.

The funding of cross-border projects in tourism has to be seen in this context, too. In many European border regions, cross-border co-operation in tourism is subsidised by the European Union solely and has thus only taken place since the late 1980s, early 1990s. It has often been asked how many projects would have been realized at all without Interreg-funds, the most applied regional development tool of the EU to support cross-border development. Without doubt, the Lake Constance area also benefited from EU subsidies. The IBT office (Internationale Bodensee Tourismus GmbH) was financed by Interreg for example. Nevertheless, the continuation of the co-operation does not only depend on these funds, which is a further condition for a successful long-term co-operation.

With regard to the implementation of a destination management strategy, the project-orientated cross-border co-operation as it took place within the scope of Interreg-programs has further disadvantages. Firstly, the co-operation only takes place during the project itself. There are generally no long-lasting structures, which would allow co-operation after the project has been finished and setting up such structures is not part of the projects, either. Furthermore, there is often an uncoordinated promotion of smaller projects with no or only a small impact on the entire area. For the coordination and realisation of touristic projects and measures in the entire destination, a certain form of institutionalisation is needed. Therefore, an adaptation of existing organisational structures is necessary for efficient destination management.

Organisation

At Lake Constance, the IBT (Internationale Bodensee Tourismus GmbH) was founded on the sound basis of 100 years of association work. As initiator and facilitator, the IBT is supposed to bring a new quality into co-operation in order to improve the development, organisation, and marketing of the destination. Being a private company, the IBT has the advantage of being able to act independently from traditional association structures or even political decisions and more closely to the market. Thus certain fields, especially marketing, can be handled more professionally. However, experience shows that there is no patent remedy for the organisation of cross-border co-operation. Decisions must be taken individually by every region and depend among other things on the current standard of cross-border co-operation. An organisation has to be established which is widely accepted and responsible for the following tasks concerning the entire area:

- Implementation of cross-border co-operations;
- Definition of the core-businesses for the entire area;
- Coordination of task distribution between the entire area, the partial areas and the communities;
- Coordination of marketing activities for the entire area;
- Establishment of a cross-border development and marketing concept;
- Supervision of the common activities.

In many border regions, cross-border co-operation in tourism has not yet developed enough to make the founding of a private umbrella organisation an option. Furthermore, on a national level more or less powerful destinations have been established (e.g. the destination in the German/Austrian Euregio Salzburg/Berchtesgadener Land/ Traunstein or the destination 'Münsterland' in the German/Dutch EUREGIO). An adaptation of structures in favour of the entire area does not make sense in these cases and cannot be enforced anyway. Rather, an already existing organisation, e.g. a regional tourism organisation, has to take the responsibility for coordination as well as for marketing tasks. This in turn requires a perfect co-operation of the partial areas. This will most probably slow down processes and make them less binding. Confidence issues and rivalry will also play a more important part.

Definition of image

Only a few border regions are perceived as a cultural and natural unity. Euregios (and other forms of institutionalisation) are synthetic objects, even if they represent a unity in a historic, cultural, and natural sense and show multifarious functional linkage. Neither the population nor outsiders perceive them as a coherent entity. Thus, they are no tourist destination. This is partly caused by the names of the cross-border regions that are often unsuitable for touristic (marketing) purposes (e.g. EUREGIO, Saar-Lor-Lux-Trier/Westpfalz, Regio TriRhena).

Therefore, the commitment to a shared definition of image is a condition for a successful cross-border planning and development of the destination, for the definition of the core business and for the commercialisation of cross-border products. It is important to know which is perceived as the most central and binding element by tourists and by the players within the destination (e.g. a geographic feature, the natural attractivity in general or certain product fields). The definition of image is simplified if the partial areas have the same or at least similar business fields. But a commitment to image can also be achieved by the definition of common themes (e.g. wine, water) or certain marketable product bundles such as cross-border 'Roads through Vineyards' ('Weinstrassen') for example.

It is essential that the definition of image be widely accepted within the destination. It must carry a content most of the players can identify with. Based on this shared definition of image, strategic provider networks for certain core business fields can be set up, presuming adequate products and a given minimum capacity.

Core business and provider networks

Successful destinations concentrate on certain core businesses. Therefore, the creation of cross-border offers like bicycle- or hiking trails do not yet make a destination. 'In comparison with informally coordinated and organised product selection (...), core businesses can be seen as services and offers that are organised in a specific way for the main forms of vacation' (Bratl/Schmidt 1998, p. 6/3).

At Lake Constance, a concentration on core business such as in the field of excursion tourism has only taken place since the founding of the IBT. A basic condition for this to happen was the formation of strategic provider networks, consisting of the hotel industry, the main providers of leisure activities, and the tourism management. These provider networks can have a positive influence on cross-border co-operation by organizing binding and lasting co-operations and creating a common identity. Furthermore, they develop products that are market-orientated and offer a certain economic security (securing the competitive capacity), and they include partners on the lowest level of the tourism supply chain (provider of principal touristic services).

Compared to the project-specific cross-border co-operation as it especially occurs within the scope of INTERREG-programs, provider networks have different advantages. They create sustainable structures for cross-border co-operation, i.e. the co-operation continues even after the projects have been finished. They integrate the touristic service providers on the lowest level, i.e. a bottom-up-strategy in order to create marketable products, and the geographic extension of provider networks depends on the respective core-business, i.e. the co-operation refers to market fields instead of political areas.

A concentration on 'classic' core business (main forms of vacation) is only possible if the different partial areas have similar business fields. Otherwise, provider networks should be worked out based on the definition of image mentioned above ('product bundles' such as 'Roads through Vineyards' or common themes) in order to achieve a common market position. An example is the border region 'Saar-Lor-Lux-Trier/Westpfalz'. As the entire area hardly shared any products, five top themes were created in co-operation with touristic and political decision makers. Those can be made an issue of and commercialized as central cross-border themes.

The provider networks and their products are an ideal marketing instrument on which a common marketing strategy can be based. These products are based on the shared strong points of the cross-border region and can be commercialized with a common definition of image. They can be used for branding and in order to create cross-border destinations in people's minds. With new cross-border products, new target groups can be reached and the operational area for traditional guest groups can be extended. Thus, the border is no longer an obstacle, but becomes an adventure for tourists and can be seen as unique selling proposition, distinguishing this destination from its national competitors.

Marketing

The basis of a successful marketing strategy is the development of the core business in product development. This can be achieved by uniting the individual key players to a cross-border provider network. A great challenge as well as a great chance, particularly in border regions, is the electronic linking of these partners. Cross-border reservation- and booking systems or electronically linked services such as the 'BodenseeErlebniskarte' have different advantages for the destination management. It is possible to have an active pricing policy, according to the respective utilization and to have influence on the partial outputs, particularly prices and quality. Furthermore, packages, developed from the core business can be commercialized in co-operation with communication and distribution partners and the offers can be placed in international reservation systems.

For visitors, the search for information is simplified. Information becomes available from one single source, and guests can make bookings for the entire cross-border co-operation via Internet. This also strengthens the perception of the border region as an entire touristic destination.

However, also on the national level, the electronic linking of the partners can only be realised in well-organised destinations. For international destinations, additional legal and technical problems have to be considered (e.g. incompatible national reservation and booking systems). Furthermore, marketing networks are of little value if the most important regional key players do not contribute.

For the performance of the marketing tasks and especially for the realisation of an effective branding, extensive financial funds are necessary. The 'Internationale Bodensee Tourismus GmbH' records in the marketing Masterplan that at least 40 per cent of the local, regional and sub-destinational marketing funds should be spent on the level of the entire area (IBT 2001). However, this is still a long way to go. The opportunity for fundraising without a specified intended use is very little. Therefore, in the future there will be attempts to raise more funds with the help of campaigns and initiatives in which the partners can participate.

Conclusions

The implementation of a destination management strategy is a difficult task, not only in border regions. The concept demands intensive stakeholder collaboration. Even on a national level, numerous psychological barriers have to be overcome, which are even more intense in international co-operations. However, destination management also provides an opportunity to achieve a better market position and to distinguish oneself from the national competitors by valuing the overall border region. The implementation of a destination management strategy can arrange the cross-border co-operation more effectively, more professionally, closer to the market and make it more sustainable and more binding. Furthermore, the co-operation on the lowest level, between the immediately affected players, is facilitated. This improves the economic situation of every single player, strengthens the regional awareness, and maybe gets Europe a step closer to the fulfilment of the often-mentioned desire of a 'Europe of regions'.

References

Bieger, T. (1997), *Management von Destinationen und Tourismusorganisationen*, (2nd edn) (München/ Oldenburg).

Bieger, T. (1998), 'Tourismusmarketing im Umfeld der Globalisierung: Aktuelle Herausforderung, innovative Lösungen und neue Strukturen', *THEXIS*, 3-98, 2–13.

Bieger, T. (2002), *Management von Destinationen*, (5th edn) (München/ Wien/ Oldenburg).

Bleile, G. (2000), 'Marktorientiertes Destinationsmanagement erfordert neue Organisationsformen des Tourismus', in M. Fontanari and K. Scherhag, (eds) *Wettbewerb der Destinationen*, (Wiesbaden: 101–109).

Bratl, H. and Schmidt, F. (1998), *Destination Management*, (Wien: Institut für regionale Innovation).

Fontanari, M. (2000), 'Trends und Perspektiven für das Destinationsmanagement – Erste empirische Ergebnisse', in M. Fontanari and K. Scherhag (eds) *Wettbewerb der Destinationen*, (Wiesbaden: 73–93).

Grashoff, C. (2000), *Destinationsmanagement als Steuerungsinstrument regionaler Tourismusentwicklung – dargestellt am Beispiel Ostfriesland*, (Germany: University of Dortmund, Diplomarbeit an der Fakultät Raumplanung).

Internationale Bodensee Tourismus GmbH (IBT) (2001), 'Marketingmasterplan 2002–2006. Für die internationale Destination Bodensee', (4, 7 November 2001).

INTERREG.CH (n.d.), 'Alpenrhein/Bodensee/Hochrhein Transeuropäische Zusammenarbeit zwischen der Schweiz und der Europäischen Union', <http://www.interreg.ch/ir2reg_bodens_d.html>, accessed 22 August 2004.

INTERREG IIIA (n.d.), 'Alpenrhein/Bodensee/Hochrhein Das Programmgebiet im Überblick', <http://www.interreg.org/>.

Iwersen-Sioltsidis, S. and Iwersen, A. (1997), *Tourismuslehre*, (Bern: Paul Hauptverlag).

Kühne, P. (2000), *Destinationsmarketing im schweizerischen Tourismus. Analyse, Entwicklung neuer Ansätze und Anwendung im Falle der Region Nidwalden – Engelberg – Vierwaldstättersee*, (Switzerland: Dissertation, Wirtschafts- und Sozialwissenschaftlichen Fakultät, University of Freiburg).

Müller, H. (1998), 'Widerstände respektieren, Widerstände überwinden – Theorie und praktische Erfahrungen bei der Bildung von Destinationen', (Switzerland: gehalten am Tourismusforum der Alpenregionen (TFA), Flims, Referat vom 9. Februar 1998).

Pechlaner, H. (2000), 'Tourismusorganisationen und Destinationen im Verbund: Produktentwicklung, Markwahrnehmung und Organisationsgestaltung als potentielle Konfliktfelder', in M. Fontanari and K. Scherhag (eds), *Wettbewerb der Destinationen*, (Wiesbaden: 27–40).

Scherhag, K. (1999), 'Destinationsmarken', in *Tourismus Jahrbuch, 1999*, 1, 39.

Spieß, S. (1998), *Marketing für Regionen*, (Wiesbaden: Deutscher Universitätsverlag).

WTO (2001), 'World-Tourism-Organization 2001 Facts and Figures – Statistics', <http://www.world-tourism.org/>, accessed n.d.

Further recommended reading

Bieger, T. (2002), *Management von Destinationen*, (5th edn), (München/ Wien/ Oldenburg).
Bratl, H. and Schmidt, F. (1998), *Destination Management*, (Wien: Institut für regionale Innovation).
DMMA (2004), 'Destination Management Monitor Austria dmma – Innovationsprojekt für ambitionierte Tourismusregionen', <http://www.dmma.at> accessed 22 August 2004.
Fontanari, M. and Scherhag, K. (eds) (2000), *Wettbewerb der Destinationen* (Wiesbaden).
Wachowiak, H. (1997), 'Tourismus im Grenzraum – Touristische Nachfragestrukturen unter dem Einfluß von Staatsgrenzen am Beispiel der Grenzregion Deutschland-Luxemburg', *Materialien zur Fremdenverkehrsgeographie* 38, (Trier).

Part III
Communication
and Information

Chapter 7

Collaborative Stakeholder Planning in Cross-border Regions

The Case of the Great Limpopo Transfrontier Park in Southern Africa

Maha Doppelfeld

Friends of Hwange National Park Trust, Zimbabwe

Introduction

In recent years, trans-boundary parks have become more significant and all parts of the world have parks that cross over international borders, and southern Africa is no exception. The following chapter focuses on Zimbabwe's Gonarezhou National Park, which is being joined with protected areas in South Africa and Mozambique, forming the Great Limpopo Transfrontier Park. The stakeholders in this venture and their collaboration efforts to successfully implement the union of the park to the neighbouring area will be focused on. This first and largest conservation area crosses international borders in Southern Africa and includes numerous stakeholders in each country, who need to cooperate to successfully manage the park. It is therefore of great importance to form a successful partnership, for the three countries involved in the park, which are Zimbabwe, Mozambique, and South Africa.

Specific issues of trans-boundary National Parks

National Parks and nature reserves have been formed since the nineteenth century (Timothy 2000a). The role of National Parks and protected areas is very important, as 'they provide virtually the only areas on the Earth's surface for natural ecosystems to occur with minimum negative human impact' (Eagles 2002:22).

Many National Parks were originally designated without giving much thought on the impact on local people or natural boundaries (Eagles 2002). Eagles (2002) states, that the Park planning challenge is not only to consider the protection of the park from development and unmanaged use, but needs to consider the community adjacent to or within it. National Parks planning and the need for improved sustainable development are intrinsically linked. Eagles (2002) identified two major challenges that are critical in successful National Parks planning. The planning process must be *scientifically and technically complete, inclusive of all stakeholders affected or included in the park, conducted in a timely and efficient manner and results in specific [...] goals for the area* (Eagles 2002:84). Secondly, the outcomes and actions identified must be placed on a firm understanding of the values of the park and their role in society (Eagles 2002).

The formation of political boundaries has caused many ecosystems to be separated and disrupted. This problem has been identified and today many countries with bordering protected areas, have decided to join these areas in order to reverse adverse impacts and improve protected areas. The formation of trans-boundary parks does not only benefit the environment, but also encourages friendly and cooperative relations between countries.

Since the early twentieth century, international parks have been created, mostly in Europe and North America (Timothy 2000a). Poland and Czechoslovakia were the first countries in 1925 to collaborate internationally by creating a border park. Today there are over 136 protected areas in 98 countries that meet across borders and have some form of co-operation (Zbiez 1999, see also Table 7.1). Zbiez (1999) states that nearly ten percent of the world total protected areas are trans-border areas. Thorsell (1990) has identified three primary possible functions of international parks: firstly to promote peace, secondly to protect and manage resources and environments and thirdly to preserve and enhance cultural values, especially of trans-boundary people (Thorsell 1990:21).

Many international parks are formed by the simple fact that two or more protected areas meet at international boundaries. Borderlands have always been favoured for designating protected areas as they often have similar attributes, such as being peripheral, marginal, sparsely populated, underdeveloped, and isolated (Timothy 2001). Recently the improvement of international relations has change the image of many borders as a place 'where people meet and not keep them apart' (Timothy 2001). Borders have become important tourism destinations and international parks are one of the most prominent attractions in border regions (Timothy 2000b). For some tourists to experience the destinations from different sides of the borders is the main appeal (Timothy 2001).

Timothy (2000b) identified three different types of international parks (Figure 7.1). Timothy's first type includes parks that lie directly on the border and function as one unit. The second type includes parks, which lie entirely in one country but are adjacent to a border. These parks can be classified as international when they are managed or financed by two or more adjacent countries. The third type is where two bordering protected areas lie adjacent to each other but are divided by a border and are managed as separate and individual units.

Table 7.1   Internationally adjoining protected areas – 1998

Region	Adjoining Protected Area Complexes	Protected Areas	Proposed complexes	Complexes with 3 countries
North America	8	42	4	0
Central and South America	24	93	15	6
Europe	45	154	26	6
Africa	34	123	12	9
Asia	25	76	12	3
TOTAL	**136**	**488**	**69**	**27**

Source: Zbiez, D. (1999)

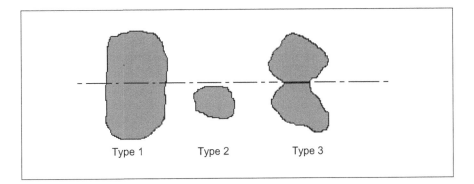

Figure 7.1   Spatial types of international parks

Source: Timothy, D. (2000)

Trans-boundary parks promote conservation, balance, harmony, integration and equity and therefore a more sustainable environment. The other benefit is that it decreases costly and needless duplication of facilities and services (Timothy 2000a). Infrastructure development, marketing and promotional benefits can be shared through joint management (Thorsell 1990, Hamilton *et al* 1998). These efforts can bridge political, cultural, and social gaps (Timothy 2000b), therefore reduce international tension, and promote peace (Thorsell 1990). But co-operation among borders can also have negative effects such as political opportunism, reinforcement of existing power structures, harmful competition, and rivalries among local authorities and bureaucratisation of the partnership process (Timothy 2002a).

Trans-boundary parks and their co-operation have created some very successful partnerships but they also face serious challenges that need to be overcome. Trans-boundary parks face the same problems as other protected areas; such as law enforcement, funding, research, and poaching (Eagles 2002), but due to their location other unique obstacles need to be tackled. Cultural and political differences can create gaps, which are difficult to bridge (Timothy 2000a, Timothy 2000b). This is made even more difficult if language barriers exist. The issue of sovereignty and territoriality is a major obstacle, when countries fiercely protect every fraction of national space (Timothy 2000b). Border disputes have been the cause for many wars and conflicts throughout history (Timothy 2001, Zbicz 1999). In order for a park to be international, the parties' involved, need to give up some degree of sovereignty and be managed by a bilateral body (Timothy 2000b). This is hard to achieve even under friendly conditions. Formalities, such as easy border crossing and customs procedures, need to be negotiated in order for trans-boundary parks to operate smoothly (Timothy 2000a, Timothy 2000b). The level of development on either side of the border, especially if a developing country is adjacent to a developed country, can cause serious obstacles (Timothy 2000a, Timothy 2000b). Disproportion occurs when one country has the resources and knowledge to manage a park but the neighbouring country does not. The location of most trans-boundary parks is marginal and these areas are often neglected in their modernisation, development and infrastructure, which can lead to lack of funding and administrative support (Timothy 2000a, Timothy 2000b).

Through the description of trans-boundary parks above, it becomes evident that the management of these parks is extremely complex and involves many stakeholders. The World Conservation Union (IUCN) and the World Commission on Protected Areas (WCPA) recognise that protected areas should be managed in co-operation, by different stakeholders, such as government agencies, local communities, indigenous peoples and private sector (Hamilton *et al.* 1998:15). Many trans-boundary parks have therefore adopted a process of collaboration, co-operation, or partnership as a model for planning and management of the area. Timothy (2000b) identified five levels of cross border partnerships in tourism (Figure 7.2), ranging from alienation to total integration of the areas.

At the International Symposium on Parks for Peace, which was held in Bormio (Italy) in 1998, the IUCN and WCPA prepared 'Good Practise Guidelines', which included guidelines for collaboration, to assist stakeholders and managers of these parks. These guidelines suggest that stakeholders of bordering countries should collaborate in wildlife, fire, customs and immigration, data and marketing management. Pest control, environmental education, and the monitoring of ecological and socio-economic factors are also important in successful partnerships (Hamilton *et al.* 1998).

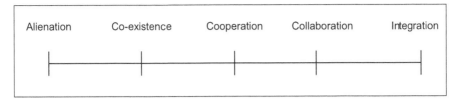

Figure 7.2   Levels of cross border partnership in tourism

Source: Timothy, D (2000b) in Bramwell and Lane (2000:23)

Eagles (2002) states, that the complexity of National Parks planning has been influenced by the accelerated awareness and knowledge of the 'intricate relationship between humans and the environment' and by the 'expanding diversity of groups with interests in parks' (Eagles 2002:74). The importance of stakeholders that are affected and influenced by National Parks has been recognised, and this is even more present in the planning of trans-boundary parks. Therefore, stakeholder collaboration as a major planning tool has emerged for trans-boundary and National Parks planning and management. The issue of stakeholder collaboration and trans-boundary Parks have not been examined in detail, despites its apparent importance in the successful management of trans-boundary parks.

Theories of stakeholder collaboration

Over the last decade, tourism planners and developers have found a new approach to more sustainable planning in the form of stakeholder collaboration, by trying to incorporate the multiple stakeholders involved in the complex and fragmented nature of the tourism industry. Wood and Gray (1991:146) define the collaboration process as:

> Collaboration occurs when a group of autonomous stakeholders of a problem domain engage in an interactive process, using shared rules, norms, and structures, to act or decide on issues related to that domain.

Collaboration involves the interaction of stakeholders, which has the 'potential to lead to dialogue, negotiations and mutually acceptable proposals' for sustainable tourism development (Bramwell & Lane 2000:1). Collaboration can gain competitive advantage as stakeholders combine their knowledge, expertise, capital, and other resources (Bramwell and Lane 2000), which is essential for the planning and management of trans-boundary parks.

In order for collaboration to occur a common issue or 'problem domain' that can only be solved by involving all stakeholders, has to be recognised. Stakeholders initially may have common or differing interest but these interests may change or be redefined during the collaboration process (Wood and Gray 1991). Collaboration often occurs where complex problems are identified and cannot be solved by a single organisation. This process involves joint decision-making approaches to solve the issue or problem, for which all stakeholders take collective responsibility for their action (Selin and Chavez 1995a). Freeman's (1984) definition of stakeholders is used, which is as follows:

> Group(s) or individual(s) that can affect or are affected by an achievement of an organization's objectives (Freeman 1984).

Different requirements, interest, beliefs, principles, and roles characterise stakeholders and the aim of collaboration is to find consensus, which is acceptable to all stakeholders. This can only take place if the applicable and valid participants have been included in the collaboration. Bramwell and Medeiros de Araujo (2000) emphasise the magnitude of correct identification by stating that 'being identified or not being identified, as a relevant stakeholder is an essential first step that affects the whole process of involving participants in collaborative planning as well as the likely outcomes of this planning'. Collaboration in tourism often involves many stakeholders and Sautter and Leissen (1999) have adopted the Stakeholder Map (Figure 7.3) from Freeman (19984) to present some main stakeholders commonly present in a destination.

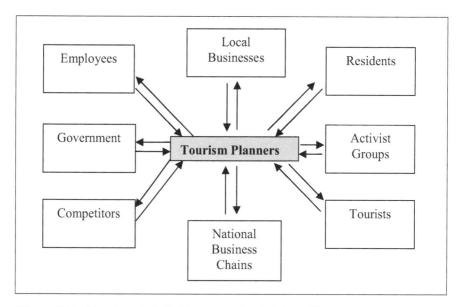

Figure 7.3   Tourism stakeholder map adopted from Freeman

Source: Sautter and Leisen 1999

It is necessary to conduct intensive audits and analysis in order to identify relevant stakeholders. Models to assist and support identification have been explored for this purpose (Freeman 1984, Hall and McArthur 1998, Weiss 1998). Stakeholder audits identify the stakeholders perceived stakes (Freeman 1984), their interest, priorities, and values (Hall and McArthur 1998). An estimate of the relative power of each stakeholder should be measured (Hall and McArthur 1998, Weiss 1998, Mikalsen and Jentoft 2001) in order to identify the stakeholders' relevance within the decision making-process.

The size of the group is of vital importance, as some form of understanding and trust must develop among stakeholders in order to achieve consensus (Medeiros de Arayo and Bramwell 2000). Therefore, the number of working group member need to be limited in numbers and as a result some relevant stakeholders may be excluded. Even with the assistance of stakeholder audits and analysis to achieve inclusion of all relevant stakeholders is often difficult due to power relations, lack of resources and time.

There are problems concerned with collaboration, both in principle and in practice. The perceptions and misconceptions that partners hold about one another, especially when environmental problems related to tourism are a concern, can cause copious conflicts. In collaboration, all stakeholders are assumed to be on equal terms with equal rights to express their views, which are to be considered and respected. The issue of power is probably one of the greatest threats to collaboration (Timothy 2000b, Parker 1999, Mason et al 2000, Hall 2000, Jamal and Getz 2000, Bramwell and Sharman 1999, Reed 2000). Power in the collaboration process is related to the extent to which a group or representative can enforce their will through 'coercion, access to material or financial resources, or normative pressure' (Bramwell and Medeiros de Araujo 2000:276). Power discrepancies make the success of the collaboration process less likely (Parker 1999, Gray 1985). Clegg and Hardy (1996) in Hall (2000) state 'power can be hidden behind trust and the rhetoric of collaboration and can be used to promote vested interests through manipulation of and capitulation by weaker parties' (Hall 2000 pp. 281:282). Power plays an issue right from the commencement of collaboration, when the relevant stakeholders are identified. Some groups or individuals may find it difficult or impossible to gain access to partnerships, even though they have legitimate interests. These differences and inequalities are most visible in cases of cultural extremes, between 'developing and developed economies, urban and rural settings, indigenous and non-indigenous peoples, and between have and have-nots' (Robinson 1999). Robinson (1999) stresses that disparity of power is similarly evident in developed countries where development can be imposed onto marginalised cultural groups without their approval. More problems related to collaboration are stated in Table 7.2.

Table 7.2   Potential problems of collaboration and partnerships in tourism planning

---

- In some places and for some issues there may only be a limited tradition of stakeholders' participation in policy making.
- A partnership may be set up simply as 'window dressing' to avoid tackling real problems head on with all interests.
- Healthy conflict may be stifled.
- Collaboration efforts may be under-resourced in relation to requirements for additional staff time, leadership and administrative resources.
- Actors may not be disposed to reduce their own power to work together with unfamiliar partners or previous adversaries.
- Those stakeholders with less power may be excluded from the process of collaborative working or may have less influence in the process.
- Power within collaborative arrangements could pass to groups or individuals with more effective political skills.
- Some key parties may be uninterested or inactive in working with others, sometimes because they decide to rely on others to produce the benefits resulting from a partnership.
- Some partners might coerce others by threatening to leave the partnership in order to press their own case.
- The involvement of a democratically elected government in collaborative working and consensus building may compromise its ability to protect the 'public interest'.
- Accountability to various constituencies may become blurred as the greater institutional complexity of collaboration can obscure who is accountable to whom and for what.
- Collaboration may increase uncertainty about the future as the policies developed by multiple stakeholder are more difficult to predict than those developed by a central authority.
- The vested interests and established practices of the multiple stakeholders may block innovation.
- The need to develop consensus, and the need to disclose new ideas in advance of their introduction, might discourage entrepreneurial development.
- Involving a range of stakeholders may be costly and time-consuming.
- The complexity of engaging all stakeholders makes it difficult to involve them all equally.
- There may be fragmentation in decision-making and reduced control over implementation.
- The power of some partnerships may be to great, leading to the creation of cartels.
- Some collaborative arrangements may outlive their usefulness, with their bureaucracies seeking to extend their live unreasonably.

Source: Bramwell and Lane 2000:7

Nevertheless, collaboration can have numerous benefits, as the interaction promotes the equity of decision-making and involves the stakeholders equally, which promotes more sustainable development, as all views are attempted to be accommodated. Stakeholders and destinations may achieve a collaborative advantage through realising a successful collaboration process (Bramwell & Lane 2000, Huxham 1996). Due to collaboration, participants can learn from other stakeholders' knowledge, which increases the consideration of other views, thereby creating attitudes that are more constructive.

Table 7.3   Potential benefits of collaboration and partnerships in tourism planning

- Stakeholders may be affected by the multiple issues of tourism development and may well be placed to introduce change and improvement.
- Decision-making power and control may diffuse to multiple stakeholders that are affected by the issues, which is favourable for democracy.
- The involvement of several stakeholders may increase the social acceptance of policies, so that implementation and enforcement may be easier to effect.
- More constructive and less adversarial attitudes may result in consequence of working together.
- The parties who are directly affected by the issue may bring their knowledge, attitudes and other capacities to the policy-making process.
- A creative synergy may result from working together, perhaps leading to greater innovation and effectiveness.
- Partnerships can promote learning about the work, skills and potential of the other partners, and also develop the group interaction and negotiating skills that help make partnerships successful.
- Parties involved in policy-making may have greater commitment to putting the resulting policies into practice.
- There may be improved coordination of the policies and related actions of the multiple stakeholders.
- There may be greater consideration of the diverse economic, environmental and social issues that affect the sustainable development of resources.
- There may be greater recognition of the importance of non-economic issues and interests if they are included in the collaborative framework, and this may strengthen the range of tourism products available.
- There may be pooling of the resources of stakeholders, which might lead to a more effective use.
- When multiple stakeholders are engaged in decision-making the resulting policies may be more flexible and also more sensitive to local circumstances and to changing conditions.
- Non-tourism activities may be encouraged, leading to a broadening of the economic, employment and social base of a given community and region.

Source: Bramwell and Lane 2000:7

Once a partnership, through understanding and learning, has reached consensus, the policies constructed can benefit everyone and therefore create sustainable development. Table 7.3. provides additional benefits resulting from stakeholder collaboration.

Numerous frameworks of collaboration in tourism have been produced to assist practitioners to plan and manage partnerships more effectively. It is surprising that this form of planning has not been explored in greater depth, in relation to trans-boundary partnerships, as it is evident that in such cases stakeholder collaboration and co-operation is inevitable. Furthermore, the creation of trans-boundary parks is becoming more popular all over the world. Timothy (2000a, 2000b) examines in detail the significance of partnerships on an international level.

Many frameworks and models exploring different stages of the collaboration process have been developed. The most prevalent framework found in existing literature is a model developed by Jamal and Getz (1995). Jamal and Getz (1995) based on Gray and Wood (1991) state that collaborative partnerships evolve over time and have identified a three-stage model. The first stage is the problem setting stage, in which key stakeholders and issues are identified. During the second stage – 'direction setting stage', the future collaborative understandings are pooled and recognized. All stakeholders should develop a sense of appreciation for the common purpose of their project. The implementation stage is the last stage, in which the policies or 'shared meanings' of the collaboration process are institutionalised (Jamal and Getz 1995:189). Gray (1996:61–62) argues that these stages are not 'necessarily separate and distinct in practice' but may overlap and recycle back to earlier issues (Gray 1996, Parker 2000).

Caffyn (2000) argues that most authors 'focus on the processes involved in collaboration rather than how they may change as partnerships develop' (Caffyn 2000:202) and aims to create a framework to trace the performance of collaboration over time. She analyses and compares different existing models of life cycles from various fields in the tourism industry. Caffyn (2000) criticises that the final stages of collaboration have not received sufficient attention and proposes that there is either continuation or several 'after-life' options that may be approached by stakeholders.

Parker (2000) identifies that a life cycle approach needs to consider and include regional economic factors, market forces and the overall competitive situation but not as variables that can be controlled by the stakeholders but as elements that can directly influence the collaboration process (Parker 2000:257).

Selin (2000) seeks to accommodate some of the issues that affect the type of collaboration that occurs and has developed a typology for sustainable tourism management. Five primary dimensions are identified, which are the 'geographic scale, legal basis, locus of control, organisational diversity and size and time frame', which were measured through representative partnerships (Selin 2000: 133).

Jamal and Getz (2000) and Ritchie (2000) both examined collaboration cases, which used roundtables composed of different stakeholders with interest and concern for the project. Ritchie's (2000) researched collaboration process used the Interest Based Negotiating technique to assist and ease the procedure. This technique focuses on separating stakeholders from the issue, interests rather than positions, devising alternatives for shared benefits and using objective criteria rather than the force of will of each stakeholder (Ritchie 2000). Jamal and Getz (2000) utilize an interpretative approach. Here the rhetorical, ideological, and power-based facets of interaction of participants involved in the roundtable can provide helpful insight to the collaboration process.

Several authors have come to conclude that the opportunity for learning from the collaboration process has not been utilised and has caused for many collaboration efforts not to reach consensus (Bramwell and Sharman 1999, Reed 2000, Tremblay 2000). The learning process more suitably tackles and deals with power relations that exist. Tremblay (2000) states, that 'participants are partly ignorant of the impacts of their actions' and that a learning approach would minimise this (Tremblay 2000:314). Reed (2000) identifies an adaptive approach towards collaboration that considers the 'turbulent, uncertain, complex and conflicting' factors in tourism planning (Reed 2000:247). This approach uses focused intervention from which unanticipated results supply opportunities for learning. Tremblay (2000) considers an evolutionary interpretation, which requires stakeholders to be adaptable to unpredictable changes during the collaboration efforts. This can be realized by learning about situations and premeditated attempts to discover a range of possibilities for development that may occur. Learning involves a certain amount of experimenting, which implies that mistakes might transpire, but from which the stakeholders will gain a better understanding and more knowledge of the situation (Tremblay 2000). Through learning about and acknowledging the different views, interest and positions of stakeholders, the process of reaching consensus might be eased (Bramwell and Sharman 1999, Reed 2000, Tremblay 2000).

The model of Bramwell and Sharman, is an 'analytical framework to assess whether collaborative arrangements are inclusionary and involve collective learning and consensus-building' (Bramwell and Sharman 1999:392). Three sets of issues are considered in the model.

The first one being the scope of collaboration and is made up of several levels of stakeholder relationships and representation. The extent to which the ranges of participating stakeholders are representative of all relevant stakeholders is considered and to which these see collaboration as a positive benefit. Whether the collaboration has included a facilitator and the stakeholders responsible for the implementation is measured. Another factor concerning the scope is whether the individuals, representing a stakeholder group, are fully representative of that group. The participation technique, which involves the stakeholders, is identified. Finally, the extent to which there is initial agreement among participants about the intended general scope of the collaboration is considered (Bramwell and Sharman 1999:395–397).

The second issue considers the intensity of collaboration among the stakeholders. The degree to which participants, recognize that collaboration is likely to construct qualitatively diverse outcomes and that their own approach to development might have to be adapted, is measured. The importance of regular stakeholder meetings and the extent to which stakeholders receive information and are consulted about the activities of collaboration is considered. Whether information is only disseminated to stakeholders or whether the interaction is direct needs to be acknowledged. Further, the interaction among participants is also essential. The interaction is supposed to 'reflect openness, honesty, tolerant and respectful speaking and listening, confidence and truth' (Bramwell and Sharman 1999:398). The degree of understanding, respect, and learning among participants from their different views, arguments, forms of knowledge, systems of meaning, values and attitudes, in addition are measured. Finally, the influence and control the facilitator exerts over the decision-making process (Bramwell and Sharman 1999:397–399) is considered.

Thirdly, the degree to which consensus emerges is calculated; by assessing when collaboration takes place. An important consideration is whether participants accept that some participants will not agree and embrace eagerly all the resulting policies. The extent to which stakeholders achieve consensus about issues and policies and how the consequences of the policies are assessed and reviewed, is acknowledged. Considerations are made about the extent to which consensus and 'ownership' materialises across the inequalities between stakeholders or reflects these discriminations. Stakeholders need to accept that there are constraints concerning the feasibility of some issues. Furthermore, stakeholders need to be committed to implement the resulting policies (Bramwell and Sharman 1999:399–401).

Background to tourism in Zimbabwe, the Gonarezhou National Park and the Great Limpopo Transfrontier Park

Since the early 1980's, tourism has been the largest revenue and foreign currency earner in Zimbabwe. Tourism planning has been conducted for many years and over the last few years, planning has changed using a more sustainable approach. Especially after the war of independence, the desire evolved to empower poverty-stricken rural communities. A community approach to sustainable tourism development was attempted, in order to improve the life style of the communities, while preserving the environment. The Communal Areas Management Programme for Indigenous Resources (CAMPFIRE) was the forerunner for Community Based Natural Resource Management in southern Africa and the concept was first practiced in 1982. CAMPFIRE programmes educated the local communities on the importance of sustaining the environment, while still receiving benefits through sport hunting and tourism. The stakeholders stated that the difficulties to implement sustainable tourism development were not as severe as in other developing country, due Zimbabwe's tourism market being more exclusive and eco-based. This today, is more difficult to implement due to the current turmoil Zimbabwe is experiencing, which has resulted in many communal areas experiencing increased poverty. This has amplified poaching, deforestation and other detrimental factors to the environment. Zimbabwe's major tourist attraction is its wildlife; therefore, its survival is of utmost importance.

There are 13 major National Parks, varying in size and location to ensure the conservation of Zimbabwe's flora and fauna. Additionally, there are many more Safari Areas locate all over the country, were wildlife management takes place, but controlled sports hunting is permitted. Furthermore in recent years, many farmers have joined their lands together to form Conservancies, were wildlife and environment is preserved with the inclusion of communities, through tourism and sport hunting.

The latest development in conservation has been the creation of trans-boundary parks with neighbouring countries. Eight of Zimbabwe's National Parks are located on or near borders and in total four trans-border parks have been proposed. On the northern border, a park with Zambia and Mozambique is under discussion. The proposed 'Four Corner Park', is located close to the Victoria Falls in Western Zimbabwe and includes areas from Zambia, Namibia, Botswana, and Zimbabwe. Another park is proposed, which includes areas of Mozambique and Zimbabwe, on the eastern border of Zimbabwe. The only trans-boundary park already in existence is the GLTP in the south of Zimbabwe.

South Africa's Kruger National Park (KNP), Mozambique's Limpopo National Park (LNP; formerly known as Coutada 16) and Zimbabwe's Gonarezhou National Park (GNP) have been united by the three countries to form Africa's largest conservation area called the Great Limpopo Transfrontier Park (GLTP). The trans-boundary Park encompasses a total area of 3,577,144 hectares of superb wildlife territory, with stunning geological splendour and abundance in wildlife, fauna, and flora (Great Limpopo Transfrontier Park website 2003). The Park aims to be 'a world-class eco-tourism destination, with extensive private sector involvement, but managed to optimise benefits for biodiversity conservation and economic development of local communities' (Great Limpopo Transfrontier Park website 2003). The GLTP is supported by umbrella areas called Great Limpopo Transfrontier Conservation Areas (GLTFCA) and includes surrounding protected and non-protected areas, which may one day, be included into the park (Integrated Tourism Development Plan, 2002).

A partnership of three unique countries proposes many challenges as they differ in terms of available funds, infrastructure, capacity, and other resources, but also in their approach towards the project. The collaboration and joint management of the park by the three countries is satisfactory, even though disparities and differences exist. Zimbabwe severely lacks funds and infrastructure, but has the advantage of having capacity and a history of exceptional wildlife management. Another problem in Zimbabwe that requires intensive funding is the removal of landmines, which are still present from the independence war more than 20 years ago. These propose severe danger to people and wildlife. Although there are many obstacles that need to be resolved, all stakeholders are confident that they will be deciphered, though it might take longer than expected.

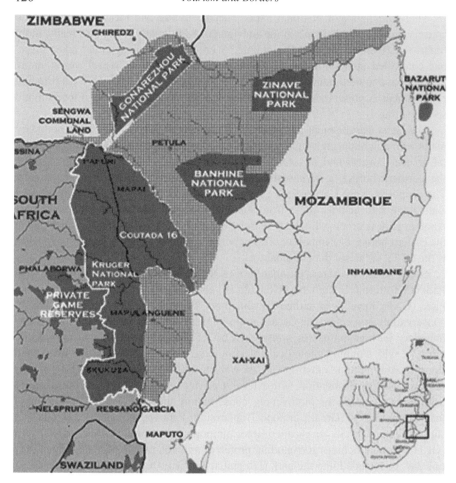

Map 7.1   The Great Limpopo Transfrontier Park and surrounding areas

Source: Great Limpopo Transfrontier Park website, 2003

A meeting between President Chissano of Mozambique and Dr. Rupert (former Southern African President of the WWF) in 1992 was an important catalyst in establishing the Park. The Southern African Development Co-operation (SADC) through the SADC Wildlife Policy and SADC Protocol on Wildlife Conservation and Law Enforcement provided a framework to forming trans-frontier parks. On the 9 of December 2002, after years of planning, meetings, and negotiations, the presidents of the three countries officially signed the GLTP into existence (Great Limpopo Transfrontier Park website 2003).

A vision statement was produced, which is as follows: *To achieve inter-state collaboration in the conservation of trans-boundary ecosystems and their associated biodiversity, promoting sustainable use of natural resources to improve the quality of life of the people of Mozambique, South Africa and Zimbabwe* (Great Limpopo Transfrontier Park website 2003). The aim of the park is to re-establish historical animal migration routes and other ecosystem functions, which have been disrupted. Another major concern is to generate jobs and revenue for the poverty-stricken local communities.

The Gonarezhou National Park is situated in the South Eastern Lowveld of Zimbabwe bordering Mozambique and covers an area of 505.300 hectares. The area offers a broad range of wildlife, exceptional geological features and is renowned for its large herds of elephant and the Chilojo Cliffs. The GNP is less known and overshadowed by more popular visitor destinations in Zimbabwe, such as Victoria Falls and Hwange National Park. Therefore, the attention given to the park for development has been less significant over the last few decades. The infrastructure, such a road networks and accommodation is poor. The GNP does not directly border any of its two neighbouring protected areas in South Africa and Mozambique and therefore a corridor (Sengwe Corridor) linking the GNP to KNP needs to be developed.

Stakeholder collaboration in Zimbabwe and the Great Limpopo Transfrontier Park[1]

The stakeholders expect and perceive to gain numerous benefits from the creation of the trans-border park, as the park ensures better conservation and greater movement of wildlife, which is especially relevant for Mozambique, as its wildlife population is nearly extinct after years of civil war. One of the greatest benefits will be the retention of southern Africa's flora and fauna. The park is the single largest conservation area in southern Africa and is expected to become a major tourist attraction. International exposure through joint marketing concepts is expected to increase tourist arrivals in Zimbabwe and the GLTP regions. From the formation of the park, large economic and financial benefits should be derived for the Zimbabwe and the region. Through the development of and involvement in the park and its surrounding areas, the local communities bordering the areas should derive benefits in the form of revenue and employment. Ample opportunities for private sector investment are present, especially in Mozambique and Zimbabwe, as these areas have very little infrastructure, which must be developed in order to successfully manage the park.

---

[1] This chapter is based on personal interviews with five relevant stakeholders involved on the Zimbabwean side of the Great Limpopo Transfrontier Park. The stakeholders represent four major sectors – the private, the public, the local community sector and the Department of National Parks and Wildlife Management (National Parks), involved in the process.

Although the benefits derived from the park are expected to be numerous, some problems and difficulties will be experienced by Zimbabwe and its partners. The local community sector and National Parks stated that the management method of the GLTP and other trans-border parks under discussion has been a top-down approach. All major decisions have been made at government and international level, which has left the local communities, in particular, as secondary stakeholders and beneficiaries. The local communities and their local authority representative should have been involved from the commencement of the project. However, this problem is being dealt with and the interest and expectations of the communities are now being considered. Research into the awareness of local communities about the GLTP has been conducted. Eighty percent of the local people were knowledgeable about the park and 70 per cent were aware of the community representative committee and its plans for the park. It was asserted that these were very positive numbers, as the area is very marginal and communication often non-existent.

Several problems need to be solved through community involvement to reduce conflict. The corridor that links the GNP and KNP is communal land and inhabited by people, which will have to be resettled to allow free movement of wildlife. The undergoing land distribution in Zimbabwe has resulted in indigenous people claiming sections of and settling in the GNP. In the past, many local communities have been displaced to make way for National Parks and other protected areas, which have caused animosity and suspicion of the National Parks department. This needs to be dealt with more sensitively and it must be ensured that the people are not moved to inferior areas. The local community sector suggests that the land should remain communal land, which will be managed in partnership with National Parks, to ensure the involvement and benefit of and for the people.

Collaboration is not a new theory to Zimbabwe and has been utilised as a planning tool in tourism and other industries before and this form of planning is believed to be the best for this project, as it is participatory and many expertises are utilised and involved, making it a more sustainable development. The public sector asserts that, 'it (collaboration) is an effective form of planning, which can be seen by the results already produced'.

It becomes apparent that the project is in the direction setting stage but 'recycles back' (Gray 1996) to the problem setting stage. Many issues have been resolved and some policies completed, but especially the issue of people living within the park needs to be resolved urgently. No policies are being implemented yet, due to a serious lack of funds. The major obstacle of a lack of funds and resources is later discussed in detail.

The scope of collaboration

Some of the relevant stakeholders involved in the park were identified through workshops, consultation, and other meetings. The problem of not all relevant stakeholders being identified from commencement of the project has caused major obstacles. This supports the statement that workshops pose a threat of exclusion of stakeholders and sustains the need to reduce the group size to a manageable number and reduce power discrepancies. This is especially the case with the local community. Most decisions, such as the formation of the treaty and Joint Management Plan only took place at the ministerial level. Stakeholders confirm that this has been corrected and that all relevant stakeholders are now properly represented. The Community Sector and National Parks pointed out, that a total representation is difficult, as only democratically elected representatives are participating or others are excluded due to being rival stakeholders. The representatives involved, however try to represent their groups views objectively and no individual views are 'being looked after at this stage'.

The government and public sector initiated the project, after being approached by South Africa. The community representative points out that the initiation has been an uncoordinated effort and therefore very time consuming. The relevant stakeholders for implementing policies are involved in the decision-making process.

The stakeholders are confident that benefits will be derived from participating in the project, even though many obstacles and uncertainties need to be overcome. The benefits and problems concerning the establishment and management of the park especially for Zimbabwe were discussed earlier. There is general agreement about economic, social, and environmental factors, which are supported by the aims and objectives of the treaty for the park. The private sector admits that their expectations were unrealistic at first, but this was reversed by a site visit, through which it became evident that there were more, pressing issues to be dealt with first before any private investment could occur. The National Parks representative pointed out that there was agreement about environmental issues, followed by economic and socio-economic issues. The community sector states that agreement was achieved through dialogue and information exchange.

Intensity of the collaboration

Stakeholders participating in collaboration should be aware that it is likely that qualitative different outcome may be reached and that participants may have to modify their approach towards the project. The private sector modified its views, after conducting a site visit and the public sector states that 'here and there you have to give in'. Through consultation, other issues may be become evident, which stakeholders have not considered before and will therefore modify their views. This confirms that collaboration is an ongoing and an always changing process and that interest may change or be redefined.

There is some division among stakeholders' opinions about the involvement of all stakeholders from the inception of the project, as only the private sector believes that this was the case. Not all stakeholders were involved from the beginning but that some were involved as the project progressed. However, at this stage of the project all stakeholders perceive that the relevant stakeholders are involved. This has caused some problems, which supports Bramwell and Meideiros de Aurojo's (2000) statement that the involvement of the correct stakeholders affects the whole process

Table 7.4   Identifying the scope, intensity, and degree of collaboration

Field	Issues considered	Private Tourism Sector	Public Tourism Sector
Scope of collaboration	Extent to which the range of participating stakeholders is representative of all relevant stakeholders	- The local people have been left out at the high level of decision making - Otherwise all other relevant stakeholders are represented	- The public sector states that no stakeholders have been left out and that all participants are relevant and critical - States that communities are involved and are represented by NGOs such as CESVI, SAFIRE and CAMPFIRE
	Extent to which relevant stakeholders see there are positive benefits to entice their participation	- Believes there are great investment opportunities but many problems concerning basic infrastructure and people residing in the area need to be overcome. - Perception that other stakeholders with other interest also see that there are positive benefits	- The stakeholders have sectional interests in which they see benefit but not in other sectors or stakeholders
	Whether the collaboration includes a facilitator and the stakeholders responsible for implementation	- The private sector suggested to Public Tourism Sector, who is the parent ministry of the private sector, to coordinate with other stakeholders - Coordination from this high level is better due to many stakeholders being government or quasi-government agencies	- The government was the initiator, who was assisted by the public sector. - The main stakeholders, being public and public sector, the local community and National Parks are all responsible for implementing
	To extent to which individuals representing a stakeholder group are fully representative of that group.	- At this point of the collaboration there is good representation of the various groups - No individuals looking after their own interest at this stage	- Due to the overall aims and objectives, which give guidance, the representatives represent the interest of the stakeholder group well and no individual interests are represented
	The number of stakeholders involved through the selected participation techniques	- Stakeholders have been identified through workshops and other forms of meetings - The number of total stakeholders involved is unclear	- Stakeholders have been identified through workshops and other forms of meetings - These meetings often take place in Chiredzi which is close to the GNP
	The extent to which there is initial agreement among participants about the intended general scope of the collaboration	- Through a site visit it became clear to the stakeholders that the problems concerning the lack of infrastructure and the local communities still living in the area needs to be resolved first before any sort of economic benefit will be derived - The problem of the lack of financial resources was identified	- Agreement of resulting policies since the inception of the project - Eased as there were aims and objectives, which were laid down in the treaty signed by the three presidents.
Intensity of the Collaboration	The degree to which participants accept that collaboration is likely to produce qualitative different outcomes and they are likely to have to modify their own approach	- As mentioned above some stakeholders modified their views after visiting the GNP - Private sector realised that other issues need to be resolved before they can realise any benefits.	- The public tourism sector states that "here and there you have to give in" to derive overall benefits and satisfy the objectives - Example given that Sengwe people did not want to move to create the corridor but consensus reached through lengthy consultation
	When and how often the relevant stakeholders are involved	- Private sector was involved from the beginning of planning - Private sector stated that all stakeholders are involved equally - No dominating stakeholders	- Not all stakeholders were involved from the inception, some were involved as the project progressed
	The extent to which stakeholder groups receive information and are consulted about the activities of the collaboration	- Stakeholders pass on information frequently to other participants in the partnerships - Private sector states that local community should not be represented at this stage due to the government not having decided on its policy towards them. To include them now would cause more confusion and might cause panic among the local communities	- Information is passed on usually through consultations and workshops - Some policies and issues only get discussed at country level and between government official therefore sometimes unclear how much information they disseminate to other stakeholders - Communities receive information through their representatives (Rural District Councils, chiefs etc)
	Whether the use of participation techniques only disseminate information or also involves direct interaction among the stakeholders	- The stakeholders mainly pass on information through meeting and workshops therefore there is direct interaction among the participants	- Most information gets disseminated through direct interaction, such as consultations and workshops
	The degree to which the dialogue among participants reflects openness, honesty, tolerant and respectful speaking and listening, confidence, and trust.	- Dialogue is respectful, tolerant and candid - Dialogue is open and honest as there are many issues that need to be dealt with	- The public tourism sector states that the interaction among participants is generally smooth but there are some hard feeling as some participants believe their interests are not really taken abroad - Some participants dominate and are more powerful due to having more funding or due to their status
	The extent to which the participants become to understand, respect and learn from each others' different interests, forms of knowledge, systems of meaning, values, and attitudes	- Private sector is hoping for the area to be declared a Tourism Development Zone, which provides tax incentives for tourism investment but has realised that there are more pressing issues that the government needs to deal with first - Therefore through dialogue the stakeholders respect the responsibilities others have	- Participants learn from each other as every stakeholder has an area of specialisation which obviously benefits other players - Each stakeholder has their own area of influence, it is therefore difficult to ignore other stakeholders as the project requires everyone to work together
Degree to which consensus emerges	Whether participants who are working to build a consensus also accept that some participants will not agree or embrace enthusiastically all the resulting policies	- Private sector acknowledges that there may be disagreement but believes the partnership is strong and will find consensus - Participants also have the example of the South African side where policies are working well, through which benefits can be seen	- Through consultation consensus will be derived even if there is disagreement
	Extent to which there is consensus among the stakeholders about the issues, the policies, the purpose of the policies, and how the consequences of the policies are assessed and reviewed.	- Participants try to find solutions to all issues to create policies that are acceptable to all - As no policies have been implemented yet the consequences can not be analysed - Lack of finance has been a major problem, making it difficult for participants to meet, to form policies that are acceptable and to implement the policies	- There are many differences, as decisions have large repercussion on the people involved and some policies may have negative impacts on other people - There is diplomacy of trying to find an agreement that is acceptable to all
	Extent to which consensus and "ownership" emerges across inequalities between stakeholders or reflect inequalities	- The private sector states that there is a high degree of ownership - Participants would be willing to finance the project from their own budget but the financial constraints are to large	- Judging from the response of participants at workshops it has emerged that they "really want it to work". - Therefore if the financial situation would improve would probably be prepared to use the stakeholders own budget
	Extent to which stakeholders accept that there are systemic constraints on what is feasible	- Participants acknowledge that there are constraints and Zimbabwe's case at the moment these are mostly financial - Views are now realistic and consensus can be reached easier as expectations are objective	- Some views may be unrealistic, but through consultation and workshops these have been identified - Therefore today expectations have been balanced through dialogue - Financial resources are a major constraint

Field	Community Representatives	National Parks
Scope of collaboration	- At the beginning consultation only took place at the ministerial level who came up with the treaty and Joint Management Plan - Now there is sufficient involvement and interest	- All stakeholders have been included and have regular meetings. - The coordinator for the transfrontier park visits the communities to consult with them about their role in the park.
	- At this stage all stakeholders perceive a benefit in the park even though there stakeholders are sometimes unsure on how these will be derived	- The stakeholders believe that there is a positive benefit by participating in the partnership
	- Many initiation had been made by various parties with very little common purpose and efforts have been uncoordinated - Therefore, it has been time consuming to come up with a consolidated, clearly defined and accepted management plan. - Government and local authorities are responsible for implementation, who are fully represented - Another community representative states that the government was the initiator	- Project started many years ago. Main initiation came from South Africa and Zimbabwe. - A number of consultancies were conducted to join Mozambique and South Africa with Zimbabwe - National Parks with the help from the government are responsible for implementation
	- Not really possible to include everyone directly and provide total representation - Some important players maybe excluded as there are seen as rival stakeholders - Development structure and management in Zimbabwe, is one of representative democracy - Sufficient representation of the community by NGOs, Rural District Councils, District Committees and chiefs	- It is difficult to achieve total representation especially with communities - In Zimbabwe, most representatives are elected democratically. These representatives should consult with the people they represent.
	- Stakeholders were identified through workshops, which now also take place at local level and include all key stakeholders but not one has taken place were all have been present	- Stakeholders were identified through workshops and consultations between stakeholders
	- At this stage now there is agreement and understanding which was achieved through dialogue as there were many disputed agendas (esp. the Sengwe corridor)	- There was general agreement about the environmental issues, with more agreement about economic issues and socio-economic factors The stakeholders representing the environmental issues have started communicating and planning who then realised that the local people should be included, too.
Intensity of the Collaboration	- The community representative is aware and believes that other stakeholders are too, that some views need to be moderated during the process of collaboration - This has already occurred as the Sengwe local community communicated with other authorities about the management of the corridor and their expectations	- Through consulting with other stakeholder, one might realise other issues and therefore there is a need to modify one approach to the project. - See progress as a maze, which the stakeholders work towards changing into something more formal
	- Not all stakeholder involved from the beginning but now all relevant stakeholders are represented - Time wasted as private sector and communities were involved late - Especially more dialogue needed between the National Parks and the local communities	- The project has a top-down approach and some people had not been informed but this is being corrected now.
	- Not much information disseminated amongst stakeholders. Interviewee gave credit to a NGO supporting the Sengwe community to have passed on info on a regular basis - Otherwise info received at various workshops	- The information disseminated from the top to the lower levels gets filtered. - Due to the top-down approach the info considered necessary only gets passed on to other stakeholders
	- Information mainly exchanged through meetings and workshops - Communities receive information during meetings with local representatives who inform communities on decisions made at higher levels	- Information is mainly exchanged through workshops at the higher level and information is given to the local communities through the Rural District Council representatives
	- Now stakeholder groups are working together to make the Joint Management Plan work - Understanding of each others views as there is sufficient dialogue and communication	- There is still mistrust among stakeholders due to having different means and agendas - The coordinator is putting in every effort to achieve consensus and agreement
	- The participants are learning from each other through dialogue - Before the project stakeholders were ignorant of each others views but now views and interest have been modified to make the project an overall success - There is no dominant stakeholder but some organisations have remained peripheral due to their capacity of resources and status	- Stakeholders learn from each others perceptions, which are expressed through dialogue - The stakeholders with more resources have more power and dominate the discussion and resulting policies
Degree to which consensus emerges	- Stakeholders acknowledge that not all policies will be supported fully, especially between communities and government - Trying to find a common solution when solving problems about policy issues	- If the stakeholders perceive that there is an overall benefit from the project they will support it
	- The stakeholders agree to a large extent as they have a common interest to derive benefits from the park - The policies try to accommodate everyone interest and views - There is wide tolerance and acceptance among participants	- There are still difficulties trying to find consensus for all issues and policies - Policies that are agreeable to all are trying to be achieved, as long as the expectations meet the 3 criteria for setting up the park; ecological, financial and social-economic benefit
	- The stakeholders have shown great interest in the success of the project and work hard for it - Problem of sourcing funds due to the current situation in the country - There is no money to start improving infrastructure, remove mines etc. and has resulted in the private sector not investing in the area - Some expectations especially that of the communities may be unrealistic, which need to be resolved through education, communication and dialogue, in order to balance interests and expectations to achieve consensus	- Stakeholders representing ecological interest often are in disagreement with agricultural representatives. - Financial issues are not so disagreed upon as much - The people in the Sengwe corridor who are to be moved have certain criteria's by which they accept their removal. - The views and expectations of stakeholders at times can be unrealistic often due to unrealistic info provided to them, therefore a large amount of education is needed to provide the correct expectations - Lack of financial resources is a major obstacle

Source: Doppelfeld, 2003, p. 42–44

and the outcomes of the planning process. In order for collaboration to be successful, representatives need to communicate with each other and the groups they represent. Here again are contradictory opinions whether this occurs on a regular basis. The private sector states that information is passed on frequently, but believes that the inclusion of the local community at this stage would cause confusion and panic. The other stakeholders believe that the process has had top-down connotations and most discussion takes place at national level, by government officials. It is unclear what and how much information is disseminated to other stakeholders. However, the community representative gives credit to a NGO, involved in the Sengwe Corridor, for distributing all information regularly and to all participants. Information is mainly distributed when workshops or other meetings take place. The benefit of this form of communication is that it is interactive and participatory, where participants have the chance to present their views. The local community receives their information from Rural District Council representatives, involved in the process. The issue of power becomes evident, as there are different levels of decision-making. Higher-level decision-making stakeholders appear to restrict access to information and therefore restrict the involvement of all stakeholders in collaboration.

The character of dialogue will influence the outcome of the collaboration process, as the style might shut out less powerful participants. Through high-quality dialogue, stakeholders may learn and gain a better understanding of each other views and expectations, which in turn simplifies reaching consensus. There is evidence that mistrust and hard feelings are present, as some stakeholders believe that their interests are not taken seriously, as their agendas differ to the more powerful stakeholders. This may suggest that the stakeholders are not involved at the different levels of the decision-making process and that power discrepancies are in force. The public sector states that interaction during workshops is usually smooth and in addition, the private sector perceives the dialogue to be respectful, tolerant, candid, open, and honest. The National Parks representative, however, asserts that the coordinator for the GLTP 'puts in every effort to achieve consensus and agreement'. Through collaboration and dialogue, stakeholders learn from each other's different interests, forms of knowledge, values, and attitudes, which improve understanding of each other. The private sector acknowledged that their views needed modification and realised that there are more pressing issues that need to be solved before their issues can be considered. This supports Tremblays (2000) statement that learning involves experimenting, during which mistakes might transpire, but from which stakeholders will gain a better understanding and more knowledge of the situation. Before the project started, stakeholders were often ignorant of each other's views, which used to cause conflicts among the different stakeholders. Further, there are some dominating stakeholders due to their status in the decision making process and provide more funds and resources.

The degree to which consensus emerges

The overall objective of the collaboration process is to reach consensus and formulate policies that are agreeable to all stakeholders involved. In Zimbabwe participants are determined to find policies that are acceptable to all and from which benefits are derived. National Parks point out that interest and expectations of all stakeholders try to be accommodated, as long as they meet the criteria for setting up the park. The criteria are ecological, financial, and social-economic benefit for and from the park and its surrounding areas. Some policies have been agreed upon, while other issues still need resolving. The resulting policies, however, have not been implemented yet due to the lack of resources in Zimbabwe. The lack of resources has also hindered frequent meetings among stakeholders to discuss issues, reach consensus and develop policies. The stakeholders understand that not all policies will be agreed upon by all participants, but believe the partnership to be strong enough to reach consensus. Furthermore, the stakeholders perceive that there are overall benefits, which will be derived from the project and therefore support it fully. Due to the very strong support by the stakeholders, a sense of ownership of the project has evolved. The stakeholders 'really want it to work' and are willing to use their own resources to make the park successful.

The views and expectations on what is feasible are often realistic, but impracticable expectation that exists are due to unrealistic information provided to the stakeholder and therefore consultation and education is needed to modify these views. There are constraints on what can be implemented and these are mainly of financial nature. The economic situation has weakened over the last five years and Zimbabwe is unable to fund the project without outside donors. Zimbabwe is, not receiving, however, any foreign funding and South Africa and Mozambique have their own resource issues. This supports Parkers (2000) statement that external factors, such as economic factors, market forces and the overall competitive situation, influence the collaboration process seriously. The lack of funds could become a major threat to the collaboration process, as it can cause a dissatisfaction and frustration among stakeholders. However, the stakeholders are willing, endeavour to implement policies, and will continue to support the project with all efforts.

Many problems and obstacles that make the process of co-operation more difficult have been identified. However, the stakeholders perceive that the benefits, which will be derived from this project, are immense and are therefore determined to make this project and the first trans-boundary Park in Zimbabwe a success.

Conclusions

The major problem of funding for the Great Limpopo Transfrontier Park project exists in Zimbabwe, which hinders the stakeholders to carry out the collaboration process successfully. This obstacle, due to the current political and economical situation of the country, will be difficult to solve and may pose a threat to the success of the park. Zimbabwe is receiving no foreign funding or help in form of resources from South Africa, Mozambique, or other countries. Therefore, the stakeholder need to submit all their efforts to source financial support and will require their own input in the form of funding, staff and any other resources that are needed, to contribute to the success of the project.

The lack of sufficient communication is highly evident, which is not only pointed out by the stakeholders, but also through very contradicting statements concerning several issues. Information is mostly distributed through workshops and meetings, which occur infrequent and should be increased. Further minutes of meetings or information of developments, maybe in the form of a newsletter and minutes, should be considered, in order to keep all stakeholder updated regularly. Especially communication with the local community should be improved, through regular town and village meetings held by the Rural District Council. Here information about any development can be distributed and feedback and input can be received from the local community.

Power relations and equity issues are a great threat to the collaboration and sustainable development to this project. The higher levels of decision-making should interact more frequently and openly, and should reduce the top-down approach to a minimum. This again, is mainly the case concerning the local communities. Issues and problems relating to all people affected by the creation of this park should be solved as early as possible before deciding on and implementing policies, as these are hard to correct if they are detrimental. If the policies are disadvantageous, further problems will be created. This can only be avoided by achieving consensus of all stakeholders and therefore participants should be treated equally in regard to decision-making.

Collaboration as a planning tool is significant and mostly appropriate in trans-boundary park planning and management, due to the many stakeholders involved. In order for policy making to be sustainable and acceptable to all parties affected and involved in trans-boundary parks, all views and interests need to be considered equally. The Great Limpopo Transfrontier Park has recognised the need for stakeholder collaboration and endeavours to include all views and interest in the planning and management of the park. In the case of the Zimbabwe stakeholders, this too, seems to be the case. The stakeholders believe that collaboration as a form of planning to be 'the way forward' for this project. However, there are several obstacles that the GLTP and Zimbabwe needs to overcome, in order to successfully and in a sustainable manner, unite the Gonarezhou National Park to the Great Limpopo Park. The main obstacles identified for the Zimbabwean efforts to collaborate successfully are power and equity issues, lack of funding, and a need to improve communication among stakeholders.

References

Adams, W.M. and Mulligan, M. (eds) (2003), *Decolonizing Nature: Strategies for Conservation in a Post-colonial Era*, (London, UK: Earthscan Publications).

Belrán, J. (ed.) (2000), *Indigenous and Traditional Peoples and Protected Areas: Principles, Guidelines and Case Studies*, (Gland, Switzerland and Cambridge, UK and WWF International, Gland, Switzerland, IUCN).

Brunner, R. and Sovinc, A. (ed) (1999), *Parks for Life: Transboundary Protected Areas in Europe*, (Ljubljana, Slovenia: IUCN/WCPA).

Buckley, R. and Pannel, J. (1990), 'Environmental Impacts of Tourism and Recreation in National Parks and Conservation Reserves' *The Journal of Tourism Studies*, 1, 24–30.

Butler, R.W. and Boyd, S. (eds) (2000), *Tourism in National Parks: Issues and Implication*, (Chichester: John Wiley).

Clarke, M. et al. (1998), *Researching and writing a dissertation in Hospitality and Tourism*, (London: International Thomson Business Press).

Daconto, G. (ed) (2003), 'Proceedings of the Seminar on the Development of Collaborative Management for the Sengwe Corridor: 20–21 April 2003', (Harare: Unpublished Report by CESVI).

Doppelfeld, M (2003), *Collaborative stakeholder planning in cross border regions: The Case of Zimbabwe's Great Limpopo Tansfrontier Park*, University of Brighton, (Eastbourne: unpublished thesis).

Dunbar, Jr. C. et al (2002), 'Race, subjectivity and the interview process', in J. Gubrium and J. Holstein (eds) 2002, *Handbook of Interview Research: Context and Method*, (London: Sage Publications).

Eagles, P.F.J. et al (2002a), *Sustainable Tourism in Protected Areas: Guidelines for Planning and Management*, (IUCN, Gland, Switzerland and Cambridge, UK).

Eagles, P.F.J. (2002b), *Tourism in National Parks and Protected Areas: Planning and Mangement*, (Wallingford: CABI).

Fontana, A. and Frey, J.H. (2000), 'The Interview: from structured questions to negotiated text', in N. Denzin and Y. Lincoln (eds) 2000, *Handbook of Qualitative Research*, (2nd edn) (London: Sage Publications).

Gallusser, W.A. in collaboration with Bürgin, M. & Leimgruber, W. (eds) (1994), *Political Boundaries and Coexistence. Proceedings of the IGU Symposium Basle/Switzerland 24–27 May 1994*, (Berne: European Academic Publisher).

Goodwin, H. (2000), 'Tourism, National Parks and Partnerships', in R.W. Butler and S. Boyd (eds) 2000, *Tourism in National Parks: Issues and Implication*, (Chichester: John Wiley, 245–261).

Goodwin, H. (2002), 'Local Community Involvement in Tourism around National Parks: Opportunities and Constraints' *Current Issues in Tourism*, 5, 3–4, 338–360.

Goodwin, H. and Roe, D. (2001), 'Tourism, Livelihoods and Protected Areas: Opportunities for Fair-trade Tourism in and around National Parks' in *International Journal of Tourism Research*, 3, 377–391.

Great Limpopo Transfrontier Park (2002), 'GLTP Joint Management Plan – January 2002 Draft Document', (Unpublished).

Great Limpopo Transfrontier Park (2003), 'The Great Limpopo Transfrontier Park', <:http://www.gkgpark.com>, accessed 27 January 2003 (Johannesburg).

Griffin, J. et al (1999), *Study on the Development of Transboundary Natural Resource Managment Areas in Southern Africa*, (Washington D.C., USA: Biodiversity Support Program).

Gunn, C.A. (1994), *Tourism Planning: Basics, Concepts and Cases*, (Bristol: Taylor and Francis).

Hall, M.C. & McArthur, S. (1998), *Integrated Heritage Management: Principles and Practices*, (Norwich: Stationary Office).

Hall, M.C. (2000), *Tourism Planning: Policies, Processes and Relationships*, (Harlow: Prentice Hall).

Hamilton, L.S. (1998), 'Guidelines for effective transboundary co-operation: Philosophy and Best Practice, in Parks for Peace Conference Proceedings 1998', (South Africa: International Conference on Transboundary Protected Areas as a Vehicle for International Co-operation held Somerset West from 16–18 September 1997, WCPA).

Hannabuss, S. (1996), 'Research interviews', *New Library World*, 97, 1129, 22–30.

Harrison, D. and Price, M.F. (1996), 'Fragile Environments, Fragile Communities An Introduction', in M.F. Price, (ed.) 1996, *People and Tourism in Fragile Environments*, (John Wiley & Sons Chichester 1–16).

Haywood K.M. (1988), 'Responsible and responsive tourism planning in the community' *Tourism Management*, 9, 105–118.

InWent – IUCN – EUROPARC (2002), *Transboundary Protected Areas: Guidelines for Good Practices and Implementation*, (Zschortau: Unpublished Report by InWent).

IUCN and WCPA (1998), *International Symposium on Parks for Peace in Bormio*, (Gland, Switzerland: Italy 18-21 May 1998, IUCN).

IUCN – The World Conservation Union (2003), *Principles, Criteria and Indicators for Sustainability of Community-Based Natural Resource Management Programmes in Southern Africa*, (Harare, Zimbabwe: IUCN).

Jamal, T.B. and Getz, D. (1994), 'The Environment-Community Symbiosis: A Case for Collaborative Tourism Planning' *Journal of Sustainability*, 2, 3, 152–173.

Jamal, T.B. and Getz, D. (1995), 'Collaboration Theory and Community Tourism Planning' *Annals of Tourism Research*, 22, 1, 186–204.

Joint Working Group and Technical Committee (2002), 'Great Limpopo Transfrontier Park; Background Information Document', (Unpublished).

Jones, B.T.B and E. Chonguica (2001), 'Review and Analysis of Specific Transboundary Natural Resource Management Initiatives in the Southern African Region', *IUCN-ROSA Series on Transboundary Natural Resource Management*, 2.

Katerere,Y. and Hill, R. and Moyo S. (2001), 'A Critique of Transboundary Natural Resource Management in Southern Africa' *IUCN_ROSA Series on Transboundary Resource Management*, 1.

Metcalfe, S. C. (1999), *Study on the Development of Transboundary Natural Resource Management Areas in Southern Africa – Community Perspectives*, (Washington, D.C., USA: Biodiversity Support Program).

Michler, I. (2003), 'Bring down the fences' *Africa Geographic*, March, 2003, 78–85.

Mohamed-Katerere, J. (2001), 'Review of the Legal and Policy Framework for Transboundary Natural Resource Management in Southern Africa', *IUCN-ROSA Series on Transboundary Management*, 3.

Mumma, A. et al (2001), 'Conflict Management and Resolution in Transboundary Natural Resource Management in Southern Africa', (Cape Town, South Africa: Report on the IUCN Regional Workshop held at Cape Manor Hotel, December 12–13, 2001, IUCN, Gland Switzerland and Cambridge, UK).

Murphy, P.E. (1985), *Tourism – A Community Approach*, (London: Routledge).

Nelson, J.G. & Serafin, R. (eds) (1997), *National parks and Protected Areas: Keystones to Conservation and Sustainable Development*, (Berlin: Springer Verlag).

Peace Parks Foundation (2003), 'Africa's Biggest Park take shape', <http:// www.peaceparks. org/content/newsroom/news_pop.php?id=59>, accessed 6 May. 2003 (Johannesburg).

Peace Parks Foundation (2003), 'Great Limpopo Transfrontier Park – current status', <:http:// www.peaceparks.org/content/newsroom/news_pop.php?id=55>, accessed 6 May. 2003 (Johannesburg).

Prentice R. (1993), 'Community-driven tourism planning and residents' preferences', *Tourism Management*, 14, 3, 218–226.

Price, M.F. (ed.) (1996), *People and Tourism in Fragile Environments*, (Chichester: Wiley).

Reed, M.G. (1997), 'Power Relations and Community-based Tourism Planning', *Annals of Tourism Research*, 24, 3, 566–591.

Robinson, M. (1999), 'Collaboration and Cultural Consent: Refocusing Sustainable Tourism', *Journal of Sustainable Tourism*, 7, 3–4, 379–397.

Ryen, A. (2002), 'Cross-cultural Interviewing', in J. Gubrium and J. Holstein (eds) 2002, *Handbook of Interview Research: Context and Method*, (London: Sage Publications).

Sandwith, T. et al (2001), *Transboundary Protected Areas for Peace and Co-operation*, (IUCN, Gland, Switzerland and Cambridge, UK).

Saunders, M. and Lewis, P. and Thornhill, A. (2003), *Research Methods for Business Students 3rd Edition*, (Harlow: Prentice Hall).

Silvermann, D. (2000), 'Analyzing talk and text', in N. Denzin, and Y. Lincoln, (eds) (2000), *Handbook of Qualitative Research 2nd Edition*, (London: Sage Publications).

Simmons D.G. (1994), 'Community participation in tourism planning', *Tourism Management*, 15, 2, 98–108.

Thorsell, J. (1990), *Parks on the Borderline: Experience in Transfrontier Conservation*, (Cambridge UK: IUCN).

Timothy, D.J. (1998), 'Cooperative Planning in a Developing Destination' *Journal of Sustainable Tourism*, 6,1, 52–68.

Timothy, D.J. (2000a), 'Cross-Border Partnership in Tourism Resource Management: International Parks along the US-Canada Border', in B. Bramwell and B. Lane (eds) (2000), *Tourism Collaboration and Partnerships: Politics, Practice and Sustainability*, (Clevedon: Channel View Publications).

Timothy, D.J. (2000b), 'Tourism and international parks', in R. Butler and S. Boyd, (eds) 2000, *Tourism National Parks: Issues and Implications*, (Chichester: John Wiley & Sons).

Timothy, D.J. (2000c), 'Tourism planning in Southeast Asia: Bringing Down Borders through Co-operation', in K.S. Chon (ed) 2000, *Tourism in South-East Asia: A new Direction*, (Birmingham: Haworth Hospitality Press).

Timothy, D.J. (2001), *Tourism and Political Boundaries*, (London: Routledge).

Tosun, C. (2000), 'Limits to community participation in the tourism development process in developing countries', *Tourism Management*, 21, 613–633.

Van der Linde, H. et al (2001), 'Beyond Boundaries: Transboundary Natural Resource Management in Sub-Saharan Africa', (Washington, D.C., USA: Biodiversity Support Program).

Wollmer, W. (2003) 'Transboundary Conservation: The Politics of Ecological Integrity in the Great Limpopo Transfrontier Park', *Journal of Southern African Studies*, 29, 261–278.

Zbicz, D.C. (1999), Transfrontier Ecosystems and Internationally Adjoining Protected Areas, (Durham, USA: IUCN).

Zimbabwe Tourism Authority (2003), 'Gonarezhou – Greater Limpopo Transfrontier Park: Tourism Development Framework', (Harare: Unpublished, Draft Working Document).

Zimbabwe Tourism Authority (2003), 'Report on the Visit to the Gonarezhou National Park (GLTP)' (Harare: Unpublished 18–19 March 2003).

Further recommended reading

Bramwell, B. and Lane, B. (eds) (2000), 'Tourism Collaboration and Partnerships: Politics, Practice and Sustainability', (Clevedon: Channel View Publications).

Bramwell, B. and Sharman, A. (1999), 'Collaboration in Local Tourism Policymaking', *Annals of Tourism Research*, 26, 2, 392–415.

Goodwin, H. (2000), 'Tourism, National Parks and Partnerships', in R.W. Butler and S. Boyd (eds) (2000), *Tourism in National Parks: Issues and Implication*, (Chicheste: John Wiley, 245–261).

Timothy, D.J. (2000b), 'Tourism and international parks', in R. Butler and S. Boyd, (eds) (2000), *Tourism National Parks: Issues and Implications*, (Chichester: John Wiley & Sons).

Chapter 8

Tourism Information and Communication Systems in Border Areas

Technical Issues, Restrictions, and Outlook

Holger Faby

European Institute of Tourism, Germany

Introduction

In tourism, heterogeneous services are primarily spread and commercialized through the communication of information. Herein it is applicable that in everyday life information concerning the issue of »travelling« is almost permanently received in either a conscious or unconscious way and – above all – it is the immediacy, approachability and reliability of these pieces of information that influence the successful commercialization of products in tourism in a positive or negative way (Faby 2004a).

Because of the great advantages of the World Wide Web (WWW) in the tourism business, the customers are enabled to get an easier and more direct access to a very large number of up-to-date information and to do reservations on web clients (Kraak 2003). Not only the global players in tourism recognized this advantage but also the Small and Micro Tourism Enterprises (SMTEs). Since the 1990s, also the SMTEs in Europe and its border areas increasingly make use of the WWW as means of marketing. Thus detailed (= inherent and geospatial decomposition) and actual (= temporal decomposition) information relating to tourism as well as the aspects of leisure can obviously be spread on a high level and comfortably commercialized through additional e-commerce-functions. The WWW opens theoretically the possibility to communicate via borders and to offer tourist information and services. Tourism web services offer the SMTEs in European bordering regions the chance to bridge disadvantages which arise from known bordering situations, periphery structures and rural rooms: The border regions in Europe have long borne the stamp of the continent's history – conventionally isolated from the capital cities, these regions often symbolised the divisions of the old continent and were at the centre of territorial disputes that tore Europe apart.

Almost all of the SMTEs in the European border areas offered tourism information services on the WWW are ending at the borders. This includes in the first way barriers in language and culture.

Proceeding on the assumption that the WWW develops more and more towards a medium of mass-communication, whose users' structure increasingly adjusts to the structure of the total population, this medium will take a continual growing in the tourist branch. By this way, the WWW is also the most recent new medium to present and disseminate geospatial data. Web maps are an ideally additional medium of communication, as information in tourism and leisure – nearly without exception – presents intelligible conception of geospatial information: cartographic information systems (web-maps) in different situations of action serve as means of information and/or as a pursuit in itself, furthermore as a pool of possible solutions concerning problems and actions. This is valid for traditional maps (paper) and – above all – for those cartographic materials applied in multi-media environments. However, for the time being, in practice those Internet applications relating to tourism prevail that integrate maps only on a secondary level of interest. Accordingly, they seize the interactive-communicative possibilities of actual Internet-technologies only rudimentary and often do not match the users' requirements (Faby 2004b).

Meaning of information in tourism

Products in tourism are characterized generally as services. At this, these are immaterial goods in first line, which are judged by the criterion of the service quality and not of the product quality (Corsten 1997, 20 pp. and Moser 1993, 464). These services contain as contact goods performances, which are sold in front of their construction. Because of the complexity and specifications, there are special information and uncertainty problems during the process of the purchase decision.

Tourist purchase decisions are always based on the knowledge of information. Consumers are inclined to procure themselves and process with high risk more information, than with low risk (Chromrik & von der Reith 1997, 21). The information intensity is in addition dependent on the journey way, journeyed distance and the complexity of the payment bundle.

The topicality, the accessibility, and the reliability of tourist information represent altogether a challenge for the tourism: potential travellers must be able to inform individually about tourism services and general basic conditions (infrastructure, culture etc.). For this the customer must be able to query easily attainable, reliable, and understandable information around the clock (and this as economically as possible, in different languages and from different places).

Information represents one of the most essential quality parameters in the service industry »tourism«. The quality of the offered information affects clearly the quality of the relationship between achievement suppliers and customers. The quality and the availability of this information as well as the put media determine as heading parameter the service value of the payment suppliers. These parameters are the most essential success basis of the tourism (Rita 2000; Brown, Emmer & van den Worm 2001, 62).

Meaning of information- and communication systems in Small and Micro Tourism Enterprises (SMTEs)

Information- and communication systems in tourism can be used for two application fields which complete each other: The first field means the use of networked communication Systems with the goal to support tasks in the back office of small and micro tourism enterprises (the communication with neighbouring regions for example to investigate information about tourist offers or booking services, E-mail communication).

As regional variant a special form of the CRS, Information- and Reservations Systems (IRS) are occurred by Small and Micro Tourism Enterprises (SMTEs). The main goal is the storage, representation, and booking of services in tourism. There is the possibility of building up various sale channels, the integration of the information- and booking services into applications in the front- and back office (Faby 2004b): by this way, IRS offer the chance of discharging the staff from tasks in the back office for favour's the tasks in the front office and conception activities. IRS-tools make it possible to work in a real network by using common IRS modules, document presentations, application programs etc. Another aspect is the support of paperless work, dynamic data combination, and real-time communication between all partners in the network and the support of statistical data preparation. The implemented tools allow uniform and automated invoices and are the buttresses of the (cooperative) product development.

The second application field does mean the use of IRS as multiple information platforms for tourists and costumers. They offer the chance to represent and communicate so far fragmented information and offers uniformly and comprehensively. Additional IRS are tools for a modern and cross-regional destination management to offer up to date as well as individually combined information and products by passing on of tourist information and services to the customer with Internet and mobile information services (including Location Based Services – LBS) as well as sale platforms on the WWW. Finally, yet importantly IRS supports the operation of as many as possible distribution platforms and a maximum of (potentially) customers and users can be reached.

The way to the future in tourism information systems: Internet goes mobile

Within the last years, the Internet was discovered by the SMTEs as sale channel. Today there is hardly any SMTE, which does not make tourist information accessible over the Internet. This does not mean that contents and the information quality of all existing information systems on the Web are adapted to the needs of the users. These can be taken quickly to the point: The Information must be presented by a high quality content, in a high temporal and spatial resolution. Further, the applications have to support safe and comfortable on-line booking services to the users.

The use of the Internet in tourism as marketing and booking tool does mean to support the users to prepare journeys (for example from at home or from the job). During the journey or the day trip, most travellers are dependent on traditionally folders, maps and travel literature. Locally up-to-date information is almost only personally available by asking employees of SMTEs.

However, tourism information can also be given as multimedia presentation on different digital platforms. These are able to offer the opportunity for comprehensive information services. On the base of the rapid developments of the telecommunication industry, several handheld devices are available which allows the use of mobile information including Location Based Services (LBS). Mobile information systems in tourism needs – in order to the user desires – to provide information and services (Luley et al. 2004). These are in detail actual tour description and guide functions, the display of route- and tour characteristics and infrastructure, refuges-locations, facilities and open time. Additional there is interest in weather forecast, actual regional weather (including warnings), information about safety and emergency and tools for help for orientation, positioning, and location. These services are called 'Location Based Services' (LBS).[1]

To overcome user desires and to gain the costumers and travellers amazing it is necessary, that the mobile systems have to provide the following information additionally: general tourism information, Points Of Interests (POIs), natural environment, entertainment and events, information about restaurants, hotels, special offers etc. and schedule and transportation.

In the meantime, numerous research projects deal with LBS and telephone providers started commercial services. Most of these projects and services deal with tourism information. The research projects refer for the most part to spatially very restricted areas; frequently they are set up cross-bordered and if they are in the European area, often they are sponsored by EU founds. After conclusion of the promoted project running time prototypes are build – but very often they are not used further because financial and staff resources are missing mostly.

Some of the started commercial services fail to become an economic success. The categories of pitfalls can be overwritten by technical possibilities, legal restrictions, and usability. The failure of commercial LBS can be justified from two perspectives (Navratil & Grum 2005): *Users* define the failure of LBS as 'costs are higher than the benefits'; low or missing benefits of the services include too expensive hardware and/or wrong, bad quality or too expensive information. *Service Providers* define the failure of LBS as 'not profitable': costs for the services are higher than the revenue (this includes indirect revenues like advertisement).

---

[1]   LBS are fully location aware applications. They deliver services through client-server or peer-to peer architectures over cellular, wireless and satellite communication networks (Raper 2005).

Special issues of tourism information systems in border areas

Already the co-operation of SMTEs in different tourist regions within a country does not always work faultlessly. The creation of a coherent information supply, which is provided over the Internet, joins at the cross-border co-operation for example further challenges. In addition, cultural differences and language barriers join next to other legal determinations (data laws etc.); these parameters aggravate the cross-border co-operation.

The simplest result of cross-border marketing in tourism should be to offer common Internet services. Here it applies to arise professionally in several languages. The next step includes offering tourism information for mobile devices and car navigation systems. The example of existing car navigation shows the barriers of cross-border-co-operation. Data – both navigating and the POIs – end at the borders. If the application-provider and the hardware support offboard navigation,[2] the user is enabled to receive new data, which does mean the end of the 'end of data set' (see Table 8.1).

Table 8.1   Advantages and disadvantages of offboard navigation systems

Advantages	Disadvantages
• Maps that are available permanently: central servers can represent whole Europe as navigation net	• Less simple putting into operation
• Current maps which are actualized at least twice in the year	• Costs for the data transmission (roaming costs of the Provider are added in addition in bordering regions), regular costs
• No additionally costs for CD-ROMs/DVDs	• Simple, rudimentary map representation
• Access to current traffic information and POIs etc	
• Low investment costs for the users	

Source: Esters 2005

The respective advantages of the present on- and offboard systems become brought together with this step. Nevertheless, these are sooner technical obstacles or aspects of the mobile communication of tourism information. It is just as important that

---

[2]   Unlike conventional navigation systems (= onboard navigation systems), the route calculation is not carried out at the offboard navigation on the user's terminal but on a central data- and application server. Maps, POIs and traffic information are provided currently and transmit the optimal route to the user's terminal in real-time.

high qualitative tourism POIs and services are integrated into the car and pedestrian navigation systems. The business models must be interesting for the SMTEs and the providers of the services. To this it requires a more extensive cross-border co-operation of the SMTEs.

Conclusions

The success of building up cross-bordered co-operation and networked based services like services on the Internet and mobile information services (including LBS) needs to fulfil two types of measures.

The first type does mean necessary *organizational measures* of the SMTEs in border regions. To let cross-border co-operation arise, particularly in bordering regions the SMTEs have to give up the monopoly and competition thinking. To be able to bridge language barriers the staff have to establish linguistic competences and to get an understanding for cultural differences. Co-operation in border regions must not fail because of restricted horizons of the SMTEs but co-operation should define about the formation and use of networks (also PPP) and supply and call of public conveyances, too. The lasting success of the cross-border co-operation justifies itself in developing and validating business models for e-Commerce and e-Work and solutions and tools for traffic, navigation services, and LBS. To be able to convert these aspects, long-term and cross-border training measures must be offered for the staff of the involved SMTEs. Openness altogether is generated for 'tourism value added services' and a consistent image building.

Besides these measures and the results just mentioned the second type should be labeled as necessary *technical measures* of the SMTEs in border regions. A very important step contains the construction and use of real cross-bordered networked IRS. This presupposes on the one hand side real working interfaces between the various software systems (here: IRS). On the other hand side the creation and the continuous care of high cross-bordered qualitative, heterogeneous, large-scaled and georeferenced information data pools (by including Points Of Interests – POIs) is necessary to be able to offer high-quality information services as argued above.

To not miss the connection to the future, existing and future communication technologies and communication standards must be taken into account by using several information platforms and media just as like the definition of heterogeneous user groups.

This presents a vast field of research for tourism, market research and cartography: Cartographic-scientific aspects ought to be considered highly for the scientific findings drawn from the observations made on the reaction of the users, their expectations and of the users' technical affinities in the examined field. This counts for cartography to be regarded as both the science and the technique of the graphic, communicative, cognitive, and technologic treatment of geospatially relevant information.

References

Brown, A. and Emmer, N. and van den Worm, J. (2001), 'Cartographic Design and Production in the Internet Era: The Example of Tourist Web Maps', *The Cartographic Journal*, 38:1, 61–72.

Chromik, R. and von der Reith, S. (1997), 'Das Reisebüro der Zukunft: Eine technologieunterstützte Kooperation. Modellbetrachtung und empirische Untersuchung' *Trends – Forschung – Konzepte im strategischen Tourismusmanagement*, 10, (Trier).

Corsten, H. (1997), *Dienstleistungsmanagement*, (München).

Esters, D. (2005), 'Offboard-Navigation und Fußgängernavigation auf mobilen Endgeräten', in S. Bruntsch & B. Löcker (ed.) (2005) *Beiträge des Spezialforums Offene Plattformen für Reiseinformationsmanagement und Mobilitätsdienste auf der AGIT 2005 – Symposium und Fachmesse für Angewandte Geoinformatik in Salzburg*, (Seibersdorf: 06. Juli 2005).

Faby, H. (2004a), 'Individuelle Reisevorbereitung mit Internetkarten – Status Quo und Potenziale' *Kartographische Nachrichten*, 54:1, 3–10.

Faby, H. (2004b), 'Untersuchung von kartographischen Medien und Nutzerbedürfnissen als Basis für zielgruppenorientierte touristische Internet-Anwendungen', *Kartographische Bausteine Band* 27, (Dresden).

Luley, P. M. et al. (2004), 'Geo-Data Presentation on Mobile Devices for Tourism Applications', in G. Gartner, (ed.) (2004) *Location Based Services and TeleCartography*, (Vienna: Proceedings of the symposium January 2004, 171–178).

Moser, K. (1993), 'Tourismuswerbung', in H. Hahn and H. J. Kagelmann (ed.) (1993), *Tourismuspsychologie und Tourismussoziologie*, (München: 463–468).

Navratil, G. and Grum, E. (2005), 'What makes Location-Based Services fail?', in G. Gartner (ed.) (2005) *Location Based Services and TeleCartography*, (Vienna: Proceedings of the symposium November 2005, 221–228).

Raver, J. (2005), 'Design constrains on operational LBS', in G. Gartner (ed.) (2005), *Location Based Services and TeleCartography*, (Vienna: Proceedings of the symposium November 2005, 21–26).

Part IV
International
Research

Chapter 9

Academic Contributions on Cross-border Issues in Tourism around the World

A Commentary International Literature Bibliography

Helmut Wachowiak
International University of Applied Sciences Bad Honnef • Bonn, Germany

Daniel Engels
International University of Applied Sciences Bad Honnef • Bonn, Germany

The following chapter intends to provide a structured overview of the existing literature on cross-border issues in tourism. Therefore, it can be used as a convenient starting point for secondary research in this field. Targeting interested scholars, students, and practitioners in tourism, this compilation of academic contributions on cross-border issues in tourism covers destination issues around the world, as well as specific issues from the point of view of different disciplines.

Most literature items deal with the opportunities as well as the problems related to border situations in tourist destinations. In addition, experiences of cross-border, inter-regional, and intra-regional co-operation between tourism stakeholders are covered. Furthermore, theoretical and applied issues are focussed upon, as well as strategic and operational issues in parallel.

To enable the reader to quickly access the desired information and easily skip through passages of minor interest, the chapter is further divided into six sections. The first section of the chapter contains literature that enables the reader to gain a general insight into types, scales, scope and functions of different borders, the social, economic and environmental importance and characteristics of border regions, the emergence of regionalism and the politics of cross-border co-operation.

The following five sections introduce the literature on Europe, the Middle East, Asia, America, and Africa. Each section is furthermore split into subsections on the literature about sub-regions and specific fields of interests (such as politics, socio-cultural affairs, environmental issues, stakeholder co-operation, and more).

A complete bibliography, sorted alphabetically according to the author's surname, is available at http://www.ashgate.com/subject_area/downloads/sample_chapters/

Tourism_and_Borders_Bibliography.pdf, enabling the reader to quickly find a title written by a desired author. In order to maintain the international character of this overview, each title is listed in the language in which it was originally published.

As already highlighted in the Introduction chapter, this bibliography is meant as an introduction to literature around the world on the broad field of border regions as tourist destinations as well as cross-border co-operation between countries to support tourism development. Obviously, it cannot list all titles available and it consequently does not consider itself to be a finite compendium but rather a convenient starting point for secondary research in this academic area. Therefore, readers interested in further studies on cross-border co-operation are advised to make use of the 'snowball effect': most titles and articles listed within this chapter offer various types of extended references themselves, enabling the reader to multiply or even hyphenate the possible resources for investigation into the desired field of interest.

Every commentary literature overview requires a detailed discussion of how it was established and the way in which relevant titles were identified, analysed, and integrated to form the final compilation. In this case, an initial interest in the topic of cross-border co-operation in tourism resulted in continuous secondary research on existing literature over several years. A key finding was the identification of a general gap in international and holistic compilations discussing this topic, which this bibliography aims to narrow. Considering the coherent relationship of cross-border issues in tourism to academic studies related to the geography of leisure, tourism and recreation, an initial search into relevant literature was initiated in the year 2004 in libraries of German universities internationally recognised for their geography and tourism related programs. Those universities and universities of applied sciences were (in alphabetical order) Aachen, Bad Honnef, Bayreuth, Berlin, Dresden, Greifswald, Innsbruck, Kiel, Lueneburg, München, Münster, Regensburg, Saarbruecken, and Trier.

After having identified, analysed, and chosen the initial set of relevant literature and studies, the search was broadened to libraries in the United Kingdom to browse the English literature available on the topic, and included the British Library in London and the University of Brighton's School of Service Management.

Having broadened the initial stock of literature, the so-called 'snowball effect' was employed in a further step to identify the international scope of the topic. Hereby, references and further readings published within the literature already identified were analysed and thus further relevant literature overviewed and evaluated.

At the same time, an international search for organisations and publications relevant to the topic was conducted with the help of the Internet using a set of key words developed from the reading of the literature during the first stages. The key words have been employed in English[1] as well as in German[2] to identify relevant articles, web pages, organisations, and project descriptions around the world.

The same key words were consequently used to search the databases of supranational political entities such as the European Union (EU), the World Tourism Organisation (UNWTO), the Association of South-East Asian Nations (ASEAN), the North-American Free Trade Association (NAFTA) and the Closer-Economic Relations between Australia and New Zealand (CER) for relevant projects and publications relating to cross-border co-operation in international tourism initiatives.

A final step of the bibliographical research for this compilation was based upon academic journals which were scanned in 2004 and 2005 by employing the key words listed above on the search engines embedded in the academic databases of Science Direct and Emerald Insight. Articles incorporated in the final bibliography were mainly taken from papers published in the Annals of Tourism Research, Tourism Geographies, Current Issues in Tourism, the International Journal of Heritage Studies, the International Journal of Tourism Research, International Tourism Reports, the Journal of Hospitality and Tourism Research, the Journal of Applied Recreation Research, the Journal of Sustainable Tourism, the Journal of Tourism Studies, the Journal of Transport Geography, the Journal of Travel Research, the Journal of Travel and Tourism Marketing, Leisure Studies, Tourism Analysis, Tourism Economics, the Tourism and Hospitality Review, Tourism Management, the National Tax Journal, the Journal of Leisure Research, the Asia Pacific Journal of Tourism Research, the Journal of Marketing, the Canadian Tax Journal, the Journal of Borderland Studies, the Journal of Hospitality Management, the Journal of European Public Policy, the International Journal of Urban and Regional Research, the Journal of Cultural Geography, the Journal of the Middle East Studies Society, and the Journal of Asian Economics.[3]

---

[1]   Key words in Englisch: international tourism co-operations, cross-border co-operations in tourism, tourism industries cross-border efforts, cross-border relationships in tourism, tourism cross-border co-operations, cross-broder relationships in tourism, tourism and borders, cross-border co-operations, international cross-border co-operations, tourism across international borders, tourism cross-border studies, cross-border studies, international tourism border studies, tourism border studies, cross-boundary tourism, cross-boundary efforts in tourism, cross-boundary initiatives, cross-boundary tourism initiatives and cross-boundary effects in tourism.

[2]   Tourismus im Grenzraum, Grenzüberschreitende Kooperationen im Internationalen Tourismus, Grenzüberschreitender Tourismus, Kooperationen im Internationalen Tourismus, Europäische Kooperationen im Tourismus, Tourismus im Grenzraum, Grenzen im Tourismus, Touristische Grenzräume, der Grenzraum im Tourismus, Tourismuskooperationen im Grenzraum.

[3]   This list represents the most important academic journals identified for the bibliography as an exhaustive list of all journal reviewed does not appear feasible. All journal contributions can be found in the alphabetical bibliography of all authors at http://www.ashgate.com/subject_area/downloads/sample_chapters/Tourism_and_Borders_Bibliography.pdf.

The searches together resulted in the following compilation of literature related to academic studies, research, and business projects conducted on cross-border issues in tourism. This should create a convenient access point for interested persons and organisations conducting future research on this area of tourism.

Literature on the relationship between borders and tourism

Before investigating specific aspects and geographically unique features of cross-border co-operation in international tourism, it is important to understand the importance of researching and closely defining the relationship between borders and tourism. Based on a conceptual framework developed by Matznetter (1979), Timothy (2001) was able to identify three aspects that facilitate the investigation of this relationship by elaborating on the influence borders impose upon tourism and tourist activities. He identified borders in relation to tourism as being either barriers, destinations themselves, or modifiers of existing tourism landscapes, and consequently capable of influencing tourism development in a positive as well as a negative manner according to their relative function in specific situations.

Tourism has become a major economic factor in a global sense. Smaller countries, in particular, that have poor natural resources and economies dependent on agricultural rather than industrial production, hope to prosper from the manifold economic benefits tourism development promises (WTO, 2005). With the increasing relevance of tourism as an economic activity, border disputes might arise between adjacent sovereign nations due to the desire of individual sovereignties to constantly increase their own share of tourist spending and economic benefits over those received by their neighbours. This applies especially to environmental attractions that are generally not contained within political borders. Border disputes in contested areas, such as the Himalayas, the Amazon rainforest or the Island of Cyprus, that, in general, might prove highly valuable for tourism development, rather counteract economic improvements, though, by halting or even diminishing tourism development.

Even though in this case the border region might be capable of being a destination in itself, the dispute creates a psychological barrier and consequently negatively influences the tourist landscape in the surrounding area. Tourists might feel unsafe through militarised borderlines, or hindered in their movements through extensive bureaucratic border-crossing procedures (Timothy, 2001) and consequently avoid visiting such areas. Obviously, tourism and borders have a very close relationship, and it is this that needs to be carefully examined, analysed, interpreted and positively shaped.

Especially suitable for highlighting the importance of close co-operation between adjacent legislative powers over areas surrounding borders are maritime spaces used for tourism. While, on the one hand, all seas belong to the international community, coastal regions always belong to the coastal zones' sovereign governments who control any economic activities within these maritime areas. Since any tourism operations are ultimately subject to sovereign national policies, which that can differ tremendously from one nation to another, tourism industries operating on open seas might find themselves subject to numerous different national policies while on the same trip. This consequently impedes smooth operations and minimises potential benefits.

Border disputes, therefore, always harm tourism development and can only be overcome through co-operation in combining and sharing existing assets in order to create a single border destination capable of benefiting all participants. This may be just one rather vague example, but it already indicates the importance of co-operation and sustainable development in international cross-border relations with regard to tourism. More detailed examples of successful and less fruitful co-operational issues will be discussed in the literature listed below.

Such co-operation initially requires a thorough understanding of tourism industries, tourism stakeholders and their individual relationships to the various types of borders, before becoming valuable to tourism planners and local populations. It appears to be self-evident, that cross-border co-operation in tourism is easy to comprehend, but rather signify a multi-disciplinary segment that requires detailed studies of stakeholder collaboration theories, sustainable tourism development in rural and peripheral areas, and the geographical and political characteristics of the relationship between tourism and international borders. Considering that especially the latter aspect of tourism studies has not been defined very well to date (Timothy, 2001), an overview of appropriate literature seems to be necessary. Therefore, the compilation within this chapter starts with conceptual contributions: generally explaining border characteristics and functions, the chances they offer and threats they pose to international co-operation in tourism industries.

General contributions of cross-border co-operation to international tourism

Understanding the relationship between borders and international tourism requires a detailed understanding of the concept of borders and their various different shapes, functions and scales. The literature in the first section deals with general descriptions of the various functions of borders, the importance and characteristics of border regions, the social implications related to borderlands and the legal and political peculiarities pertinent to such areas. Universally applicable characteristics and features of international cross-border co-operation are identified and discussed, in order to build a profound basis for understanding the specific regional, national or continental differences applicable to international trans-boundary tourism co-operation.

Firstly, it is important to mention that all political borders are human creations (Timothy, 2001) that through their form and type are not only able, but specifically designed to indicate the relationship a nation-state has with its neighbours. It could be argued that the less visible and enforced a borderline, the friendlier the relationship, making the existence of co-operation across borders more likely. The literature listed in the first sub-section below, therefore, intends to give a general insight into the manifold aspects of cross-border co-operation in tourism. The relationship between borders and international tourism is investigated, best practices and worst case scenarios are discussed, and general considerations about border regions and their importance are examined. Since any form of co-operation requires collaboration between different stakeholders and participants, some literature additionally deals with stakeholder and collaboration theory in tourism (see especially Bramwell and Lane, 2000), as well as sustainable tourism development (see especially Pezolli, 1997). In borderlands that are often signified by their rural and peripheral character and where tourism is regarded as a primary tool for economic development, such theories and concepts are necessary in order to understand the characteristics of tourism planning and development (see especially Butler, 1996; Krakover, 1985; Timothy, 95b; Timothy, 2000a).

Table 9.1   Selected literature: General explanations in cross-border co-operation in tourism

ANZÁLDUA, G. (1987), *Borderlands/la frontera: the new mestiza*, (San Fransisco: Aunt Lute Books).
ASCHER, B. (1984), 'Obstacles to international travel and tourism', *Journal of Travel Research*, 22:1, 2–16.
BLAKE, G. (1993), 'Transfrontier collaboration: A worldwide survey', in A.H. Westing (ed.), *Transfrontier Reserves for Peace and Nature: A Contribution to Human Security*, (Nairobi: United Nations Environment Programme, 35–48).
BLAKE, G. (1994), 'International transboundary collaborative ventures', in W.A. Gallusser (ed.), *Political Boundaries and Coexistence*, (Bern: Peter Lang, 359–371).
BLATTER, J. (1997), 'Explaining crossborder co-operation: A border-focused and border-external approach', *Journal of Borderlands Studies*, 7:1/2, 151–174.
BLATTER, J. (2000), 'Emerging Cross-Border regions as a step towards sustainable development? Experiences and Considerations from examples in Europe and North America': International Journal of Economic Development, <www.spaef.com/IJED_PUB/v2n3/v2n3_4_blatter.pdf>, accessed on 25 March 2003.
BRAMWELL, B. AND LANE, B. (EDS) (2000), *Tourism Collaboration and Partnerships: Politics, Practice and Sustainability*, (Clevedon: Channel View Publications).
BRESLIN, S. AND HUGHS, C. W. AND PHILIPS, N. AND ROSAMOND, B. (2002), *New regionalisms in the global political economy*, (London/NewYork: Routledge).
BUTLER, R.W. (1996), 'The development of tourism in frontier regions: Issues and approaches', in Y. Gradus and H. Lithwick (eds), *Frontiers in Regional Development*, (Lanham, MD: Rowman & Littlefield, 213–229).
BUURSINK, J. (2001), 'The Binational Reality of Border-Crossing Cities', *GeoJourna*, 54:1, 7–19).
BOYLE, R. (1989), 'Partnership in practice', *Local Government Studies*, 15:2, 17–27).
CAFFYN, A. (2000), 'Is There a Tourism Partnerships Life Cycle', in B. Bramwell and B. Lane (eds) *Tourism Collaboration and Partnerships: Politics, Practice and Sustainability*, (Clevedon: Channel View Publications).

CAMBRIDGE SYSTEMATICS, INC. AND TSI CONSULTANTS (2001), *Cross-Border Trade and Travel Study – Final Report and Analysis Results*, <http://www.wcog.org/library/imtc/travelstudy. Pdf>, accessed on 2 February 2003.

CHARPENTIER, J. (1988), *La co-operation transfrontaliere interregionale: Vortrag vor dem Eurpa-Institut in Saarbrücken am 23. Oktober 1987*, (Saarbrücken: Vorträge, Reden und Berichte aus dem Europa-Institut, 123).

CUTLER, B. (1991), 'Welcome to the Borderlands', *American Demographics*, 13:2, 44–57.

DAVIS, G. (1961), *My country is the world: The adventures of a World Citizen*, (London: McDonald's).

DAVIS, G. AND GUMA, G. (1992), *Passport to Freedom: A Guide for World Citizens*, (Washington: Seven Locks Press).

DEMKO, G.J. AND WOOD, W.B. (eds) (1994), *Reordering the World: Geopolitical Perspectives on the Twenty-first Century*, (Boulder, CO: Westview Press).

DEUTSCHER VERBAND FÜR ANGEWANDTE GEOGRAPHIE (1986), *Fremdenverkehr und Freizeit: Entwicklung ohne Expansion. – Ergebnisse der 3. Fachtagung der Regionalen Arbeitsgruppe Saar / Mosel / Pfalz am 21. Juni 1985 in Trier*, (Bochum: Deutscher Verband für angewandte Geographie).

DO AMARAL, I. (1994), 'New reflections on the theme of international boundaries', in C.H. Schofield (ed.), *World Boundaries,Global Boundaries*, (London: Routledge, 1, 16–22).

DONNAN, H. AND WILSON, T.M. (1999), *Borders: Frontiers of Identity, Nation and State*, (Oxford: Berg).

ESKELINEN, H. AND LIIKANEN, I. AND OKSA, J. (1999), *Curtains of iron and gold: Reconstructing borders and scales of interaction*, (Aldershot: Ashgate).

FORSCHUNGSSTELLE FÜR REGIONALE LANDESKUNDE (1983), *Grenzüberschreitende Zusammenarbeit. Pragmatisch oder institutionalisiert? Bericht und Kommentare*, (Flensburg: Dokumentationen zur deutsch-dänischen Grenze, 3)

FRAAZ, K. (1988), *Bewertung der Fremdenverkehrsförderung im Rahmen der Gemeinschaftsaufgabe*, in *Akademie für Raumforschung und Landesplanung* (ed.), *Fremdenverkehr und Regionalpolitik*, (Hannover: ARL, 131–153).

GALLUSSER, W.A. AND BÜRGIN, M. AND LEIMGRUBER, W. (eds) (1994), *Political Boundaries and Coexistence. Proceedings of the IGU Symposium Basle/Switzerland 24–27 May 1994*, (Bern: European Academic Publisher).

GEENHUIZEN, M. V. AND RATTI, R. (eds) (2001), *Gaining advantage from open borders: An active space approach to regional development*, (Aldershot/Burlington: Ashgate).

GETZ, D. (1987), *Tourism planning and research: Traditions, models and futures*, in *Australian Travel Workshop Proceedings of the Australian Travel Workshop*, (Bunburry: Australian Travel Workshop, 407–448).

GETZ, D. (1992), 'Tourism planning and destination life cycle', *Annals of Tourism Research*, 19, 752–770.

GIROT, P.O. (ed.) (1994), *World Boundaries, The Americans*, (London: Routledge, 4).

GRADUS, Y. AND LITHWICK, H. (EDS) (1996), *Frontiers in Regional Development*, (Lanham, MD: Rowman & Littlefield).

GREVEN, H. AND MEYER, B. AND GABBE, J.-D. (1980), *EUREGIO: Modell grenzüberwindender Zusammenarbeit*, (Hannover: Niedersächsische Landeszentrale für politische Bildung).

GRUBER, G. AND LAMPING, H. AND LUTZ, W. AND MATZNETTER, J. AND VORLAUFER, K. (eds) (1979), *Tourism and Borders: Proceedings of the Meeting of the IGU Working Group – Geography of Tourism and Recreation*, (Frankfurt a.M.: Institut für Wirtschafts- und Sozialgeographie der Johann Wolfgang Goethe Universität).

GRUNDY-WARR, C. (ed.) (1994), *World Boundaries, Eurasia* (London: Routledge, 3).

GUNN, C.A. (1994), *Tourism Planning: Basics, Concepts, Cases – 3rd Edition*, (Washington D.C.: Taylor and Francis).

HALL, S. (1991), 'Old and new identities, old and new ethnicities', in A.D. King (ed.) *Culture, globalization, and the world-system: Contemporary conditions for the representation of identity*, (Binghamton: State University of New York at Binghamton).

HALL, S. (1996), 'Introduction: Who needs "identity"?', in S. Hall and P. du Gay (eds) *Questions of cultural identity*, (London: Sage Publications).

HANDELMAN, S. (2000), 'Two nations, indivisible Time', (published on 10 July 2000), pp. 20–27.

HANNERZ, U. (1997), 'Borders', *International Social Science Journal*, 154, 537–548.

HAMILTON, L.S. (1998), *Guidelines for effective transboundary co-operation: Philosophy and Best Practice, in International Conference on Transboundary Protected Areas as a Vehicle for International Co-operation held Somerset West*, (Somerset: WCPA, South Africa from 16–18 September 1997. Parks for Peace Conference Proceedings).

HANSEN, N. (1983), 'International co-operation in border regions: An overview and research agenda', *International Regional Science Review*, 8:3, 255–270.

HANSEN, N. (1986), 'Border region development and co-operation: Western Europe and the US-Mexico borderlands in comparative perspective', in O.J. Martinez (ed.) *Across Boundaries: Transborder Interaction in Comparative Perspective*, (El Paso: Center for Inter-American and Border Studies, University of Texas, 31–44).

HARTSHORNE, R. (1936), 'Suggestions on the terminology of political boundaries', *Annals of the Association of American Geographers*, 26, 56–57.

HASSON, S. AND RAZIN, E. (1990), 'What is hidden behind a municipal boundary conflict?', *Political Geography Quarterly*, 9:3, 267–283.

HAYWOOD, K.M. (1988), 'Responsible and responsive tourism planning in the community', *Tourism Management* 9, 105–118.

HOIVIK, T. AND HEIBERG, T. (1980), 'Centre-periphery tourism and self-reliance', *International Social Science Journal*, 32:1, 69–98.

HOLDICH, T.H. (1916), *Political Frontiers and Boundary Making*, (New York: Macmillan).

HOUSE, J.W. (1981), 'Frontier studies: An applied approach', in A.D. Burnett, and P.J. Taylor, (eds) *Political Studies from Spatial Perspectives*, (New York: Wiley, 291–312).

HUDMAN, L.E. (1978), 'Tourist impacts: The need for planning', *Annals of Tourism Research*, 5, 112–125.

HUDSON, B.M. (1979), 'Comparison of current planning theories: Counterparts and contradictions', *Journal of the American Planning Association*, 45, 387–398.

INSKEEP, E. (1991), *Tourism Planning: An Integrated and Sustainable Development Approach*, (New York: Van Nostrand Reinhold).

JAMAL, T.B. AND GETZ, D. (1995), 'Collaboration theory and community tourism planning', *Annals of Tourism Research* 22, 186–204.

KABES, V. (1987), 'Der Griff über die Grenze – Tourismus und Völkerverständigung', *Zeitschrift für Fremdenverkehr*, 42:3, 8–13.

KALRA, V.S. AND PUREWAL, N.K. (1999), 'The strut of the peacocks: Partition, travel and the Indo-Pak border', in R. Kaur and J. Hutnyk (eds) *Travel worlds: Journeys in contemporary cultural politics*, (London: Zed Books, 54–67).

KAUR, R. and Hutnyk, J. (eds) (1999), *Travel worlds: Journeys in contemporary cultural politics*, (London: Zed Books).

KRAKOVER, S. and Gradus, Y. (eds) (2002), *Tourism in frontier areas*, (Lanham, MD: Lexington Books).

LANG, W. (1989), 'Die normative Qualität grenzüberschreitender Regionen: Zum Begriff der "soft institution"', *Archiv des Völkerrechts*, 27, 253–285.

LEIMGRUBER, W. *(1998),* 'Defying political boundaries: Transborder tourism in a regional context' *Visions in Leisure and Business*, 17:3, 8–29.

LOPEZ, B. (1989), *Crossing Open Ground*, (New York: Vintage Books).

LORENZ, T. AND STOKLOSA, K. (N.D.), 'Bibliographie zur Grenzregion' http://www.wsgn.uni-ffo.de/bibliographie.pdf, accessed on 23 February 2005.

LUNDBERG, D.E. AND LUNDBERG, C.B. (1993), *International Travel and Tourism – 2nd Edition*, (New York: Wiley).

MARTINEZ, O.J. (ed.) (1986), *Across Borders: Transborder Interaction in Comparative Perspective*, (El Paso: Texas Western Press).

MARTINEZ, O.J. (1994), 'The dynamics of border interaction: New approaches to border analysis', in C.H. Schofield (ed.) *World Boundaries, Vol. 1, Global Boundaries*, (London: Routledge, 1–15).

MATZNETTER, J. (1979), 'Border and tourism: Fundamental relations', in G. Gruber and H. Lamping, and W. Lutz and J. Matznetter and K. Vorlaufer (eds), *Tourism and Borders: Proceedings of the Meeting of the IGU Working Group – Geography of Tourism and Recreation*, (Frankfurt a.M.: Institut für Wirtschafts- und Sozialgeographie der Johann Wolfgang Goethe Universität, 61–75).

MINISTRY OF THE INTERIOR (1974), *Instructions Concerning Movement and Stay in Frontier Zone*, (Helsinki: Ministry of the Interior, Headquarters of the Frontier Guards).

MORETTI, G. (1983), 'Die Dynamik des Fremdenverkehrsflusses in Europa', in author unknown, *Fremdenverkehr in Europa*, (Göttingen, 81–87).

NORTHERN DIMENSION FORUM (n.d.), 'Regional and Cross-border co-operation', <http://www.ndforum.net/?cat=3&sub=15> , accessed on 11 February 2003.

O'BYRNE, D.J. (2001), 'On passports and border controls', *Annals of Tourism Research*, 28:2, 399–416.

O'MAOLIN, C. (2000), *North-South co-operation on Tourism*, (Armagh: The Centre for Cross Border Studies).

OHMAE and KENICHE (1993), 'The rise of the region state', *Foreign Affairs*, 72, 78–87.

PERKMANN, M. and SUM, N. (eds) (2002), *Globalization, regionalization and cross-border regions*, (Hampshire: Palgrave Macmillan).

PEZOLLI, K. (1997a), 'Sustainable Development Literature: A Transdisciplinary Bibliography', *Journal of Environmental Planning and Management*, 40:5, 575–601.

PEZOLLI, K. (1997b), *Sustainable Development: A Transdisciplinary Overview of the Literature*, (Journal of Environmental Planning and Management. Vol. 40, No. 5, pp. 549–574).

PRESCOTT, J.R.V. (1987), *Political Frontiers and Boundaries*, (London: Allen and Unwin).

RATTI, R. and REICHMANN, S. (1993), *Theory and Practice of Transborder Co-operation*, (Frankfurt am Main).

RICHARD, W.E. (1993), 'International planning for tourism' *Annals of Tourism Research* 20, 601–604).

RINGER, G. (1998), 'Introduction', in G. Ringer (ed.) *Destinations: Cultural Landscapes of Tourism*, (London: Routledge, 1–13).

RUMLEY, D. AND MINGHI, J.V. (1991), 'Introduction: The border landscape concept', in D. Rumley and J.V. Minghi (eds), *The Geography of Border Landscapes*, (London: Routledge, 1–14).

RUMLEY, D. AND MINGHI, J.V. (EDS) (1991), *The Geography of Border Landscapes*, (London: Routledge).

SCHOFIELD, C.H. (ed.) (1994), *Global boundaries, world boundaries*, (London/NewYork: Routledge, 1).

SLUBICE REGIONAL GOVERNMENT (2002), 'Cross-Border Co-operation', <http://www.slubice.pl/english/wspolpraca.html>, accessed 11 February 2003.

SMITH, G. and MALKIN, E. (1997), 'The border' *Business Week*, (published on 12 May 1997), pp. 64–74.

STEINECKE, A. (ed.) (1981), *Interdisziplinäre Bibliographie zur Fremdenverkehrs- und Naherholungsforschung. Beiträge zur allgemeinen Fremdenverkehrs- und Naherholungsforschung. Berliner geographische Studien. Band 9*, (Universität Berlin).

SUPPER, F. (1983), 'Der Beitrag der Städtepartnerschaften zur Entwicklung des lokalen Fremdenverkehrs', *Fremdenverkehr in Europa*, 89–96, (Göttingen).

TENHIÄLÄ, H. (1994), 'Cross-border co-operation: Key to international ties', *International Affairs*, 6, 21–23.

TIMOTHY, D.J. (1995a), 'International boundaries: New frontiers for tourism research', *Progress in Tourism and Hospitality Research*, 1:2, 141–152.

TIMOTHY, D.J. (1995b), 'Political boundaries and tourism: Borders as tourist attractions', *Tourism Management* 16, 525–532.

TIMOTHY, D.J. (1998a), 'Collecting places: Geodetic lines in tourist space', *Journal of Travel and Tourism Marketing*, 7:4, 123–129.

TIMOTHY, D.J. (1998b), 'Cooperative tourism planning in a developing destination', *Journal of Sustainable Tourism*, 6,:1, 52–68.

TIMOTHY, D.J. (1998c), 'International boundaries and tourism: Themes and issues', *Visions in Leisure and Business*, 17:3, 3–7.

TIMOTHY, D.J. (2000a), 'Borderlands: An unlikely tourist destination?', *Boundary and Security Bulletin*, 8:1, 57–65.

TIMOTHY, D.J. (2001), 'Tourism in the borderlands: Competition, complementarity and cross-frontier co-operation', in S. Krakover and Y. Gradus (eds), *Tourism in Frontier Areas*, (Baltimore, MD: Lexington Books, 233–258).

TIMOTHY, D.J. (2002), 'Tourism in borderlands: Competition, complementarity, and cross-frontier co-operation', in S. Krakover and Y. Gradus (eds), *Tourism in frontier areas*, (Lanham, MD: Lexington Books, 233–258).

TIMOTHY, D.J. (2003), *Tourism and Political Boundaries*, (London/New York: Routledge Advances in Tourism, Routledge Taylor & Francis Group).

TIMOTHY, D.J. AND MAO, B. (1992), *Tourism in international exclaves*, (Paper presented at the annual meeting of the Canadian Association of Geographers, Ontario Division, Scarborough, Ontario, 31 October).
TIMOTHY, D.J. AND TEYE, V.B. (2004), 'Political boundaries and regional co-operation in tourism', in A.A. Lew and C.M. Hall and A.M. Williams (eds), *A Companion to Tourism*, (London: Blackwell, 584–595).
TIMOTHY, D.J. AND PRIDEAUX, B. AND KIM, S.S. (2004), 'Tourism at borders of conflict and (de)militarised zones', in T.V. Singh, (ed.) *New Horizons in Tourism: Strange Experiences and Stranger Practices*, (Wallingford, UK: CAB International, 83–94).
VON BÖVENTER, E. (1969), 'Walter Christaller's central places and peripheral areas: The central place theory in retrospect', *Journal of Regional Science*, 9:1, 117–124.
WALLASCH, K. (2002), 'Das Trinser Konzept grenzüberschreitender Kooperationen, in TRANS', Internet-Zeitschrift für Kulturwissenschaften. Nr. 12/2002, <www.inst.at/trans/12Nr/wallasch12.htm>, accessed on 19 October 2003.
WEST, J.P. AND JAMES, D.D. (1983), 'Border tourism', in E.R. Stoddard and R.L. Nostrand and J.P. West (eds), *Borderlands Sourcebook: A Guide to the Literature on Northern Mexico and the American Southwest*, (Norman: University of Oklahoma Press, 159–165).
WEST, R. (1973), 'Border towns: What to do and where to do it', *Texas Monthly* 1, 62–73 and 109.

Social aspects of border regions and cross-border co-operation

As will be discussed in detail later on, borders, today, often function as divisions between the cultural entities of commonly shared values, norms, languages and identities that signify an innate community. As such, they enclose an area generally inhabited by a specific people that are constantly surrounded by peers sharing similar attributes of a national identity. Such identities are clearly visible in central areas far away from any boundaries, where anthropologists and sociologists are easily able to examine sovereignties' common histories and understand an individual's relationships with peers and foreigners. Nevertheless, what about those national citizens living in more remote and peripheral areas such as border regions? Living conditions there are influenced not only by people's own national features, but also heavily by those of their neighbours. This, in turn, might influence the formation of identity for a citizen, as well as value perception. Even though the perception of borders has changed over the years, it is still possible to define specific common geographical features and attributes of borders.

According to Timothy (2001), it is important to investigate the scale of borders and elaborate on the level of impact a specific border imposes upon human interactions. He identifies three larger scales, namely international borders, sub-national borders, and third-order, or lower-level borders. Whereas international borders indicate a language and cultural barrier and consequently are capable of imposing a high impact upon human interactions, sub-national borders rather divide specific folklore existing within a common national identity and are rather signified by legal barriers. As such, sub-national borders impose less impact upon human interactions since, in general, crossing such a border only imposes very little change in a perceived environment, and languages and customs primarily re main identical. The lowest impacts upon human interactions impose third-order boundaries that divide cities or counties where people are often unaware of the actual crossing of such borderlines. With increasing global literacy and education and ongoing globalisation, such impacts might change in the future, and the actual impact of international borders might weaken due to increased familiarity with other customs and languages. This situation is visible in part, already, on investigation of the social attributes of living in border regions in which two or more cultures merge to form a distinct border identity.

The unique social aspects that are perceivable among people inhabiting frontier zones, form the topic of the literature listed within the following sub-section. This attempts to investigate, identify and examine the existence of a border identity, common to all those people living in and near frontier zones, that is applicable generally. Psychological determinants and perceptions of geographical as well as more psychological forms of borders (see Scott, 1995), their perception in the minds of the local populations, their influence on spatial behaviour and the importance of territoriality for an individual's identity formation (see especially Wilson and Donnan, 1998) are just a few examples of issues related to borderland tourism and currently being discussed and investigated in international academia.

Table 9.2   Selected literature: Social aspects of cross-border co-operation in tourism

ATLAS OF POPULATION AND ENVIRONMENT (2000), 'Migration and Tourism', <http://www. ourplanet.com/aaas/pp./population05.html> , accessed on 3 April 2003.
BACHLEITNER, R. and SCHIMANY, P. (eds) (1999), *Grenzenlose Gesellschaft – grenzenloser Tourismus*, (München: Profil Verlag Tourismuswissenschaftliche Manuskripte, 5.).
BECKER-SCHULTES, J. (1983), 'Auswirkungen einer "partiell-offenen" Grenze auf das Verhalten der Bevölkerung – das Beispiel der Gemeinden Lichtenberg und Schirnding in Oberfranken', *Arbeitsmaterialien zur Raumordnung und Raumplanung* 26, (Bayreuth).
BOYD, S.W. (1997), *The impact of human disaster on tourism destination regions*, (Northern Cyprus: Paper presented at GAU International Tourism Conference on Challenged Tourism, Girne, 3–7 December).
D'AMORE, L.J. (1983), 'Guidelines to planning in harmony with the host community', in P.E. Murphy (ed.) *Tourism in Canada: Selected Issues and Options*, (Victoria, BC: University of Victoria, Department of Geography, 135–159).
DE KADT, E. (1979), 'Social planning for tourism in the developing countries', *Annals of Tourism Research* 6, 36–48.
DI MATTEO, L. (1999), 'Using alternative methods to estimate the determinants of cross-border trips', *Applied Economics* 31:1, 77–88.
EADE, J. (ed.) (1997), *Living the Global City*, (London: Routledge).

EDWARDS, J.N. AND FULLER, T.D. AND VORAKITPHOKATORN, S. (1994), 'Why people feel crowded: An examination of objective and subjective crowding', *Population and Environment*, 16:2, 149–160.

KNIGHT, D.B. (1982), 'Identity and territory: Geographical perspectives on nationalism and regionalism', *Annals of the Association of American Geographers*, 72, 514–531.

KNIGHT, D.B. (1994), 'People together, yet apart: Rethinking territory, sovereignty, and identities', in G.J. Demko and W.B Wood (eds), *Reordering the World: Geopolitical Perspectives on the Twenty-first Century*, (Boulder, CO: Westview Press, 71–86).

KORTEN, D.C. (1981), 'Management of social transformation', *Public Administration Review*, 41, 609–618.

LEIMGRUBER, W. (1989), 'The perception of boundaries: Barriers or invitation to interaction?', *Regio Basiliensis*, 30, 49–59.

MAIER, M.A. (1992), 'Border jumpers', *Hispanic* 5:9, 70.

MARKS, I.M. (1987), *Fears, Phobias, and Rituals: Panic, Anxiety, and Their Disorders*, (New York: Oxford University Press).

MASLOW, A. (1954), *Motivation and Personality*, (New York: Harper and Row).

MATHIESON, A. AND WALL, G. (1982), *Tourism: Economic, Physical and Social Impacts*, (London: Longman).

MATLEY, I.M. (1977), 'Physical and cultural factors influencing the location of tourism', in E.M. Kelly (ed.) *Domestic and International Tourism*, (Wellesley, MA: The Institute of Certified Travel Agents, 16–25).

MINGHI, J.V. (1963b), 'Television preference and nationality in a boundary region', *Sociological Inquiry* 33:2, 165–179.

MINGHI, J.V. (1991), 'From conflict to harmony in border landscapes', in D. Rumley and J.V. Minghi (eds), *The Geography of Border Landscapes*, (London: Routledge, 15–30).

RYAN, C. (1991), *Recreational Tourism: A Social Science Perspective*, (New York: Routledge).

RYDEN, K.C. (1993), *Mapping the Invisible Landscape: Folklore, Writing, and the Sense of Place*, (Iowa City: University of Iowa Press).

SACK, R.D. (1986), *Human Territoriality: Its Theory and History*, (Cambridge: Cambridge University Press).

SCHILLING, H. (1986), 'Leben an der Grenze', *Forschung Frankfurt*, 3:4, 28–33.

SCOTT, J. (1995), 'Sexual and national boundaries in tourism', *Annals of Tourism Research*, 22, 385–403.

SOBOL, J. (1992), 'Life along the line', *Canadian Geographic*, 112, 46–56.

TAVRIS, C. AND WADE, C. (1995), *Psychology in Perspective*, (New York: Harper Collins).

THORPE, G.L. AND OLSON, S.L. (1997), *Behavior Therapy: Concepts, Procedures, and Applications, 2nd Edition*, (Boston: Allyn and Bacon).

WACKERMANN, G. (1979), 'Projection socio-culturelle du tourisme et isochrones moyens en espace frontalier', in G. Gruber and H. Lamping and W. Lutz and J. Matznetter and K. Vorlaufer (eds), *Tourism and Borders: Proceedings of the Meeting of the IGU Working Group – Geography of Tourism and Recreation*, (Frankfurt a.M.: Institut für Wirtschafts- und Sozialgeographie der Johann Wolfgang Goethe Universität, 295–307).

WILSON, T.M. AND DONNAN, H. (EDS) (1998), *Border Identities: Nation and State at International Frontiers*, (Cambridge: Cambridge University Press).

Geographical features of cross-border regions and their influence on tourism

Historically, borders were concrete structures, such as the Great Wall of China or Hadrian's Wall in Great Britain, that indicated a specific people's military territory and frontiers to sovereignty and national legislative power. However, Timothy (2001) found out that, surprisingly, the most concrete borders often turned out to be the weakest since they were the ones most heavily fought over and constantly shifted. So, it should not be surprising that the form of borders evolved throughout the centuries and only very few visible concrete borderlines remain in a modern world of globalisation, which "is often depicted as boundary-melting process" (Yang, 2004: 4).

With increasing globalisation, cross-border regional development became a focus for economists and geographers (Sparke and Sidaway and Bunnel and Grundy-Warr, 2004) and regionalisation soon was argued to be potentially regarded as a counterpart to globalisation (Yang, 2004). Modern borders were considered to function as divisions between national entities that comprised shared laws, norms, values and beliefs rather than sovereign military areas, and thus became stronger, the more invisible they were. Increasingly, borders have merged into distinctively unique cultural landscapes, thereby forming border regions that do not necessarily adhere to the geodetic lines imposed by national governments. Because the focus here is on the issue of trans-boundary co-operation in international tourism, the specific aspects of creating and investigating such borderlands are only peripherally touched upon, and the interested reader is advised to check Lorenz and Stoklosa (not dated) for a profound compilation of literature dealing with the specific situations of such border regions in general. Nonetheless, it is important to highlight that researchers, such as Herzog (1990), House (1980), Minghi (1994a) and Rumley and Minghi (1991), soon began to realise this development towards bi- sometimes even multinational border regions and investigated what became known as borders-of-the-mind, a concept which is further elaborated later.

Geographically, it is important to closely define different types of borders and their specific functions in order to understand the resulting relationships between an individual border and tourism activities in its surrounding areas. There are two distinctly different types of borders, namely physiographical and geometric borders (Timothy, 2001). Physiographical borders are oriented along geographical features, such as mountain ranges or rivers, and therefore indicate natural divisions that often adhere to historically established social patterns without artificially dividing cultural groups that naturally feel they belong together. Such forced divisions might occur, though, when borders are of a geometric nature. Borderlines of this kind are created without underlying natural features and rather orient themselves along human settlements or cultural features in the best case and military take-over or occupation in the worst case. Examples of geometric borders that pose tremendous problems for tourism development are the division between North and South Korea, the Islands of Cyprus and Ireland or the formerly divided city of Berlin, examples that will be discussed in more detail later on.

Having identified the two distinct types of borders, several researchers (see especially Leimgruber, 1980; Pearcy, 1965; Prescott, 1987) set out to examine the different functions of a border. In general, they agreed that any border marks the limits of territory and sovereignty. Economically, its role is to control and monitor the flow of goods, services and people alike, in order to maintain its territorial economic prosperity, which, if necessary, may even be defended by military actions along its borderlines. In addition, any border also functions as an ideological barrier, a function that is especially important when considering the division between nation states belonging to profoundly different economic and social value systems, as is the case with North and South Korea. Which functions predominate and how much each function influences life in the surrounding areas, differs tremendously from border to border and requires individual attention and research. In terms of tourism, it is important to mention that tourism landscapes around borders are manifold in nature due to the extent of development on either side of the border. One of the most important features in tourism research in border areas, therefore, is the distribution and frequency of border crossing facilities, the ease of doing so and the importance of tourism and related activities for the different sides. Such border gateways create funnels which consequently impact and shape the spatial developments of border regions and action spaces of their populations (Timothy, 2001).

Timothy and Tosun (2003) point out that identifying and discussing barrier functions of borders is the essence of tourism borderland research, since physical and psychological barriers erected at the border crossings by the host and home countries heavily impact travel and tourism activities. Such barriers, at least in their physical form, mainly relate to geographical features and man-made divisions that are erected to influence spatial developments and action spaces of different forms of tourist and non-tourist activities. Such issues are discussed in detail in the literature listed below. In addition, some literature deals with tourism geographies in general and the geography of borderlands in particular and offers a broad base for secondary research on geographic features influencing the capability and potential success of cross-border co-operation in international tourism.

Table 9.3   Selected literature: Geographic aspects of cross-border co-operation in tourism

BECKER, CH. (1982a), *Die Raumwirksamkeit unterschiedlicher Fremdenverkehrsformen,* (Vermessung, Photogrammetrie, Kulturtechnik, 146–153).
BECKER, CH. (1982b), 'Aktionsräumliches Verhalten von Urlaubern im Mittelgebierge', *Materialien zur Fremdenverkehrsgeographie* 9, (Trier).
BOGGS, S.W. (1932), 'Boundary functions and the principles of boundary making', *Annals of the Association of American Geographers* 22:1, 48–49.
BOGGS, S.W. (1940), *International Boundaries: A Study of Boundary Functions and Problems,* (New York: Columbia University Press).

CHRISTALLER, W. (1955), 'Beiträge zu einer Geographie des Fremdenverkehrs', *Erdkunde* 9:1, 10–19.

COMEAUX, M.L. (1982), 'Attempts to establish and change a western boundary', *Annals of the Association of American Geographers* 72, 254–271.

Framke, W. (1982), 'Grenze und Kulturlandschaft – Überlegungen zum Basler Symposium, 5.–8. Oktober 1981', *Erdkunde* 36:3, 207–209.

Germann, R. E. (1983), 'Grenzen einer nationalen Raumordnungspolitik', in M. Lendi, (ed.) *Elemente zur Raumordnungspolitik*, (Zürich: 103–112).

GLASSNER, M.I. (1996), *Political Geography, Second Edition*, (New York: Wiley).

GOODALL, B. (1987), *Dictionary of Human Geography*, (New York: Facts on File Publications).

GOTTMAN, J. (1973), *The Significance of Territory*, (Charlottesville: University Press of Virginia).

GROTH, P. AND BRESSI, T.W. (eds) (1997), *Understanding Ordinary Landscapes*, (New Haven: Yale University Press).

GRUNDY-WARR, C. AND SCHOFIELD, R.N. (1990), 'Man-made lines that divide the world', *Geographical Magazine* 62:6, 10–15.

HANSEN, N. (1977), 'Border Regions: A critique of spatial theory and empirical case-study', *Annals of Regional Science* 11, 1–14.

HARD, G. (1987), '„Bewußtseinsräume" – Interpretationen zu geographischen Versuchen, regionales Bewußtsein zu erforschen', *Geographische Zeitschrift* 75, 127–148, (Stuttgart).

HAUBRICHS, W. AND SCHNEIDER, R. (eds) (1994), *Grenzen und Grenzregionen*, (Saarbrücken: Veröffentlichungen der Kommission für saarländische Landesgeschichte und Volksforschung, 22).

HEIGL, F. (1978), *Ansätze einer Theorie der Grenze*, (Wien: Schriftenreihe der Österreichischen Gesellschaft für Raumforschung und Raumplanung).

HEIGL, F. (ed.) (1973/74), *Probleme grenznaher Räume*, (Innsbruck: Schriftenreihe des Instituts für Städtebau und Raumordnung, 1–2).

ISTEL, W. (1983), 'Auswirkungen nationaler Grenzen auf die grenzüberschreitende Raumplanung', in J. Maier (ed.) *Staatsgrenzen und ihr Einfluß auf Raumstrukturen, Teil 1*, (Bayreuth: Arbeitsmaterialien zur Raumordnung und Raumplanung, 23, 20–51).

JONES, S.B. (1943), 'The description of international boundaries', *Annals of the Association of American Geographers* 49, 241–255.

JONES, S.B. (1945), *Boundary Making*, (Washington, D.C.: Carnegie Endowment).

KIRBY, K.M. (1996), *Indifferent Boundaries: Spatial Concepts of Human Subjectivity*, (New York: Guildford Press).

KIRK, W. (1963), 'Problems of geography', *Geography* 48, 357–371.

KOLOSSOV, V. AND O'LAUGHLIN, J. (1998), 'New borders for new world orders: Territorialities at the fin-de-siecle', *GeoJournal* 44:3, 259–273.

KRISTOFF, L.K.D. (1959), 'The nature of frontiers and boundaries', *Annals of the Association of American Geographers* 49, 269–282.

LEFEBVRE, H. (1991), *The Production of Space*, (Oxford: Basil Blackwell).

LEIMGRUBER, W. (1980), 'Die Grenze als Forschungsobjekt der Geographie', *Regio Basiliensis* 21, 67–78.

LOWENTHAL, D. (1961), 'Geography, experience, and imagination: Towards a geographical epistemology', *Annals of the Association of American Geographers* 51, 241–261.
MATLEY, I.M. (1976), *The Geography of International Tourism*, (Washington, D.C.: Association of American Geographers).
MINGHI, J.V. (1963a), 'Boundary studies in political geography', *Annals of the Association of American Geographers* 53, 407–428.
PEARCE, G.E. (1965), 'Boundary functions', *Journal of Geography* 64:8, 346–349.
RATZEL, F. (1892), 'Die politischen Grenzen', *Mitteilungen der Geographischen Gesellschaft für Thüringen zu Jena* 11, 69–73.
RATZEL, F. (1896), 'The territorial growth of states', *Scottish Geographical Magazine* 12, 351–361.
REYNOLDS, D.R. AND MCNULTY, M.L. (1968), 'On the analysis of political boundaries as barriers: A perceptual approach', *East Lakes Geographer* 4, 21–38.
ROBINSON, G.W.S. (1959), 'Exclaves', *Annals of the Association of American Geographers* 49, 283–295.
RUPPERT, K. (1972), *Regionalgliederung und Verwaltungsgebietsreform als gesellschaftspolitische Aufgabe – Geographie im Dienste der Umweltgestaltung*, (Wiesbaden: Tagungsberichte und wissenschaftliche Abhandlungen des Deutschen Geographentages in Erlangen 1971).
SMALL, J. AND WITHERICK, M. (1995), *A Modern Dictionary of Geography – 3rd Edition*, (London: Edward Arnold).
TURBEVILLE III, D.E. AND BRADBURY, S.L. (1997), *Borderlines, border towns: Cultural landscapes of the post-free trade 49th parallel*, (Forth Worth, Texas: Paper presented at the annual meeting of the Association of American Geographers).
WATERMAN, S. (1987), 'Partitioned states', *Political Geography Quarterly* 6:2, 151–170.

Determinants of environmentally oriented cross-border co-operation in international tourism

Successful cross-border co-operation in international tourism often relates to environmental protection. As already indicated above, natural landscapes do not necessarily adhere to man-made borders and thus might create a cultural landscape that extends across two or more different national territories. These areas are particularly suitable for the creation of special purpose areas such as bi- or multi-national parks or biosphere reserves, since natural areas are highly valuable for the tourism industry. In addition, borderlands, by their very nature, often prove to be predestined for the creation and formation of international nature parks due to their peripheral remoteness, relatively sparse population, and untouched scenic nature. Employing cross-border nature reserves for the development of local tourism enables participants from each side of the border to prosper economically without necessarily having to significantly change their historically established ways of life and engage in industrial or other forms of unsustainable production.

Emerging eco-trends in tourism industries have displayed an increasing demand for untouched scenery and rural ways of life, thereby offering tremendous opportunities for peripheral and remote border landscapes to gain their share of economic benefits. Studies conducted by several researchers during the 1990s (i.e. Palomäki, 1994; Tenhiälä, 1994; Timothy, 1999a; Wachowiak, 1994d) indicated the scope of partnership activities important for border regions engaging in the creation and operation of trans-frontier nature parks. According to their findings, the most important aspect is the equal protection of natural and cultural resources on both sides of the border. Infrastructure needs to be developed co-operatively and in harmony with the environmental requirements of the area. Also, it should be provided in a way that enables visitors to cross the border conveniently within the individual parks.

Last, but by no means least, issues of human resources, marketing, and promotion need to be addressed collectively, but only after possibly prohibiting border restrictions have been lifted and legal formalities jointly agreed upon. Engagement in co-operative planning, development and operations subsequently requires a thorough understanding of collaborative stakeholder co-operation in order to suit the needs and desires of all participants. If the efforts are to be lastingly beneficial, strict adherence to the principles of sustainability is necessary: according to Timothy (2001: 170), the principles of sustainability are equity, efficiency, integration, balance, harmony and ecological integrity and are naturally 'espoused and encouraged when cross-border co-operation is put into practice and when residents, who, owing to their customarily marginal role in national functions, have not been involved in planning, are given a voice in decision making'.

The literature listed below, therefore, intends to give a more in-depth discussion on these briefly introduced determinants, as well as specific aspects of environmental protection across international boundaries and its employment in tourism industries. These general discussions will be joined by detailed case studies and research reports on individual trans-frontier nature parks that are situated all over the world and discussed in their geographic contexts later on in this chapter.

Table 9.4   Selected literature: Environmental aspects to cross-border co-operation in tourism

BAYRISCHES STAATSMINISTERIUM FÜR LANDESENTWICKLUNG UND UMWELTFRAGEN (2001), 'Umweltpolitische Herausforderungen für Grenzregionen', <http://www.umweltministerium.bayern.de/aktuell/newsroom/reden/2001/030501.htm> , accessed on 24 April 2003.
BLAKE, G. AND CHIA, L. AND GRUNDY-WARR, C. AND PRATT, M. AND SCHOFIELD, C.H.(eds) (1997), *International Boundaries and Environmental Security: Frameworks for Regional Co-operation*, (Amsterdam: Kluwer).
BLATTER, J. (1994a), *Erfolgsbedingungen grenzüberschreitender Zusammenarbeit im Umweltschutz: Das Beispiel Gewässerschutz am Bodensee*, (Freiburg: EURES, EURES Discussion paper, 37).

BLATTER, J. (1994b), *Erfolgsbedingungen grenzüberschreitender Zusammenarbeit im Umweltschutz. Das Beispiel Gewässer- und Auenschutz am Oberrhein*, (Freiburg: EURES, EURES Discussion paper, 43).

BREYMEYER, A. AND NOBLE, R. (eds) (1996), *Biodiversity conservation in transboundary protected areas: Proceedings of an International Workshop Bieszczady and Tatra National Parks, Poland May 15–25, 1994*. (Washington: National Academy Press).

BRUNN, S.D. AND MUNSKI, D.C. (1999), 'The International Peace Garden: A case study in locational harmony', *Boundary and Security Bulletin* 7:3, 67–74.

DUFFEY, E. (ed.) (1983), *Naturparks in Europa: Ein Führer zu den schönsten Naturschutzgebieten von Skandinavien bis Sizilien*, (München: Christian-Verlag/Stuttgart: Deutscher Bücherbund).

FORDNEY, C. (1996), 'Boundary wars', *National Parks* 70:1, 24–29.

GRASSHOPPER (1994), 'Trans-European Protected Belt and its Cross-Border Zones Grasshopper, 5, Winter 1994', <http://www.zb.eco.pl/ gh/5/belt_e.htm>, accessed 22 February 2003.

HENDERSON, L. (1993), 'Forging a link: two approaches to integrating trade and environment', *Alternatives* 20:1, 30–36.

INWENT – IUCN – EUROPARC (2002), *Transboundary Protected Areas: Guidelines for Good Practices and Implementation*, (Zschortau: Unpublished Report by InWent).

KENNEY, J. (1991), 'Beyond park boundaries: The spoils of development and industry imperil nearby national parks', *National Parks* 65:7/8, 20–25.

LUCAS, R.C. (1964), 'Wilderness perception and use: The example of the Boundary Waters canoe area', *Natural Resources Journal* 3:3, 394–411.

MECCA CONSULTING (n.d.), 'Tourism and Environmental Plannin', <http://www.mecca-consulting.at/projekteeng.html>, accessed on 3 April 2003.

MEYER, R. (1980), 'Gemeinsame Aktionen an den Staatsgrenzen zur Förderung grenzüberschreitender Natur- und Nationalparke', *Naturschutz- und Naturparke* 97, 31–34.

NIEWIADOMSKI, Z. (1996), 'Bieszczady National Park', in A. Breymeyer, and R. Noble (eds), *Biodiversity conservation in transboundary protected areas: Proceedings of an International Workshop Bieszczady and Tatra National Parks Poland May 15–25, 1994*, (Washington: National Academy Press).

STEFFENS, R. (1993), 'Not just another roadside attraction', *National Parks*, 67:1, 26–31.

THE 2ND EUROPEAN YOUTH WATER CONGRESS (2001), 'The European Forum – Water and Tourism', <http://www.youth-water-congress.org/forum.htm>, accessed on 3 April 2003.

THORSELL, J. (1990), *Parks on the Borderline: Experience in Transfrontier Conservation*, (Cambridge: IUCN).

THORSELL, J. AND HARRISON, J. (1990), 'Parks that promote peace: A global inventory of transfrontier nature reserves', in J. Thorsell (ed.), *Parks on the Borderline: Experiences in Transfrontier Conservation*, (Gland: IUCN, 3–21).

TIMOTHY, D.J. (2000b), 'Tourism and international parks', in R.W. Butler and S.W. Boyd (eds), *Tourism and National Parks: Issues and Implications*, (Chichester: Wiley, 263–282).

VOGEL, D. (1997), 'Trading up and governing across: Transnational governance and environmental protection', *Journal of European Public Policy* 4:4, 556–571.

WACHOWIAK, H. (1995), 'Grenzüberschreitende Zusammenarbeit im umweltschonenden Tourismus', in P. Moll (ed.), *Umweltschonender Tourismus – Eine Entwicklungsperspektive für den ländlichen Raum?*, (Bonn: Kuron, 93 – 103).

WEINGROD, C. (1994), 'Two countries, one wilderness', *National Parks* 68:1, 26–31.

WESTING, A.H. (1993), 'Building confidence with transfrontier reserves: The global potential', in A.H. Westing (ed.), *Transfrontier Reserves for Peace and Nature: A Contribution to Human Society*, (Nairobi: UNEP, 1–15).

WESTING, A.H. (ed.) (1993), *Transfrontier Reserves for Peace and Nature: A Contribution to Human Society*, (Nairobi: UNEP).

WOODRUFFE, B.J. (1998), 'Conservation and the rural landscape', in D. Pinder (ed.), *The New Europe: Economy, Society and Environment*, (Chichester: Wiley, 455–476).

ZBICZ, D.C. (1999), *Transfrontier Ecosystems and Internationally Adjoining Protected Areas*, (Durham: IUCN).

Policies and political issues of cross-border co-operation in international tourism

Certainly, the most important and most difficult aspect of cross-border co-operation in general, especially with regard to the tourism industry, relates to policies and political issues. It is important to bring local, regional and national policies into a coherent line, in order to create a single tourism policy capable of capturing the desired economic benefits of a prospering tourism industry and to avoid negative threats and implications. However, tourism policies, at least as far as border regions are concerned, also have to be compatible with those policies that have been created on the other side of the border and need to consider national interests that might differ tremendously between neighbouring nations. Specifically due to the fact that economic interests make up a great part of engaging in tourism industries and that tourism is often seen as the most convenient and promising industry for peripheral, remote and rather rural border landscapes, co-operation across national borders is not easily achieved and is often accompanied by mutual suspicion and the desire to increase one's own share over that of the neighbour. To overcome such mutual suspicion, it is important to engage in participative planning and strategy formulation, which is accomplishable through the identification of the all important stakeholders and local populations' participation in policy creation and industry operations (Greer, 2002).

The literature listed below, therefore, concentrates mainly on the different aspects of national policy formation relating to cross-border co-operation with specific regard to tourism. It is commonly agreed among researchers (Greer, 2002; Timothy, 2001) that tourism, especially in border regions, increasingly requires cross-border co-operation to maintain its local competitiveness and avoid duplications. Such co-operation has been identified as the result of ongoing globalisation and has already been successful on a political, diplomatic, economic and military level, as numerous international organisations and entities such as the EU, NATO, UNO, UNWTO or NAFTA, to name just a few popular examples, have already proven. The formation of such international entities obviously changed the nature of international relations and influenced tourism by spurring and facilitating co-operation between adjacent and distant national entities. Despite the positive effects that begin to surface, Smith and Pizam (1998) highlight, though, that it is too early to understand the entire effect that such international organisations exert on tourism. Research, therefore, should continue to examine the individual effects of regional co-operation across existing boundaries. Since tourism attractions may be located nearby, along or sometimes even across political borders, cross-border co-operation in international tourism might prove capable of minimising parallel and duplicating developments (two opera houses, state libraries, zoological gardens and the like in very close proximity to border regions), thereby saving valuable resources for potential specialisation, which might prove to be an economically more valuable and sustainable strategy (Timothy, 1995; Timothy, 2001).

The fact that national entities might even discover their specific border demarcations to be attractive tourism assets, such as the former Check-point Charly in Berlin, Germany, or the 'Four Corners Monument', demarcating the border between the US states of New Mexico, Utah, Arizona and Colorado, tends to indicate the tremendous potential such borderlands create for national and bi- or even multi-national tourism development. As Timothy (2001) argues, increasing globalisation has spurred the creation of cross-boundary regions and it is these regions that have already, especially in the case of Europe, become the essence of profitable tourism development in the future, highlighting the importance of further investigations into cross-border co-operation in tourism.

Thus, the literature listed below discusses the need for established political structures to adjust to these newly emerging situations on their national borderlines and incorporates research findings and related studies discussing the general aspects of regionalisation and cross-border tourism strategy formation.

Table 9.5   Selected literature: Politics of cross-border co-operation in tourism

BEYER AND ERDOSI AND GREIF AND HAHNE AND KLEMENCIC AND MAIER AND MOLL AND RAMMING AND SPITZNER (n.d.), 'Regionalpolitik in und für periphere, grenznahe Gebiete', *Arbeitsmaterialien zur Raumordnung und Raumplanung* 86, (Bayreuth).
BLATTER, J. (1996), *Political co-operation in cross-border regions: Two explanatory approaches*, (Zurich: ERSA 36th European Congress).
BRAMWELL, B. AND SHARMAN, A. (1999), 'Collaboration in local tourism policymaking', *Annals of Tourism Research* 26, 392–415.
BURNETT, A.D. AND TAYLOR, P.J. (eds) (1981), *Political Studies from Spatial Perspectives*, (New York: Wiley).
DORSEY, J. (1990), 'Tourism council urges governments to drop barriers to travel', *Travel Weekly* 49:17, 6.

DUCHACEK, I.D. (1986), 'International competence of subnational governments: borderlands and beyond', in O.J. Martinez (ed.), *Across Borders: Transborder Interaction in Comparative Perspective*, (El Paso: Texas Western Press, 11–28).

GABBE, J. AND MALCHUS, K. AND MAHNKOPF, H. AND MARTINOS (1999), 'Institutionelle Aspekte der grenzüberschreitenden Zusammenarbeit', <http://www.aebr.net/publikationen/pdfs/inst_asp_99. de.pdf>, accessed on 24 April 2003.

GUNKEL, P. (1983), 'Öffentliche Investitionsentscheidungen im Fremdenverkehr. Überprüfung und Konzeption von Bewertungsansätzen', *Europäische Hochschulschriften*, 5:438, (Frankfurt a. M.).

HALL, C.M. (1994B), *Tourism and Politics: Policy, Power and Place*, (Chichester: Wiley).

HALL, D.R. (1990C), 'The "communist world" in the 1990's', *Town and Country Planning* 59:1, 28–30.

HANSEN, N. (1983), 'International co-operation in border regions: An overview and research agenda', *International Regional Science Review*.

HOCKING, B. AND MCGUIRE, S. (eds) (1999), *Trade Politics: International, Domestic, and Regional Perspectives*, (London: Routledge).

HOUSE, J.W. (1980), 'The frontier zone: A conceptual problem for policy makers', *International Political Science Review*, 1:4, 456–477.

KOLBE, K. (1997), 'Institutionen der grenzüberschreitenden Zusammenarbeit', <http://members.aol.com/KSUKolbe/saarlor2.htm>, accessed on 24 April 2003.

MCMILLAN-SCOTT, E. (1992), 'Die Entwicklung einer gemeinschaftlichen Fremdenverkehrspolitik', in Amt für amtliche Veröffentlichungen der europäischen Gemeinschaft (ed.), *Fremdenverkehr in Europa*, (Luxemburg: Amt für amtliche Veröffentlichungen der europäischen Gemeinschaft, 9–18).

SCHULGEN, H. (1988), 'Die regionale Fremdenverkehrspolitik braucht neue Leitgedanken', in Akademie für Raumforschung und Landesplanung (ed.), *Fremdenverkehr und Regionalpolitik*, (Hannover: ARL, 255–275).

SÖNMEZ, S.F. (1998), 'Tourism, terrorism, and political instability', *Annals of Tourism Research* 25, 416–456.

TIMOTHY, D.J. (1996), 'Small and isolated: The politics of tourism in international exclaves', *Acta Turistica* 8:2, 99–115.

TIMOTHY, D.J. (2001), *Tourism and Political Boundaries*, (London: Routledge).

WORLD TRAVEL AND TOURISM COUNCIL (1991), *Bureaucratic Barriers to Travel*, (Brussels: WTTC).

Legal aspects of cross-border co-operation

Closely related to politics but, due to its massive importance for cross-border co-operation, worthy of separate discussion, are the legal prerequisites and conditions in borderlands, as well as potential criminal results of increased border traffic and facilitated crossing procedures. Borders, as already indicated above, are not always friendly lines of cultural or national separation, but are also capable of denoting a division between strongly opposing economic, legal and jurisdictional systems. In such instances, borders tend to function as insurmountable obstacles, highly safeguarded, to separate a specific people from the rest of the world, as was the case with the Berlin Wall that separated the Capitalist World from the Socialist block, for example. Even in cases where divergences are only restricted to very few jurisdictional differences, as is the case with most borders in the modern world, economic opportunities, highly suitable for tourism businesses, have sometimes spurred the impression that tourism actually increases criminal and illegal activities. A good example of this is cross-border travel for gambling purposes, an industry specifically important for North-American Indian reservations, or simply the intent of participating in activities considered illegal or unlawful in the visitors' home countries. Borderlands have, therefore, sometimes been regarded as predestined creators of vice (see especially Bowman, 1994), hence, unsafe places of criminal and indecent activities. Additionally, global terrorist activities have raised issues of safety and resulted in increased visa regulations and immigration procedures, sometimes heavily hindering cross-border co-operation in tourism.

The literature listed within the following sub-section, therefore, concentrates on investigations and studies conducted to clarify the relationship between tourism, terrorism and other forms of criminal activities and the implications such situations potentially pose upon cross-border co-operation.

Table 9.6   Selected literature: Legal considerations in cross-border co-operation in tourism

BEYERLEIN, U. (1988), 'Rechtsprobleme der lokalen grenzüberschreitenden Zusammenarbeit: Beiträge zum ausländischen öffentlichen Recht und Völkerrecht', <http://www.virtual-institute.de/de/hp/beitr96.cfm>, accessed 23 February 2005.
Bowman, K.S. (1994), 'The border as locator and innovator of vice', *Journal of Borderlands Studies* 9:1, 51–67.
JOYCE, J. (1990), 'Court rules for border bargains', *Guardian*, 13 June 1990.
PIZAM, A. (1982), 'Tourism and crime: Is there a relationship?', *Journal of Travel Research*, 20:3, 7–10.
POMMERSHEIM, F. (1989), 'The crucible of sovereignty: Analyzing issues of tribal jurisdiction', *Arizona Law Review* 31:2, 329–363.
RYAN, C. (1993), 'Crime, violence, terrorism and tourism: An accidental or intrinsic relationship?', *Tourism Management* 14, 173–183.
SCHLOEGEL, B. (1982), *Grenzüberschreitende interkommunale Zusammenarbeit: Voraussetzungen und Rechtsgrundlagen sowie Beispiele in der Abwasserbeseitigung, Energie-und Wasserversorgung und im Nahverkehr*, (Berlin: Verlag E. Schmidt).
WORLD PRESS REVIEW (1993), 'Border crime', *World Press Review* 40:6, 32.

Economic considerations of cross-border co-operation with specific regard to international tourism industries

Considering the history of international cross-border relations, it appears that all co-operation was initiated through economic interests and the desire to prosper by creating bigger and stronger markets. This development has seen the creation of supra-national entities such as NAFTA, the North American Free Trade Association, ASEAN, the Association of South-East Asian Nations, or, CER, the Closer Economic Relations between Australia and New Zealand, to name just a few. Even though all of these entities and organisations differ in size and their level of national integration and supranational legitimacy, they all combine the economic interest of prospering on not only a national, but a supra-national regional level.

Probably the most famous of these entities is the EU. Today, it is a legitimate sovereign body, reigning over a single market and unitary currency, which also started out simply as an economic agreement between a few national entities interested in the creation of a bigger market. With tourism having become a major economic industry in recent years and expected to grow dramatically within the near future (WTO, 2005), it should not be surprising, that much of these entities' attention is put into tourism development. Economies have long realised the importance of tourism as an economic factor and its capability of prospering remote, peripheral and rural areas distributed along national boundaries. Once again, the European Union has been signposting these efforts by being highly involved in tourism policy creation and subsequent tourism development through various programs, elaborated on in detail later. These efforts in tourism development across national boundaries have arguably resulted in the creation of unique borderlands, economically independent cross-border regions that have formed through territorial integration processes, which increasingly become popular tourist destinations themselves (Blatter, 2000). But regional economic integration is not restricted to Europe. It has been observed globally and has led to increased economic interest in borderlands, where researchers have already begun to recognise the creation of specifically designed tourism and recreational business districts capable of catering specifically for visiting neighbours (Getz, 1993; Stansfield and Rickert, 1970).

The literature listed below hence intends to offer convenient access to preliminary secondary research on economic considerations underlying cross-border co-operation in tourism industries.

Table 9.7   Selected literature: Economic considerations of cross-border co-operation in tourism

AMTSBLATT der Europäischen Gemeinschaften (1999), 'Das Beschäftigungspotential der Tourismuswirtschaft', <http://europa.eu.int/eurlex/pri/de/oj/dat/1999/c_178/ c_ 17819990623de00030013.pdf> , accessed on 25 April 2003.

CROSS-BORDER Economic Bulletin (2001), 'The Cross-Border Economy After September 11', <http://www.rohan.sdsu.edu/faculty/jgerber/oct01.htm> , accessed on 3 April 2003.

EAST WEST Economic Corridor (2002), 'Regional Information', <http://www.visit-mekong.com/ewec/> , accessed on 3 April 2003.

ECONOMIST (1989), 'Bouncing the beach ball', *The Economist*, 25 February, 38.

ECONOMIST (1991), 'On the borderline', *The Economist,* 23 November, 58.

ECONOMIST (1992), 'The future, and by and large it works', *The Economist*, 27 June, 61.

ECONOMIST (1993a), 'Coming of age in Andorra', *The Economist*, 13 March, 60.

ECONOMIST (1993b), 'Propinquity pays', *The Economist*, 20 February, 67.

ECONOMIST (1995), 'Red Sea Riviera?', *The Economist,* 16 December, 58.

ECONOMIST (1997), 'Berliners see red', *The Economist*, 8 March, 56.

EL-AGRAA, A.M. (1990), 'The theory of economic integration', in A.M. El-Agraa (ed.), *The Economics of the European Community*, (New York: St Martin's Press, 79–96).

ELLIOTT, M. (1997), 'Hey, can you spare a "EURO"?', *Newsweek*, 17 February, 48–49.

EUROPEAN UNIVERSITY INSTITUTE (ed.) (1998), 'Rise in cross-border tourism seems to be biggest impact of the euro for citizens in a border region', <http://www.ieem.org. mo/publications/eurodisc/HTML/euro/html/00152/index5.htm> , accessed 13 March 2003.

GETZ, D. (1993), 'Planning for tourism business districts', *Annals of Tourism Research*, 20:3, 583–600.

HALL, C.M. (1994a), 'The closer economic relationship between Australia and New Zealand: Implications for travel and tourism', *Journal of Travel and Tourism Marketing* 3, 123–131.

HALL, D.R. (1991b), 'Introduction', in D.R. Hall (ed.), *Tourism and Economic Development in Eastern Europe and the Soviet Union*, (London: Belhaven Press, 3–28).

HANSEN, N. (1977), 'The economic development of border regions', *Growth and Change*, 8:4, 2–8.

HOSEASON, J. (1998), 'The Euro affair: How a single currency for Europe will affect tourism', *Tourism* 96, 10–11.

KLEMZ, W. (1989), *Interkommunale Zusammenarbeit im Fremdenverkehrsbereich: Zugleich ein Beitrag zur ökonomischen Analyse des Zweckverbandes*, (Frankfurt am Main/ Bern/ New York/ Paris: Verlag Peter Lang and Europäische Hochschulschriften, 5:999, Volks- und Betriebswirtschaft).

LINTNER, B. (1991a), 'Forgotten frontiers: Peace brings investors to notorious border region', *Far Eastern Economic Review*, 16 May, 23–24.

LÖSCH, A. (1954), *The Economics of Location*, (New Haven, CT: Yale University Press).

LYONS, J. (1991), 'Border merchants', *Forbes*, (published on 19 August 1991), pp. 56–57.

MASHINSKY, V. (1996), 'Border tax on the rich', *Izvestia*, (published on 18 December1996), p. 2.

McWHIRTER, W. (1992), 'A monster spending spree' *Time*, (published on 23 March 1992), p. 7.

MIKESELL, J.L. (1970), 'Central cities and sales tax rate differentials: The border city problem', *National Tax Journal* 23:2, 206–213.

MIKESELL, J.L. (1971), 'Sales taxation and the border county problem', *Quarterly Review of Economics and Business* 11:1, 23–29.

OHMAE, K. (1995), *The End of the Nation State: The Rise of Regional Economies*, (London: HarperCollins).

PAYNE, T. (1987), 'Economic issues', in C. Clarke and T. Payne (eds), *Politics, Security and Development in Small States*, (London: Allen & Unwin, 50–62).

RICHARDSON, H.L. (1993), 'NAFTA means economic growth', *Transportation and Distribution* 34:11, 42–48.

RODRIGUEZ, M. AND PORTALES, J. (1994), 'Tourism and NAFTA: Towards a regional tourism policy', *Tourism Management* 15 , 319–322.

SMITH, G. AND MALKIN, E. (1997), 'The border', *Business Week*, 12 May, 64–74.

SPEHL, Harald (1981), *Einfluß der Grenzlage auf Betriebe in peripheren Regionen: Erhebungsergebnisse. Arbeitspapier 5*, (Trier: Universität Trier, Schwerpunkt Stadt- und Regionalplanung).

STADTFELD, F. (ed.) (1989), *Chancen und Risiken eines europäischen Tourismus-Binnenmarktes*, (Worms: Schriftenreihe zur Touristik 2).

STANSFIELD, C.A. AND RICKERT, J.E. (1970), 'The Recreational Business District', *Journal of Leisure Research*, 2:4, 213–225.

WILLIAMS, A.M. AND SHAW, G. (1998), 'Tourism and economic development', in D. Pinder (ed.) *The New Europe: Economy, Society, and Environment*, (Chichester: Wiley, 177–201).

WORLD TOURISM ORGANIZATION (1988), *The Problems of Protectionism and Measures to Reduce Obstacles to International Trade in Tourism Services*, (Madrid: WTO).

WORLD TOURISM ORGANIZATION (1998), 'Tourism to Benefit from Single European Currency', <http://www.world-tourism.org/pressrel/europa.htm> , accessed on 17 March 1998.

WORLD TOURISM ORGANIZATION (2000a), *Compendium of Tourism Statistics, 1994–1998*. (Madrid: WTO).

WORLD TOURISM ORGANIZATION (2000b), <http://www.world-tourism.org/>, accessed on 26 August 2000.

Specific aspects of cross-border co-operation in international tourism

Having discussed and introduced literature on the general nature of cross-border co-operation in tourism, their social, political, legal and economic prerequisites, chances, opportunities, threats, implications, benefits and results, the following assemblage introduces literature that discusses very specific aspects of cross-border tourism relations that could not easily be classified with any issue addressed above. The literature below describes studies undertaken on day-trippers, on international cruises (Bar-On, 1988), or the unique circumstances exclusive to sovereign encapsulation (Boal, 1994), and offers first insight into the issues arising within quasi-states, partitioned states and divided cities (Butler and Mao, 1995; Butler and Mao, 1996; Dunn, 1994). Additionally, some literature deals with preliminary studies investigating frontiers to tourism yet to be explored, namely cyberspace (Rojek, 1998) and outer-space (Smith, 2000; White, 1999; Wilson, 2000).

More profound discussions, introducing current literature available on the emerging phenomenon of cross-border shopping and the particular dependencies evident in microstates and remote island nations, will finally round off the section that concentrates on particular aspects surfacing when investigating cross-border co-operation in international tourism industries. As such, it might prove a valuable first pit-stop for closely defining potential future research questions for further academic studies in the field of cross-border co-operation in international tourism industries.

Table 9.8   Selected literature: Specific aspects of cross-border co-operation in tourism

BARKER, D. AND MILLER, D.J. (1995), 'Farming on the fringe: Small-scale agriculture on the edge of the Cockpit Country', in D. Barker and D.F.M. McGregor (eds), *Environment and Development in the Carribean: Geographical Perspectives*, (Kingston: University of the West Indies Press, 271–292).
BAR-ON, R. (1988), 'International day trip, including cruise passenger excursions', *Revue de Tourisme*, 43:4, 12–17).
BOAL, F.W. (1994), 'Encapsulation: Urban dimensions of national conflict', in S. Dunn (ed.) *Managing Divided Cities*, (Ryburn: Keele, 30–40).
BORNEMAN, J. (1998), 'Grenzregime (border regime): The Wall and it's aftermath', in T.M. Wilson and H. Donnan (eds), *Border Identities: Nation and State at International Frontiers*, (Cambridge: Cambridge University Press, 162–90).
BROHMAN, J. (1996), 'New directions in tourism for third world development', *Annals of Tourism Research* 23, 48–70.
BUTLER, R.W. AND MAO, B. (1995), 'Tourism between quasi-states: International, domestic or what?', in R.W. Butler and D. Pearce (eds), *Change in Tourism: People, Places, Processes*, (London: Routledge, 92–113).
BUTLER, R.W. AND MAO, B. (1996), 'Conceptual and theoretical implications of tourism between partitioned states', *Asia Pacific Journal of Tourism Research*, 1:1, 25–34.
CHURCH, A. AND REID, P. (1996), 'Urban power, international networks and competition: The example of cross-border co-operation', *Urban Studies* 33:8, 1297–1318.
CLARK, T. (1994), 'National boundaries, border zones, and marketing strategy: A conceptual framework and theoretical model of secondary boundary effects', *Journal of Marketing* 58, 67–80.
COMITE REGIONAL DE TOURISM NORD-PAS DE CALAIS (n.d.), 'Waterway tourism in Euroregional territories', <http://www.partir-en-croisiere.com/an/decouvrir.asp>, accessed on 3 April 2003.
DUNN, S. (ed.) (1994), *Managing Divided Cities*, (Ryburn: Keele).
INTERREGIIIA IRELAND/WALES (n.d.) Priority 2 Measure 2, 'Culture, Heritage and Tourism', <http://www.interreg.ie/main.html>, accessed on 3 April 2003.
JESSOP, B. (1995), 'Regional economic blocs, cross-border co-operation, and local economic strategies in postcolonialism', *American Behavioral Scientist* 38:5, 674–715.

JOHNSTON, M. AND MAURO, R. AND DILLEY, R.S. (1991), *Trans-border tourism in a regional context*, (Kingston, Ontario: Paper presented at the annual meeting of the Canadian Association of Geographers).

KAMRAVA, M. (1993), *Politics and Society in the Third World*, (London: Routledge).

KELLER, C.P. (1984), 'Centre-periphery tourism development and control', in J. Long and R. Hecock (eds), *Leisure, Tourism and Social Change*, (Edinburgh: Centre for Leisure Research, 77–84).

KRAKOVER, S. (1985), 'Development of tourism resort areas in arid regions', in Y. Gradus (ed.), *Desert Development: Man and Technology in Sparselands*, (Dordrecht: D. Reidel Publishing, 271–284).

OLSEN, D.H. AND TIMOTHY, D.J. (1999), 'Tourism 2000: Selling the Millennium', *Tourism Management* 20, 389–392.

OPPERMANN, M. (1998), 'Tourism space revisited', *Tourism Analysis* 2, 107–118.

PEARCE, D.G. (1999a), 'Introduction: Issues and approaches', in D.G. Pearce, and R.W. Butler (eds) *Contemporary Issues in Tourism Development*, (London: Routledge, 1–12).

PRETES, M. (1995), 'Postmodern tourism: The Santa Claus industry', *Annals of Tourism Research* 22, 1–15.

RICHTER, L.K. (1992), 'Political instability and tourism in the Third World', in D. Harrison (ed.) *Tourism and the Less Developed Countries*, (London: Belhaven, 35–46).

ROJEK, C. (1998), 'Cybertourism and the phantasmagoria of place', in G. Ringer (ed.) *Destinations: Cultural Landscapes of Tourism*, (London: Routledge, 33–48).

SCOTT, J. (1995), 'Sexual and national boundaries in tourism', *Annals of Tourism Research* 22, 385–403.

SMITH, V.L. (2000), 'Space tourism: The 21st century "frontier"', *Tourism Recreation Research*, 25:3, 5–15.

WHITE, M. (1999), 'Entrepreneurs study space tourism with "reality" 15–20 years in future', *Sentinal Tribune*, (published on 23 September 1999), p. 10.

WILSON, J. (2000), 'Postcards from the moon: A lunar vacation isn't as far-out an adventure as you think', *Popular Mechanics*, June, 97–99.

Cross-border shopping

A major aspect of cross-border relations is the ever-increasing number of travellers that engage in shopping related activities across national boundaries. Whether such trips are motivated by the search for nationally unattainable goods and services, or rather by the desire to hunt for bargains and save some money, is often difficult to determine and academia has only recently begun to examine and understand the principles of cross-border shopping as a tourism and leisure activity. A national economy can lose tremendous amounts when its citizens cross their national borders to shop in neighbouring countries, leaving entire regions subject to economic decline. It should, therefore, not be surprising that such activities are seldom actively supported by generating governments and sometimes even strongly opposed as being unsuitable for sustainable regional development (Timothy, 2001; Timothy and Butler, 1995). None the less, cross-border shoppers have a profound impact on tourism industries, due to the fact that they often engage in more than just shopping. They take a day off, visit monuments or museums and dine out. Cross-border shopping activities, therefore, relate closely with tourist activities and consequently deserve to be addressed individually within this chapter.

Even though sovereign nations fear losing valuable revenues when their citizens hunt for bargains and shop across the border, they often facilitate such situations by engaging in the formation of free-trade areas and lifting the tariffs and customs on an ever increasing number of products. One of the most extensive and far-reaching of those free-trade areas is probably the so-called Eurozone, a single market with a unitary currency created for the European Union according to principles set out in the Schengen agreement. This agreement has tremendously spurred cross-border shopping activities within the EU, since any citizen in a member state is eligible to buy as much of any given good or product as is acceptable for private consumption in any given member country. This situation has amplified bargain hunting across national boundaries, especially for goods such as petrol, gas, cigarettes, or alcohol, and hence granted some countries, including Luxembourg and the Netherlands, a competitive economical advantage over others (see especially Bygvra, 1990; Gramm, 1983; Hidalgo, 1993). Even though, initially, this meant a loss in direct tax revenues for some countries, it has also stimulated the flow of people across the continent, increased awareness of national and regional produce and resulted in the creation of tourism niches such as food and shopping tourism in Belgium or France, indicating the positive impact cross-border shopping can have on international tourism industries. Consequently, the interest in this segment of recreational and leisure activities rose and several studies began to investigate cross-border shopping, its determinants, participants and impacts in more detail.

Richard (1996) was one of the first to realise that in cross-border tourism between neighbouring nations two distinct tourist groups exist, namely the traditional tourists and the shoppers. Due to the fact that shoppers are often day-trippers they have previously been shamefully neglected in tourism related studies, as this group, by definition, does not belong to the closely defined group of tourists. This might explain the relative youth of cross-border shopping related research in tourism studies. On further examination of the phenomenon of cross-border shopping on a global perspective, he identified that the exchange rate of currencies was a powerful indicator of visitation patterns, especially among the latter group. Considering that exchange rates have always been an indicator of any international tourism activities, though, such realisations have spurred more economically inspired research (see DiMatteo, 1993) which indicates that taxation schemes rather than simply the exchange rates on either side of the border, influence spending and attendance level. Such schemes thus prove to be valuable tools when attempting to control cross-border shopping activities (see especially Lucas, 2004) and employ them for regional tourism development. Other determinants identified in cross-border shopping are customs and tariffs, as well as the distance between visitors' actual place of residence and their desired shopping destination (see especially Timothy and Butler, 1995). As already mentioned, research into these issues is rather in its infancy and requires further in-depth investigations.

The literature listed below, therefore, discusses this newly emerging field of tourism research in general, offers a global as well as very detailed regional perspective concentrating on specific destinations for cross-border shopping activities, and intends to offer a convenient starting point for further research into this economically important borderland phenomenon.

Table 9.9   Selected literature: Cross-border shopping activities

ASSOCIATED PRESS (1998), 'Drop in Canadian dollar draws US shoppers across border', *The Toledo Blade*, (published on September 1998).
BOISVERT, M. AND THIRSK, W. (1994), 'Border taxes, cross-border shopping, and the differential incidence of the GST', *Canadian Tax Journal* 42:5, 1276–1293.
BONDI, N. (1997), 'Shoppers head to Windsor for deals: High exchange rate on American dollar fuels big-ticket buys', *Detroit News*, (published on 30 December 1997).
BONDI, N. (1998), 'Bargain hunters hit Canada: Loonie's record fall an economic windfall for Americans shopping across the border', *Detroit News*, (published on 30 January 1998).
CANADIAN Chamber of Commerce (1992), *The Cross Border Shopping Issue*, (Ottawa: Canadian Chamber of Commerce).
CHAMBERLAIN, L. (1991), *Small Business Ontario Report No.44: Cross Border Shopping*, (Toronto: Ministry of Industry, Trade and Technology).
CHATTERJEE, A. (1991), 'Cross-border shopping: Searching for a solution', *Canadian Business Review* 18, 26–31.
DANISH Cross Border Institute (2002), 'Cross-Border Shopping in the spring of 2001 – after the Oeresund Bridge and Schengen', <http://www.ifq.dk/en/ publications.html>, accessed on 3 April 2003.
ECONOMIC AND BUSINESS REVIEW INDONESIA (1993), 'Cross-border shopping craze', *Economic and Business Review Indonesia* 57, 34.
FITZGERALD, J.D. AND QUINN, T.P. AND WHELAN, B.J. AND WILLIAMS, J.A. (1988), *An Analysis of Cross-Border Shopping*, (Dublin: The Economic and Social Research Institute).
GETZ, D. (1993), 'Tourist Shopping Villages: Development and Planning Strategies', *Tourism Management* 14, 15–26.
GOODMAN, L.R. (1992), 'A working paper on crossborder shopping: The Canadian impact on North Dakota', in H.J. Selwood and J.C. Lehr (eds) *Reflections from the Prairies: Geographical Essays*, (Winnipeg: University of Winnipeg, Department of Geography, 80–89).
GOVERNMENT OF ONTARIO (1991), 'Report on Cross-Border Shopping', (Toronto: Standing Committee on Finance and Economic Affairs).

GRAMM, M. (1983), 'Einkaufen im belgisch-niederländisch-deutschen Dreiländereck – ein Beispiel für grenzüberschreitende räumliche Interaktionen und ihrem Beitrag zur Entwicklung eines grenzüberschreitenden Nationalbewußtseins', in J. Maier (ed.), *Staatsgrenzen und ihr Einfluß auf Raumstrukturen, Teil 1*, (Bayreuth: Arbeitsmaterialien zur Raumordnung und Raumplanung 23, 51–69).

GOVERNMENT OF NEW BRUNSWICK (1992), *A Discussion Paper on Cross Border Shopping*, (Fredericton: Department of Economic Development and Tourism).

HIDALGO, L. (1993), 'British shops suffer as "booze cruise" bargain hunters flock to France', *Times*, (published on 22 November 1993), p. 5.

JANSEN-VERBEKE, M.C. (1990), 'From leisure shopping to shopping tourism', in *Proceedings of the International Sociological Association Annual Conference*, (Madrid: International Sociological Association, 1–17).

KEMP, K. (1992), 'Cross-border shopping: Trends and measurement issues', *Canadian Economic Observer* 5, 1–13.

LEIMGRUBER, W. (1988), 'Border trade: The boundary as an incentive and an obstacle to shopping trips', *Nordia* 22:1, 53–60.

LUCAS, V. (2004), 'Cross-border shopping in a federal economy', *Regional Science and Urban Economics* 34, 365–385.

MESCHEDE, W. (1983), *Geschäftseinzugsbereiche und räumliches Einkaufsverhalten der niederländischen und deutschen Bevölkerung im Raum Enschede-Gronau*, Arbeitsberichte 2. Münster/ Westfalen, Arbeitsgemeinschaft Angewandte Geographie Münster.

MICHALKÓ, G. and Timothy, D.J. (2001), 'Cross-border shopping in Hungary: Causes and effects', *Visions in Leisure and Business* 20:1, 4–22.

MINGHI, J.V. (1999), 'Borderland "day tourists" from the East: Trieste's transitory shopping fair', *Visions in Leisure and Business* 17:4, 32–49.

NGAMSOM, B. (1998), 'Shopping tourism: A case study of Thailand', in K.S. Chon (ed.), *Proceedings, Tourism and Hotel Industry in Indo-China and Southeast Asia: Development, Marketing, and Sustainability*, (Houston, TX: University of Houston, 112–128).

OHSAWA, Y. (1999), 'Cross-border shopping and commodity tax competition among governments', *Regional Science and Urban Economics* 29, 33–51.

RICHARD, W.E. (1995), 'Segmenting cross-border tourism with a focus on the shopping component', in B.D. Middlekauff (ed.), *Proceedings of the Annual Meeting of the New England-St Lawrence Valley Geographical Society*, (Burlington, VT: New England-St Lawrence Valley Geographical Society, 183–197).

RICHARD, W. E. (1996), 'Cross Border Tourism and Shopping: The Policy Alternatives', <http://www.Usm.maine.edu/cber/mbi/winter96/ tourism.htm>, accessed on 25 April 2003.

TIMOTHY, D.J. (1999b), 'Cross-border shopping: Tourism in the Canada-United States borderlands', *Visions in Leisure and Business* 17:4, 4–18.

WASSERMANN, D. (1996), 'The borderlands mall: Form and function of an imported landscape', *Journal of Borderlands Studies* 11:2, 69–88.

WEIGAND, K. (1990) 'Drei Jahrzehnte Einkaufstourismus über die deutsch-dänische Grenze', *Geographische Rundschau* 42:5, 286–290.

Cross-border co-operation with regard to international tourism in Europe

As already indicated above, Europe, and specifically the European Union, has consecutively evolved as an international role model for cross-border co-operation between individual sovereign nations. By its very nature of being a union formed by national sovereign members, the EU has proven to be inseparably associated with cross-border relations and co-operation and so taken on a leading role in cross-border co-operation. The pure history of the EU, as well as the numerous experiences gained from co-operation of uncountable scope and scale, ranging from loosely existing joint promotions to highly interdependent cross-border entities, offer interested researchers into this subject a truly bursting spring of information that hardly ever dries up. It was only for individual nations willing to jointly co-operate and promote general similarities without necessarily having to give up national identities and sovereignties, that the EU was founded, and its expansion ever since has proven this concept right. Such wide-ranging co-operation comprising economic, social and political life within the member states naturally soon began to involve tourism industries as well, and so it should not be surprising that the EU has managed to create a single tourism market organised and supervised by centrally legitimised tourism policies.

Due to the Schengen treatment, there are no travel restrictions for travelling on European citizens within the member states, and visa and immigration procedures have been minimised to practically nihilistic levels. As a result, numerous borderlands have formed, created by co-operation across now easily surmountable national borders, and produced the Europe of the regions, a political entity less ruled by its national members but signified by its multinational regions. This development has been actively promoted, specifically in terms of tourism, through unitary European programs such as ENVIREG, INTERREG or LEADER (Leimgruber, 1998, Timothy, 2001), designed to promote cross-border regional development and economic prosperity. It was these programs that initiated the formation of new political entities all over the EU, the EUREGIOS, which engage in energetically promoting and actively supporting cross-border co-operation on social, economical, political, legal and recreational levels. These efforts have even been further stimulated since the introduction of the Euro, the unitary currency for the Eurozone, which obviously tremendously facilitated economic co-operation, business transactions, and travel across the continent. Having introduced the general background of the European Union, the following section will discuss cross-border co-operation with specific regard to tourism within the EU, in particular, and on the European continent in general.

Because much of this co-operation is situated around the German borders, a large proportion of the literature listed below is published in German. This might be because Germany geographically holds a very central position within the European continent. Since it is surrounded by European partner nations, the possibilities for engaging in cross-border relations and co-operation across national boundaries multiply, especially considering that it twice offers the possibility of co-operating with more than one partner. One of these situation is given in the BeNeLux region where there is the option of co-operating with the Netherlands and Belgium within just one co-operation, and in the south of Germany with Austria and Switzerland. None the less, various scholars have conducted research on border co-operation and published in English and French, the two official languages of the EU, so that the following chapter still legitimately claims to offer a basis for further internationally inaugurated research on cross-border relations with regard to tourism on the European continent.

Table 9.10   Selected literature: General structure of European cross-border o-operation

AGRICOLA, S. (ed.) (1994), 'Freizeitmarkt in Europa – Grenzüberschreitende Freizeitmärkte nach Eröffnung des gemeinsamen Marktes', *Schriftenreihe Themendienst Freizeit-Wirtschaft* 1, (Duisburg).

BECKER, CH. (ed.) (1992), *Perspektiven des Tourismus im Zentrum Europas*, (Trier: Europäisches Tourismus Institut GmbH an der Universität Trier).

BLACKSELL, M. (1998), 'Redrawing the political map', in D. Pinder (ed.), *The New Europe: Economy, Society and Environment*, (Chichester: Wiley, 23–42).

BOSENIUS, U. (1981), 'Grenzüberschreitende Zusammenarbeit. Europäisches Rahmenübereinkommen über die grenzüberschreitende Zusammenarbeit zwischen Gebietskörperschaften', *Der Landkreis* 51, 112–114.

BRUNN, G. (ed.) (1998), *Grenzüberschreitende Zusammenarbeit in Europa: Theorie – Empirie – Praxis, internationale Konferenz in Aachen, 18. – 19. September 1997* (Siegen).

BUCKLEY, P.J. AND WITT, S.F. (1990), 'Tourism in the centrally-planned economies of Europe', *Annals of Tourism Research* 17, 7–18.

BUNDESMINISTERIUM FÜR WIRTSCHAFT AND ADAC (1991), *Europäisches Jahr des Tourismus 1990 – Eine Idee braucht Hilfe*, (Bonn/München: Dokumentationsbericht der nationalen und paneuropäischen Projektanträge aus der Bundesrepublik Deutschland).

CATUDAL, H.M. (1979), *The Exclave Problem of Western Europe*, (Tuscaloosa: University of Alabama Press).

CHRISTALLER, W. (1963), 'Some considerations of tourism in Europe: The peripheral regions-underdeveloped countries-recreation areas', *Paper of the Regional Science Association* 12, 95–105.

CHURCH, A. AND REID, P. (1995), 'Transfrontier co-operation, spatial development strategies and the emergence of a new scale of regulations: The Anglo-French border', *Regional studies* 29:3, 297–306.

CIACCIO, C. (1979), 'L'organisation d'un espace périphérique: l'exemple de Pantellerie, limite méridional du tourisme en Italie', in G. Gruber and H. Lamping and W. Lutz and J. Matznetter and K. Vorlaufer (eds), *Tourism and Border: Proceedings of the Meeting of the IGU Working Group – Geography of Tourism and Recreation*, (Frankfurt am Main: Institut für Wirtschafts- und Sozialgeographie der Johann Wolfgang Goethe Universität, 253–265).

CLOß, H. AND GAFFGA, P. (1983), 'Die Grenze als tägliche Erfahrung, Beispiel: Raum Trier – Fachwissenschaftliche und fachdidaktische Hinweise zur unterrichtlichen Behandlung', *Geographie und Schule* 5:24, 31–40.

DEWAILLY, J.M. (1979), 'Fascination et pesanteur d'une frontière pour le tourisme et la recreation l'example Franco-Belge', in Gruber and H. Lamping and W. Lutz and J. Matznetter and K. Vorlaufer (eds), *Tourism and Borders: Proceedings of Meeting of the IGU Working Group – Geography of Tourism and Recreation*, (Frankfurt am Main: Institut für Wirtschafts- und Sozialgeographie der Johann Wolfgang Goethe Universität, 309–317).

ECHIKSON, W. (1997), 'A new map for an old continent: EU picks 11 nations to be partners', *Christian Science Monitor*, (published on 15 December 1997), p. 7.

EMS-DOLLART-REGION (1990), *Operationelles Programm für die EG-Förderinitiative INTERREG, (1991–1993)* (Leer: EDR).

ENDRUWEIT, G. (1983), 'Grenzlage als Bewußtseins- und Imageproblem – das Beispiel des Saarlandes', in LAG Hessen/Rheinland Pfalz/Saarland (ed.), *Probleme räumlicher Planung und Entwicklung in den Grenzräumen an der deutsch-französisch-luxemburgischen Staatsgrenze*, (Hannover: 137–168).

EREG – EIGENSTÄNDIGE REGIONALENTWICKLUNG IM GRENZRAUM (2001), 'Wie Alles began', <http://www.tuwien.ac.at/pr/ news/2001/010726A_EREG.pdf>, accessed on 24 April 2003.

EUROPÄISCHE KOMMISSION AND DIREKTION FÖRDERUNG DES UNTERNEHMENS UND VERBESSERUNG DES UMFELDS AND REFERAT TOURISMUS (1994), *Culture and countryside: 48 projects co-financed by the European Commission in 1992*, (Luxemburg: Office for Official Publication of the European Communities).

EUROPEAN COMMISSION (1995), *The Role of the Union in the Field of Tourism*, (Brussels: Commission of the European Communities).

FREIHERR VON MALCHUS, V. (1983), 'Möglichkeiten grenzüberschreitender Zusammenarbeit in Europa', *Der Landkreis* 53, 235–238.

FRICK, W. AND SCHALLER, C. (1990), 'Europäische Regionen und regionale Tourismusförderung', in U. Braun-Moser (ed.), *Europäische Tourismuspolitik*, (Sindelfingen: 150–167).

GABBE, J. (1992), 'EUREGIO – Regionale grenzüberschreitende Zusammenarbeit in der Praxis', in Akademie für Raumforschung und Landesplanung (ed.), *Grenzübergreifende Raumplanung – Erfahrungen und Perspektiven der Zusammenarbeit mit den Nachbarstaaten Deutschlands: Forschungs- und Sitzungsberichte der Akademie für Raumforschung und Landesplanung* 188, 187–208, (Hannover).

GARRETT, G. AND WEINGAST, B.R. (1993), 'Ideas, interests, and institutions: Constructing the European Community's Internal Market', in J. Goldstein and R.O. Keohane (eds), *Ideas in Foreign Policy. Beliefs, Institutions, and Political Change*, (Ithaca/London: Cornell University Press, 173–206).

GONIN, P. (1994), 'Régions frontalières et développement endogène: de nouveaux territoires en construction au sein de l'Union Européenne', *Hommes et Terres du Nord* 2:3, 61–70.

GRIMM, F.-D. (1995), 'Veränderte Grenzen und Grenzregionen, veränderte Grenzbewertung in Deutschland und Europa', in F.D. Grimm (ed.), *Regionen an deutschen Grenzen. Strukturwandel an der ehemaligen innerdeutschen Grenze und an der deutschen Ostgrenze: Beiträge zur Regionalen Geographie* 38, 1–16 (Leipzig: Institut für Länderkunde).

HAMM, R. (1996), *European Border Regions – Driving Force of European Integration?*, (Zürich: Paper presented at the 36th European Congress of the European Regional Science Association).

HANSEN, N. (1986), 'Border region development and co-operation: Western Europe and the US-Mexico borderlands in comparative perspective', in O.J. Martinez (ed.) *Across Boundaries: Transborder Interaction in Comparative Perspective*, (El Paso: Center for Inter-American and Border Studies, University of Texas, 31–44).

HAUSMANN, W. (1986), '3. Europäische Konferenz der Grenzregionen, Borken, Bundesrepublik Deutschland, 4. – 6. September 1984', *Schriftenreihe Landes – und Stadtentwicklungsforschung des Landes Nordrhein-Westfalen*, Sonderveröffentlichungen 0.032.

HERSTER, S. (n.d.), 'Gutachten zu Hemmnissen bei der Gründung einer grenzüberschreitenden Branchengruppe"Tourismus"imRahmendesINTERREGll-Projektes"Euregionalisierung der beruflichen Qualifizierung und der Erwachsenenbildung in der euregio rhein-maas-nord"', <http://www.euregio.krefeld.schulen.net/trias_403_II/tourismus/ tourismus.pdf>, accessed on 20 March 2003.

JENNER, P. AND SMITH, C. (1993), 'Europe's microstates: Andorra, Monaco, Liechtenstein and San Marino', *EIU International Tourism Reports* 1, 69-89.

KEARNEY, E.P. (1992), 'Redrawing the political map of tourism: The European view', *Tourism Management* 13, 34–36.

KNIGGE, D. (2001), 'Maßnahmen der Europäischen Gemeinschaft im Bereich des Fremdenverkehrs', *Europäische Hochschulschriften* 2, (Frankfurt am Main: Rechtswissenschaft).

KOMMISSION DER EUROPÄISCHEN GEMEINSCHAFT (1982a), *Etappen nach Europa. Chronik der Europäischen Gemeinschaft*, (Brüssel/Luxemburg: Kommission der Europäischen Gemeinschaft).

KOMMISSION DER EUROPÄISCHEN GEMEINSCHAFT (2001), 'Zusammenarbeit für die Zukunft des Tourismus in Europa', <http://www.alpenforschung.de/skripte/EUtourismus-2001-zusammenarbeit.pdf>, accessed on 25 April 2003.

LANG, S. and Becker-Marx, K. (1981), 'Untersuchung der Möglichkeiten und Beschränkungen einer kommunalen und regionalen Kooperation über die europäischen Binnengrenzen', in K. Becker-Marx (ed.), *Stand der grenzüberschreitenden Raumordnung am Oberrhein*, (Heidelberg: Heidelberger Geographisches Institut, 73–105).

LEIMGRUBER, W. (1981), 'Political boundaries as a factor in regional integration: Examples from Basle and Ticino', *Regio Basiliensis* 22, 192–201.

LEIMGRUBER, W. (1991), 'Boundary, values and identity: The Swiss-Italian transborder region', in D. Rumley and J.V. Minghi (eds), *The Geography of Border Landscapes*, (London: Routledge, 43–62).

LIBERDA, E. (1997), 'Regionalentwicklung in Grenzregionen: Eine Euregio als Regionalentwicklungsstrategie? Das Beispiel der Inn-Salzach-Euregio', *SIR- Mitteilungen und Berichte* 25, 79–90, (Salzburg: Salzburger Institut für Raumordnung Verlag).

MAILLAT, D. (1990), 'Transborder regions between members of the EC and non-member countries', *Built Environment* 16:1, 38–51.

MINGHI, J.V. (1994a), 'European borderlands: International harmony, landscape change and new conflict', in C. Grundy-Warr (ed.) *World Boundaries, Vol. 3, Eurasia*, (London: Routledge, 89–98).

OEDEKERK, F.C.J. (1983), 'Arbeid in een Grensgebied. Internationaal forensisme in de Regio Basel', *Geografisch Tijdschrift* 17, 39–48.

ORDEMANN, B. (1989), *Die Europäische Gemeinschaft und der Fremdenverkehr – mögliche Folgen des europäischen Binnenmarktes 1992 für den Fremdenverkehr in der Bundesrepublik Deutschland*, (Trier: Diplomarbeit).

PEARCE, D.G. (1992), 'Tourism and the European Regional Development Fund: The first fourteen years', *Journal of Travel Research* 30:3, 44–51.

PERKMANN, M. (2003), 'Cross-border regions in Europe', *European Urban and Regional Studies* 10:2, 153–171.

POLIVKA, H. (1982), 'Die Regio als Wirtschaftsraum', *Regio Basiliensis XXIII* 1/2, 117–138.

ROBINSON, G. AND MOGENDORFF, D. (1993), 'The European tourism industry: Ready for a single market?', *International Journal of Hospitality Management* 12, 21–31.

SCHLAMMERL, J. (1983), *Überörtliche Zusammenarbeit im Bereich der Fremdenverkehrsinfrastruktur – Entwicklung, Auswirkungen und Erfahrungen am Beispiel der Erholungsgebiete Obere Kyll, Bitburger Stausee und Rheinhessische Schweiz*, (Trier: Diplomarbeit).

SCOTT, J. and Sweedler, A. and Ganster, P. and Eberwein, W. (eds) (1996), *Border Regions in Functional Transition. European and North American Perspectives on Transboundary Interaction*, (Erkner: Regio Series of the IRS 9, 171–191).

SCOTT, J.W. (1989), 'Transborder co-operation, regional initiatives, and sovereignty conflicts in western Europe: The case of the Upper Rhine Valley', *Publius: The Journal of Federalism* 19, 139–156.

SCOTT, J.W. (1993), 'The institutionalization of transboundary co-operation in Europe: Recent development on the Dutch-German border', *Journal of Borderlands Studies* 8:1, 39–66.

SCOTT, J.W. (1998), 'Planning co-operation and transboundary regionalism: Implementing policies for European border regions in the German-Polish context', *Environmental and Planning* 16, 605–624.

SCOTT, J. W. (1999), 'European and North American contexts for cross-border regionalism', *Regional Studies* 33:7, 605–618.

STAATSKANZLEI RHEINLAND-PFALZ (ed.) (1992), *Handbuch der grenzüberschreitenden Zusammenarbeit in Rheinland-Pfalz*, (Mainz: Staatskanzlei Rheinland-Pfalz).

SOUTH EAST REGION IN EUROPE (2002), 'Resume of the Interreg IIIA programme based on the Operational Programme and the Programme Complement', <http://www.go-se.gov.uk/key%20business/europrogrammes/ euro_iinterreg.htm>, accessed on 3 April 2003.

SWEET, A.S. AND SANDHOLTZ, W. (1997), 'European integration and supranational governance', *Journal of European Public Policy* 4:3, 297–317.

TYKKYLÄINEN, M. AND BOND, D. (1997), 'Border as Opportunities in the Regional Development of Europe', in author unknown, *Proceedings of the 37th European Congress of the Regional Science Association*, (Rome).

WACHOWIAK, H. (1994a), 'Grenzüberschreitende Zusammenarbeit im Tourismus auf der Ebene der öffentlichen Hand entlang der westdeutschen Staatsgrenze', *Raumforschung und Raumordnung* 52:6, 397–405.

WACHOWIAK, H. (1994b), *Grenzüberschreitende Zusammenarbeit im Tourismus – eine Analyse grenzüberschreitender Maßnahmen an der westlichen Staatsgrenze der Bundesrepublik Deutschland zwischen Ems-Dollart und Baden-Nordelsaß-Südpfalz*, (Trier: Europäisches Tourismus Institut).

WACHOWIAK, H. (1995), 'Touristische Entwicklungsperspektiven für Grenzregionen im Europäischen Binnenmarkt', in Deutsche Gesellschaft für Freizeit (ed.) *Freizeit, Tourismus und Europäischer Binnenmarkt: Daten-Analysen-Perspektiven-Politik-Marketing*, (Erkrath: Deutsche Gesellschaft für Freizeit, 151–170).

WACHOWIAK, H. (1997), 'Tourismus im Grenzraum – Touristische Nachfragestrukturen unter dem Einfluß von Staatsgrenzen am Beispiel der Grenzregion Deutschland-Luxemburg', *Materialien zur Fremdenverkehrsgeographie* 38, (Trier).

WANHILL, S. (1997), 'Peripheral area tourism: A European perspective', *Progress in Tourism and Hospitality Research* 3:1, 47–70.

WU, C. (1998), 'Cross-border development in Europe and Asia', *GeoJournal* 44:3, 189–201.

Policies of European cross-border co-operation in tourism

As already discussed earlier, a major component of cross-border relations and tourism co-operation, namely the creation and authorisation of coherent policies, proves to be less of a bias on the European continent than it is in other parts of the world. The EU, being a political entity of its own, legitimised through its sovereign member states and equipped with extensive jurisdictional and legislative powers executed by a single European parliament, enabled the creation of a single European tourism market. Such a single market is arguably easier to administer through a coherent European tourism policy agreed upon in the European parliament, rather than having to go through the obstacles of deciding on every aspect individually and from co-operation to co-operation. Being able to operate in such a collaboratively created supra-national policy-framework, co-operation among EU member states hence rather resembles co-operation across sub-national borders and subsequently tends to deal with specified details rather than general policies.

Co-operation program leaders are significantly liberated from extensive bureaucracies and are consequently able to concentrate on smooth operations and constant improvements through fast decision-making even though operating in two different sovereign market places at the same time. Any co-operation within the EU, therefore, is not only able to operate but also be initiated and solely existent on a regional or even local level, instead of having to go through extensive national approval procedures, which might have been required in the case of strict visa regulations or unclear foreign affair policies. This facilitated the necessary legal and political procedures for international cross-border co-operation and supported the requirements for sustainable regional economic development, where it is highlighted that successful and sustainable co-operation needs to be regionally planned, jointly developed and collaboratively operated (Bramwell and Lane, 2000). Such a framework does not solely benefit intra-European relations and co-operation, though, it also facilitates cross-border co-operation on the edges of EU territory.

Considering the fact, that the EU functions as a political entity, co-operation along the Greek-Turkish border is accordingly subject to the same political framework as co-operation along the Finnish-Russian borderline, to name just two contrasting examples. Even though the geographic or political situation, historic relations and social composition within these regions differs tremendously, the framework applies to both, and facilitates the operation of cross-border co-operation and foreign affairs between the individual sovereign European nation-states, their co-operating partner sovereignties and the supra-national entity of the European Community, through clearly formulated policies and negotiating parties. The literature listed below further investigates the existing EU tourism policies (see especially Braun-Moser, 1991), indicates its development, and examines its implications for cross-border co-operation on the European continent.

Table 9.11   Selected literature: Policies of European cross-border co-operation in tourism

AMTSBLATT DER EUROPÄISCHEN GEMEINSCHAFTEN (1992), 'Vertrag über die Europäische Union unterzeichnet zu Maastrich am 7. Feb. 1992', <http://europa.eu.int/eurlex/de/ treaties/dat/ EU_treaty.html#0001000001>, accessed 25 April 2003.

AMT FÜR AMTLICHE VERÖFFENTLICHUNGEN (1991), 'Aktionsplan der Gemeinschaft zur Förderung des Fremdenverkehrs', *Dokumente der Kommission der Europäischen Gemeinschaften*, (Luxemburg: Amt für Amtliche Veröffentlichungen, 1–33).

AMT FÜR AMTLICHE VERÖFFENTLICHUNGEN DER EUROPÄISCHEN GEMEINSCHAFTEN (1997), 'Bericht der Kommission an den Rat, das Europäischen Parlament, den Wirtschafts- und Sozialausschuß und den Ausschuß der Regionen zu Maßnahmen der Gemeinschaft, die sich auf den Tourismus auswirken 1995–96', *Dokumente, Kommission der Europäischen Gemeinschaften*, (Luxemburg: Amt für Amtliche Veröffentlichungen).

BRAUN-MOSER, U. (1991), *Europäische Tourismuspolitik*, (Sindelfingen: Libertas Verlag).

DUPUY, P.M. (1982), 'Legal aspects of transfrontier regional co-operation', *West European Politics* 5, 50–63.

EUREGIO (ed.) (1989), *Grenzüberschreitendes Durchführungsprogramm für die EUREGIO 1989–1992*, (Gronau: EUREGIO).

EUREGIO (ed.) (1990), *Operationelles grenzüberschreitendes Durchführungsprogramm für die EUREGIO 1990–1993*, (Gronau: EUREGIO).

GRENZREGIO RHEIN-MAAS-NORD (ed.) (1978), *Übereinkommen zur Bildung der Grenzregio Rhein-Maas-Nord gemäß Beschluß des Regio-Rates vom 13 Dezember 1978*, (Roermond).

HUNTOON, L. (1998), 'Immigration to Spain: Implications for a unified European Union immigration policy', *International Migration Review* 32:2, 423–450.

INTERREG IIIa (2000), 'Programm 2000 – 2006: Programmplanungsdokument zur Förderung der grenzüberschreitenden Zusammenarbeit', <http:// www.salzburg.gv.at/pdf-programm_de.pdf>, accessed on 24 April 2003.

HELMUT, W. (1983), 'Regional Policy and International Border Areas of the EC: A Case Study Concerning the Belgian-Dutch-German Border Area', in J.S. Adams and W. Fricke and W. Herden (eds), *American-German International Seminar Geography and Regional Policy: Resource Management by Complex Political Systems*, (Heidelberg: Ruprecht-Karls Universität, 298–305).

KOMMISSION DER EUROPÄISCHEN GEMEINSCHAFTEN (1982b), 'Erste Überlegungen zu einer Fremdenverkehrspolitik der Gemeinschaft', *Bulletin der Europäischen Gemeinschaften* 4, Beilage, (Brüssel: Kommission der Europäischen Gemeinschaft, Generalsekretariat, 1–37).

KOMMISSION DER EUROPÄISCHEN GEMEINSCHAFTEN (1982c), 'Eine Fremdenverkehrspolitik der Gemeinschaft. Erste Überlegungen', *Bulletin der Europäischen Gemeinschaften* 4, Beilage.

LEZZI, M. (1993), 'Raumordnungspolitik in den 90er Jahren zwischen Regionalismus und Supranationalisierung? Untersuchung in deutschen und schweizerischen Grenzregionen', *Tagungsbericht und wissenschaftliche Abhandlungen* 49:4, (Stuttgart/Wiesbaden: Deutscher Geographentag, 66–74).

Historical background and development of tourism cross-border co-operation in Europe

As indicated earlier, within Europe, cross-border relations and the resulting co-operation are often built upon historically manifested long-lasting relationships. The literature listed below, therefore, intends to give a general overview of how such historic relations influence modern co-operation and across-the-border relations. The issues addressed within the literature comprise the influence borders have had and still exert on the social composition and identity formation within border regions and how such formerly highly fought over and militarised boundaries between nation states were overcome to create the modern EU. One specific aspect examined is the impact passport facilities potentially exert upon the creation of tourism phenomena (see Warszynska and Jackowski, 1979). It is not only the political implications such control situations create that are interesting here. Nor is it restricted to investigation of the stress and resulting potential reluctance towards crossing the border that is created by strenuous visa procedures and entry regulations (Timothy, 1995; Timothy, 2001), but it is also due to the function of a passport in general. The passport itself was identified as a means of consolidating national state identity and, according to O'Byrne (2001: 40), even "perhaps, the most important symbol of the nation-state system". Since nationhood closely relates to insider- vs. outsidership (O'Byrne, 2001) it naturally requires an 'other' as opposed to oneself. This 'other' needs to be identified and defined and passport controls and further rituals undertaken in border crossing procedures have proved highly suitable (see O'Byrne, 2001, for a detailed discussion on the history of the development and use of passports).

By controlling and monitoring who enters or leaves a country, nation-states were able to distinguish between peers and others and, by comparing and contrasting, consolidate their individual identity. With the ratification of the Schengen treatment, the need for EU citizen passports for travel within the Union vanished and supported the notion of the European Union growing closer together and creating a truly European identity.

As with numerous other issues discussed within this chapter, research into these subjects and themes is only in its infancy and requires further academic studies and investigations. For this reason, the literature listed below intends to offer a suitable starting point for examining the historic determinants of modern cross-border co-operation in Europe.

Table 9.12   Selected literature: Historical background of European cross-border co-operation

AKADEMIE FÜR RAUMFORSCHUNG UND LANDESPLANUNG (1969), 'Grenzbildende Faktoren in der Geschichte', *Akademie für Raumforschung und Landesplanung-Forschungs- und Sitzungsberichte* 48, (Hannover).

ALTMANN, J. (1955), 'Der grenzüberschreitende Reiseverkehr 1950–54', *Statistisches Monatsheft Schleswig-Holstein* 7, 285–288.

BECKER, E. (VORWORT) (1983), 'Dreiländereck Vormals. Alte Fotos aus dem Saargau' (Saarbrücken: Saarbrücker Druckerei und Verlag).

COX, H. L. (1983), 'Die Auswirkungen der deutsch-niederländischen Staatsgrenze von 1815 auf die volkstümliche Heiligenverehrung im Rhein-Maasgebiet. Ein Beitrag zur kulturräumlichen Stellung des Rhein-Maas-Gebietes', *Rheinisch-Westfälische Zeitschrift für Volkskunde* 28, 111–131.

DÜWELL, K. AND KÖLLMANN, W. (eds) (1983), Von der Entstehung der Provinzen bis zur Reichsgründung, Rheinland-Westfalen im Industriezeitalter Vol. 1, (Wuppertal/Hamm).

FEHRENBACH, E. (1982), 'Die Einführung des französischen Rechts in der Pfalz und in Baden', in F.L. Wagner (ed.), Strukturwandel im pfälzischen Raum vom Ancien Regime bis zum Vormärz (Speyer: Verlag der Pfalz, 61–71).

GLOECKNER, C. (1982), 'Von den Grenzländern zu einer Europäischen Großregion (Saar-Lor-Lux)', Wirtschaft und Standort 7, 2–6).

GRIESER, D. (1983), Historische Strassen in Europa: Von der Via Appia bis zur Avus, (Hamburg: Kabel Verlag).

KEMPEN, L. (1983), 'Unsere niederländische Sprachvergangenheit. Sprachumbruch am Niederrhein', Kalender für das Klever Land 33, 192–195.

KOLTZ, J. (1983), 'Die Geschichtlichen Verbindungen zwischen Luxemburg und Trier. III. Teil', Jahrbuch 1983, (Kreis Trier/Saarburg, 340–367).

KURZ, G. (1981), 'Die Entwicklung des grenzüberschreitenden Reiseverkehrs in der Bundesrepublik Deutschland von 1965–1979', Schriftenreihe Fremdenverkehr 4, 92–96, (Heilbronn).

LUTZ, H. AND RUMPLER, H. (ed.) (1982), Österreich und die deutsche Frage im 19. und 20. Jahrhundert: Probleme der politisch–staatlichen und sozio-kulturellen Differenzierung im deutschen Mitteleuropa, (München/Oldenbourg: Wiener Beiträge zur Geschichte der Neuzeit 9).

MAIER, J. (1990), 'Staatsgrenzen und ihr Einfluß auf Raumstrukturen und Verhaltensmuster: Grenzen gegenüber der CSSR und der DDR', Arbeitsmaterialien zur Raumordnung und Raumplanung 26 (Bayreuth).
MAIER, J. (1993), 'Staatsgrenzen und ihr Einfluß auf Raumstrukturen und Verhaltensmuster: Die Grenzöffnung zur CSFR und ihre Auswirkungen auf die Stadt Weiden/Opf. Und ihr Umland', Arbeitsmaterialien zur Raumordnung und Raumplanung 118, (Bayreuth).
MARTIN, P. (1983), *Salon Europas, Baden-Baden im 19.Jahrhundert*, (Konstanz).
MATHIEU, F. (1983), *Wasserbillig im 19. und 20. Jahrhundert 2*, (Luxemburg).
MEDVEDEV, S. (1999), 'Across the line: Borders in post-Westphalian landscapes', in H. Eskelinen and I. Liikanen and J. Oksa (eds), *Curtains of Iron and Gold: Reconstructing Borders and Scales of Interaction*, (Aldershot: Ashgate, 43–56).
MÜLLER-LANKOW, B. (1980), 'Freizeitforschung 1979–80: Dokumentation', Deutsche Gesellschaft für Freizeit 47.
TRAUSCH, G. (1983), 'Deutschland und Luxemburg vom Wiener Kongress bis zum heutigen Tage: Die Geschichte einer Entfremdung', in J. Becker and A. Hillgruber (eds), *Die Deutsche Frage im 19. und 20. Jahrhundert*, (München: 185–220).
WEBER, H. (1984), Grenzvermessung Deutschland – Luxemburg: Die Entstehung der Grenze in den Jahren 1815/16, sowie ihre Vermessung und Dokumentation in den Jahren 1980–1984, (Koblenz: Landesvermessungsamt Rheinland-Pfalz).
WARSZYNSKA, J. and JACKOWSKI, A. (1979), 'Impact of passport facilities in the passenger traffic between Poland and the German Democratic Republic (GDR) on the development of touristic phenomena', in G. Gruber and H. Lamping and W. Lutz and J. Matznetter and K. Vorlaufer (eds), *Tourism and Borders: Proceedings of the Meeting of the IGU Working Group – Geography of Tourism and Recreation*, (Frankfurt a.M.: Institut für Wirtschafts- und Sozialgeographie der Johann Wolfgang Goethe Universität, 353).

Geographical considerations of cross-border relations in Europe

Whether a borderland is capable of developing as a tourist destination after all, often depends upon geographical factors that influence the area around the boundary. Aspects to be considered when investigating these determinants are the potential duplication of tourism resources on the different sides of the border, the level of available infrastructure on either side of the border, the existence of natural or man-made resources capable of being employed in tourism industries, or the historic land-use practices, to name just a few. Geographical considerations obviously greatly influence cross-border co-operation, since it is these features that often create the basis for physiographical boundaries. Such boundaries also regularly prove capable of attracting tourists searching for untouched nature, scenic views or rural and peripheral ways of life.

Developing any economic activity within such regions highly depends upon the geographical conditions and the influence they pose upon smooth and economically viable operations. The literature combined below investigates some of these issues in more detail and, therefore, offers a profound first insight into the influence geographical structures pose upon socio-economical activities in borderlands. The focus is thereby placed upon different forms of spatial distribution and land-use as well as residents' and visitors' resulting action-spaces. None the less, some literature discusses the general aspects of European geography and introduces its ubiquity and singularities.

Table 9.13   Selected literature: Geographical aspects of European cross-border co-operation

AKADEMIE FÜR RAUMFORSCHUNG UND LANDESPLANUNG (1992), 'Grenzübergreifende Raumplanung: Erfahrungen und Perspektiven der Zusammenarbeit mit den Nachbarstaaten Deutschlands', *ARL-Forschungs- und Sitzungsberichte* 188, (Hannover).
ARBEITSGEMEINSCHAFT ALPEN-ADRIA, KOMMISSION FÜR RAUMORDNUNG UND UMWELTSCHUTZ (1982), *Gemeinsamer Raumplanungsbericht* (Klagenfurt: ARGALP).
ARBEITSGEMEINSCHAFT DER LÄNDER UND REGIONEN DER OSTALPENGEBIETE AND KOMMISSION FÜR RAUMORDNUNG UND UMWELTSCHUTZ (1987), 'Wirtschaft, Tourismus und Verkehr als Komponenten der räumlichen Entwicklung im Gebiet der Arbeitsgemeinschaft Alpen-Adria', ETI-Studien 1. (Linz: Amt der Oberösterreichischen Landesregierung, München: Bayerisches Staatsministerium für Landesentwicklung und Umweltfragen).
ARNOLD-PALUSSIERE, M. (1983), 'Die grenzüberschreitende regionale Zusammenarbeit auf dem Gebiet der Raumordnung: Rheintal, Elsaß, Pfalz, Baden, Nordwestschweiz', *ARL-Beiträge* 71, 379–393, (Hannover).
AUST, B. (1983), 'Die staatliche Raumplanung im Gebiet Saar-Lor-Lux-Regionalkommission', (Saarbrücken: Universität des Saarlandes, Arbeiten aus dem Geographischen Institut der Universität des Saarlandes, Sonderheft 4).
BALON, E. (1983), 'Strukturräume und Funktionsbereiche im salzburgisch-bayrischen Grenzgebiet', Meitteilungen und Berichte des SIR 3–4, 32–47.
BECKER-MARX, K. (1981a), 'Modelle der Organisatorischen Lösung; insbesondere: Das Modell einer "Planungsgemeinschaft Oberrhein"', in K. Becker-Marx (ed.) *Stand der grenzüberschreitenden Raumordnung am Oberrhein*, (Heidelberg: Heidelberger Geographisches Institut, 36–45).
BECKER-MARX, K. (ed.) (1981b), 'Stand der grenzüberschreitenden Raumordnung am Niederrhein', Heidelberger geographische Arbeiten 71, (Heidelberg: Geographisches Institut der Universität Heidelberg)..
BECKER-MARX, K. (1983), 'Raumordnung im Oberrheingraben: Der Versuch einer Zusammenarbeit im Grenzgebiet zweier großer europäischer Nationen', in *43. Deutscher Geographentag Mannheim 1981*, (Wiesbaden: 327–330).
BREUER, H. (1981), '"Grenzfälle" – Hemmungs- oder Anregungsfaktoren für räumliche Entwicklungen (insbesondere aufgezeigt im Belgisch-Niederländisch-Deutschen Grenzraum)', in *Festschrift für Felix Monheim zum 65. Geburtstag 2*, (Aachen: 425–438).
BULLINGER, D. (1982), 'Probleme und Ansätze grenzüberschreitender Raumkoordinierung, dargestellt an den Beispielen Bodensee, Hoch- und Oberrhein', *Mitteilungen 1982, 3–6*, 105–115, Österreichisches Institut für Raumplanung.

BULLINGER, D. (1983), 'Unterschiedliche Probleme und Ansätze der grenzüberschreitenden Raumkoordinierung an drei Beispielen: Bodensee, Hochrhein, Oberrhein', in *43. Deutscher Geographentag Mannheim 1981*, (Wiesbaden: 308–310).

BURTON, R.C.J. (1994), 'Geographical patterns of tourism in Europe', *Progress in Tourism, Recreation and Hospitality Management* 5, 3–25).

BRÜCHER, W. (ed.) (1982), 'Pilotstudie zu einem Saar-Lor-Lux-Atlas', Schriftenreihe der Regionalkommission Saarland-Lothringen-Luxemburg-Rheinland-Pfalz 8, (Saarbrücken).

BRÜCHER, W. (1983), 'Inhalte und Ziele eines grenzüberschreitenden Kartenwerkes in der Region Saar-Lor-Lux', in *43. Deutscher Geographentag Mannheim 1981*, (Wiesbaden, 324–327).

COAKLEY, J. (1982), 'National territories and cultural frontiers: Conflicts in the formation of states in Europe', *West European Politics* 5:4, 34–49.

COLE, J. and Cole, F. (1993), *The Geography of the European Community*, (London: Routledge).

DEGE, W. (1983), 'Grenzüberschreitende Zentrenbeziehungen und räumliche Integration in der Regio Basiliensis', in *18. Deutscher Schulgeographentag Basel 1982*, Tagungsband, (Basel: 138–155).

GRIMM, F.D. (ed.) (1995), 'Regionen an deutschen Grenzen. Strukturwandel an der ehemaligen innerdeutschen Grenze und an der deutschen Ostgrenze', *Beiträge zur Regionalen Geographie* 38, (Leipzig: Institut für Länderkunde).

ISTEL, W. AND ROBERT, J. (1982), 'Raumordnung beiderseits der Grenze der Bundesrepublik Deutschland zu den Nachbarstaaten der Europäischen Gemeinschaft sowie der Schweiz und Österreich – unter besonderer Berücksichtigung der Zentren und Achsen', ARL-Beiträge 59, (Hannover).

KISTENMACHER, H. (1988), 'Die Auswirkungen der Dezentralisierung auf die Verwaltungsstruktur sowie auf das Raumordnungssystem in Frankreich und Erfordernisse der grenzüberschreitenden Zusammenarbeit' *Arbeitsmaterialien der Akademie für Raumforschung und Landesplanung* 146, (Hannover).

KISTENMACHER, H. AND GUST, D. (1983), 'Erfordernisse und Probleme der grenzüberschreitenden Abstimmung bei der Raumplanung im deutsch-französichen Grenzraum am Oberrhein', in LAG Hessen/Rheinland Pfalz/Saarland (ed.) *Probleme räumlicher Planung und Entwicklung in den Grenzräumen an der deutsch-französich-luxemburgischen Staatsgrenze*, (Hannover: 41–70).

LENDI, M. (1983), 'Werdende europäische Raumordnungspolitik', in M. Lendi (ed.), *Elemente zur Raumordnungspolitik*, (Zürich: 117–130).

LEVY, F. AND FLICKINGER, N. AND SCHULZ, C. (1999), Grenzüberschreitende Raumordnungsstudie Diedenhofen: Trier, Regionalkommission Saarland-Lothringen-Luxemburg-Trier-Westpfalz, Arbeitsgruppe Raumordnung, (Merzig: Merziger Druckerei und Verlagsgesellschaft mbH).

MAIER, J. (ed.) (1983), 'Staatsgrenzen und ihr Einfluß auf Raumstrukturen und Verhaltensmuster 1, Grenzen in Europa', *Arbeitsmaterialien zur Raumordnung und Raumplanung* 23, (Bayreuth).

MAIER, J. (1994), 'Staatsgrenzen und ihr Einfluß auf Raumstrukturen und Verhaltensmuster: Auswirkungen von Grenzen (ausgewählte Fallbeispiele)', Arbeitsmaterialien zur Raumordnung und Raumplanung 126, (Bayreuth).

MAIER, J. and WEBER, J. (1979), 'Tourism and leisure behaviour subject to the spatial influence of a national frontier: The example of Northeast Bavaria', in G. Gruber and H. Lamping and W. Lutz and J. Matznetter and K. Vorlaufer (eds), *Tourism and Borders: Proceedings of the Meeting of the IGU Working Group – Geography of Tourism and Recreation*, (Frankfurt a.M.: Institut für Wirtschafts- und Sozialgeographie der Johann Wolfgang Goethe Universität, 111–127).

MOLL, P. (1983), 'Der Beitrag der Raumordnung zur Überwindung der Grenzen im Gebiet Saarland / Rheinland-Pfalz – Lothringen – Luxemburg', in LAG Hessen/Rheinland Pfalz/ Saarland (ed.), *Probleme räumlicher Planung und Entwicklung in den Grenzräumen an der deutsch-französich-luxemburgischen Staatsgrenze*, (Hannover, 71–105).

MOLL, P. (1990), 'Grundlagen und Ergebnisse der grenzüberschreitenden räumlichen Abstimmung und Planung im Raum Saar-Lor-Lux-Trier-Westpfalz', in *Saar-Lor-Lux-Trier-Westpfalz: Eine Region auf dem Weg nach Europa*, (Trier: 20.-22. September 1990, 83–91, Grenzüberschreitender Hochschulkongreß).

PEDRESCHI, L. (1957), 'L'exclave Italiano in terra Svizzera di Campione d'Italia', *Revista Geografica Italiana* 64, 23–40.

ROBERT, J. (1983), 'Räumliche Planung in Belgien, den Niederlanden und Nordrhein-Westfalen', Schriftenreihe Landes- und Stadtentwicklungsforschung des Landes Nordrhein-Westfalen 1, (Dortmund).

RUPPERT, K. (1979), 'Funktionale Verflechtungen im deutsch-österreichischen Grenzraum', in G. Gruber and H. Lamping and W. Lutz and J. Matznetter and K. Vorlaufer (eds), *Tourism and Borders: Proceedings of the Meeting of the IGU Working Group – Geography of Tourism and Recreation*, (Frankfurt a.M.: Institut für Wirtschafts- und Sozialgeographie der Johann Wolfgang Goethe Universität, 95–110).

SCHREIBER, T. (1983), 'Thematische Karten aus der Euregio Maas-Rhein: "Radwandern Kreis Aachen": Möglichkeiten zur Auswertung im Unterricht', *Informationen und Materialien zur Geographie der Euregio Maas-Rhein* 12, 1–4.

SCHWENDERMANN, F. (1983), 'Infrastrukturelle Verflechtungen über die Grenze am Hochrhein', in *43. Deutscher Geographentag Mannheim 1981*, (Wiesbaden, 310–314).

SIEMONS, S. (1989), Struktur und aktionsräumliches Verhalten von niederländischen und deutschen Feriendorfgästen – dargestellt an zwölf Feriendörfern der Region Trier, (Trier: Diplomarbeit).

WEBER, H. (1983), 'Der Wasserlauf als Staatsgrenze. Mit Berücksichtigung der besonderen Verhältnisse der deutsch-luxemburgischen Grenze', Nachrichtenblatt der Vermessungs- und Katasterverwaltung Rheinland-Pfalz 26:3, 174–196 and 4, 262–285.

ZOEBEL, CHR. (1982), 'Grenzüberschreitende Raumordnung und Landesplanung im Weserraum', *Die Weser* 56:2, 23–28.

Cross-border co-operation concerned with environmental protection

Closely related to geographical features, but, due to its importance for the tourism industry and its reluctance to adhere to man-made boundaries presented individually, the following sub-section concentrates on literature discussing cross-border co-operation concerned with environmental protection and international preservation efforts across the European continent. The creation of multinational trans-boundary nature reserves and international parks are just one aspect of such protective co-operation that additionally comprises activities related to assuring water quality protection or landscape preservation. Considering the increasing deforestation and decay in moors and wetlands across Europe, especially the latter aspect will prove to be of major importance for future preservational efforts. Several multinational nature parks have already been established and are operated across national borders. One example, is the bi-national nature park situated in the border region between Germany and Luxemburg; another one is the adjacent nature park between Germany and Belgium.

Issues examined, investigated, and academically researched in terms of their feasibility, implications and results might prove valuable for further multi-national nature and biosphere reserves developments. Besides the benefits created through environmental protection, these developments offer tremendous potential for economically viable as well as environmentally sustainable tourism businesses. Their importance for future tourism related businesses is likely to grow, and consequently requires a thorough understanding of its underlying forces to prevent overuse and negative exploitation.

Table 9.14   Selected literature: Cross-border co-operation and environmental protection

BESSLING, B. (1981), 'Grenzüberschreitende Hochmoore und Heiden zu den Niederlanden im Kreis Borken', *Mitteilungen der Landesanstalt für Ökologie, Landschaftsentwicklung und Forstplanung Nordrhein-Westfalen* 6:2, 39–44.
BORCHERDT, Chr. (1964), 'Die Veränderungen in der Kulturlandschaft beiderseits der saarländisch-lothringischen Grenzen', in *Tagungsberichte und wissenschaftliche Abhandlungen des Deutschen Geographentages 1963*, (Wiesbaden: 335–350).
BRUGGER, A. and DILLMANN, E. (1983), *Der Bodensee: Eine Landeskunde im Luftbild*, (Stuttgart: Theiß-Verlag).
DENISIUK, Z. and STOYKO, S. and TERRAY, J. (1997), 'Experience in cross-border co-operation for national park and protected areas in central Europe', in J.G. Nelson and R. Serafin (eds), *National Parks and Protected Areas: Keystones to Conservation and Sustainable Development*, (Berlin: Springer, 145–150).
DUFFEY, E. (ed.) (1983), *Naturparks in Europa. Ein Führer zu den schönsten Naturschutzgebieten von Skandinavien bis Sizilien*, (München: Christian-Verlag/Stuttgart: Deutscher Bücherbund).
EIFELVEREIN (ed.) (1961), *Naturpark Südeifel: Führer durch das Naturparkgebiet und seine Nachbarbezirke am luxemburgischen Ufer von Auer und Our*, (Düren: Eifelverein).
EUROPEAN COMMISSION (2000), 'Results of the Karelia Parks Development Project', <http://parks.karelia.ru/2.html>, accessed on 3 April 2003.
FRIEDRICHS, G. IN CO-OPERATION WITH KOMMISSION FÜR DEN DEUTSCH-LUXEMBURGISCHEN NATURPARK UND DER STAATSKANZLEI RIP (OBERSTE LANDESPLANUNGSBEHÖRDE) (1972), *Deutsch – Luxemburgischer Naturpark, Landschafts – und Entwicklungsplan*, (Luxembourg).

GRASSHOPPER (1994), 'Trans-European Protected Belt and its Cross-Border Zones', Grasshopper 5, Winter 1994, <http://www.zb.eco.pl/ gh/5/belt_e.htm>, accessed 22 February 2003.

HOMBURGER, W. (1983), 'Landschaftsentwicklung und Umweltvorsorge im Vergleich der Oberrheinländer', in *43. Deutscher Geographentag Mannheim 1981*, (Wiesbaden, 314–317).

INSTITUT FÜR LANDES- UND STADTENTWICKLUNGSFORSCHUNG (1986), *Grenzüberschreitende Umweltschutzkontakte zwischen den Niederlanden und Nordrhein-Westfalen*, (Dortmund: Institut für Landes- und Stadtentwicklungsforschung 5).

JOB, H. (1992), 'Grenzübergreifende Probleme landschaftsbezogener Erholungsformen im Deutsch-Luxemburgischen- und Deutsch-Belgischen Naturpark', in Chr. Becker and W. Schertler and A. Steinecke (eds), *Perspektiven des Tourismus im Zentrum Europas*, (Trier: ETI-Studien 1, 46–64).

KERSBERG, H. (1983), 'Naturlandschaft und Landschaftswandel im Emsland unter Berücksichtigung von Entwicklungsprogrammen im grenznahen Raum', in H. Heineberg (ed.), *Exkursionen in Westfalen und angrenzenden Regionen*, (Paderborn: Schöningh, 391–396).

KERZ, R. (1987), 'Der Deutsch-Belgische Naturpark: Ein Beispiel grenzüberschreitender Zusammenarbeit', *Naturschutz- und Naturparke* 125, 39–44.

KLIMKIEWICZ, M. (2000), 'Development of Trans-boundary Sustainable Tourism in the Carpathians: A Case of the Eastern Carpathians Biosphere Reserve', in J. Wyrzykowski, (ed.), *Conditions of the Foreign Tourism Development in Central and Eastern Europe: Changes in model of tourism in the last decade*, (Wrocaw: University of Wrocaw, Institute of Geography 6).

KLIMKIEWICZ, M. (2002), *Sustainable Tourism Development in Mountain Regions: The Case of the International East Carpathians Biosphere Reserve*, (Prague: Research support scheme, 1741/1999).

KÖHLER, T. AND SAALBACH, J. (2001), *Oberrheingebiet: Leben beiderseits der Grenze*, (Lauterbourg: Justus Perthes Verlag Gotha GmbH).

KUHN, E. L. (ed.) (1981), 'Leben am See im Wandel – der Bodenseeraum auf dem Weg in die Moderne', *Leben am See* 16, (Friedrichshafen: Landratsamt).

LEUENBERGER, TH. AND WALKER, D. (1992), 'Euroregion Bodensee – Grundlagen für ein grenzüberschreitendes Impulsprogramm', in *Gutachten im Auftrag der deutschen, österreichischen und schweizerischen Bundesländer und Kantone der Bodenseeregion*, (St. Gallen: Hochschule St. Gallen).

NFI – NATURFREUNDE INTERNATIONALE (n.d.), 'Projekte', <http://www.ecotrans.org/nfi.htm>, accessed on 24 April 2003.

SCHMITZ, G. AND WIESE, B. (1981), 'Die Luxemburger Schweiz – Berdorf und das Müllerthal im Deutsch-Luxemburgischen Naturpark', in G. Schmit and B. Wiese, *Luxemburg in Karte und Luftbild*, (Luxemburg: 79–81).

THANNHEISER, D. (1983), 'Westmünsterland. Moore im deutsch-niederländischen Grenzraum', in H. Heineberg (ed.), *Exkursionen in Westfalen und angrenzenden Regionen*, (Paderborn: Schöningh, 53–60).

THE 2ND EUROPEAN YOUTH WATER CONGRESS (2001), 'The European Forum – Water and Tourisme', <http://www.youth-water-congress.org/forum.htm>, accessed on 3 April 2003.

VARNIERE-SIMON, F. (1991), 'L'evolution contrastee de l'amanagement touristique transfrontaliere en Ardennes: Perspectives Franco-Belges', *Revue Geographique de l'Est*, 31:2, 113–121.
VIEDEBANTT, K. (ed.) (1983), *Urlaub rund um den Bodensee Heyne–Reisebücher*, (München: Heyne-Verlag).
VON MOLTKE, K. (1987), 'Handbuch für den grenzüberschreitenden Umweltschutz in der EUREGIO Maas-Rhein', *Schriftenreihe Landes- und Stadtentwicklungsforschung des Landes Nordrhein-Westfalen* 1:45, (Dortmund).
WINDISCH, W. W. (1981), *Parke und wandere rund um den Bodensee, 40 Rundwanderungen für Autofahrer auf deutscher, österreichischer und schweizer Seite*, (Mannheim: Südwestdeutsche Verlagsanstalt, Buchreihe Parke und Wandere).
YOUNG, L. AND RABB, M. (1992), 'New park on the bloc: Hungary, Czechoslovakia, and Austria must overcome obstacles to create Eastern Europe's first trilateral park', *National Parks* 66:1, 35–40.
ZENGERLING, Theo (1981), 'Region Hochrhein – Bodensee. Zweiteilige Region mit grenzüberschreitenden Verflechtungen', *Der Gemeinderat* 24:3, 16–18.
ZINKO, Y. and Kravchuk, Y. and Boguckyj, A. and Brusak, V. and Gnatiuk, R. (1993), 'Problems of the formating the Roztocza-Gologory-Krzemieniec belt of protected nature (in Polish)', in K.H. Wojciechowski (ed.), *Edukacja Ekologiczna i Ochrony Środowiska na Pograniczach*, (Lublin: Towarzystwo Wolney Wszechnicy Polskey, 14–18).

Cross-border co-operation in the Alpines

One of the biggest and probably the most popular environmental European attractions are the Alpines. Considering their sheer size and unique geographical position connecting five European nations, namely Germany, Austria, Italy, France and Switzerland, they offer uncountable opportunities for studying cross-border relations and co-operation on numerous environmental as well as political and economic subjects. Nations adjacent to these mountain ranges have to co-operate in terms of environmental protection with regard to traffic, for example. Any actions undertaken within this pristine and fragile environment need to be jointly developed in order to keep the infrastructural situation alive. Nevertheless, such infrastructure development is just one example of the need for cross-border collaboration within the Alpines. The literature listed below discusses several other issues. The main focus is placed on the Alpines' potential to function as a recreational space employed for tourism and leisure industries that adhere to principles of sustainability and environmental protection.

To co-ordinate any development within the Alpines, the Arbeitsgemeinschaft Alpen-Adria (ARGE ALP) was initiated and is jointly managed by representatives of all adjacent nation states. As a legitimate organisation entrusted with the sustainable development of any consumption patterns within these alpine regions, the Arge Alp has ratified the Alpenkonvention, a single document stating the aims and objectives of environmental protection and industrial and economic development within the mountains, and thereby set a sustainable framework for future expansion of any given industry operating within this fragile environment.

The literature listed below intends to introduce the different opinions and considerations with regard to tourism development in the Alpines, as well as the current situation of cross-border co-operation among adjacent nation states hoping to prosper from exploiting the Alpines for the tourism business.

Table 9.15   Selected literature: Cross-border co-operation in the Alpines

---

ANDREAE, C. A. (ed.) (1983), *Die Alpen als europäischer Erholungsraum*, (Innsbruck).

ARBEITSGEMEINSCHAFT ALPENLÄNDER (1982), Freistaat Bayern, Autonome Provinz Bozen-Südtirol, Kanton Graubünden, Region Lombardei, Land Salzburg, Land Tirol, Autonome Provinz Trient, Land Vorarlbert, Eine Information über die Arbeitsgemeinschaft Alpenländer, (München: Arge Alp).

BARNICH, H. (1981), 'Das Leitbild der ARGE ALP (Arbeitsgemeinschaft Alpenländer)', Berichte zur Raumforschung und Raumplanung 25: 4, 23–26.

BRAUN, J. (UNTER MITARBEIT VON KLAUS MAYR) (1998), 'Eures-T-Interalp-Projekt: Das Verhältnis von Eures-Interalp und den Euregios im bayrisch-österreichischen Grenzraum IV', (Linz: Institut für Sozial- und Wirtschaftswissenschaften).

GEBHARDT, H. (1987), 'Perzeption von Grenzen und Grenzüberschreitende Verflechtungen Beispiele aus dem Alpenraum', Revue Geographique de l'Est 27 :1, 39–51.

HERINGER, J. AND RICCABONA, S. (1999), 'Tourismus grenzüberschreitend – Naturschutzgebiete Ammergebirge, Außerfern, Lechtaler Alpen: Gemeinsame Fachtagung 23. – 24. Juni 1998 in Füssen', in Bayerische Akademie für Naturschutz und Landschaftspflege (ed.), Laufener Seminarbeiträge 3, (Tirol: Umweltanwalt, Kleine Schriftenreihe des Tiroler Umweltanwaltes 16, Laufen/Salzach).

INTERREG.CH(n.d.), 'Alpenrhein/Bodensee/Hochrhein:TranseuropäischeZusammenarbeit zwischen der Schweiz und der Europäischen Union', <http://www.interreg.ch/ir2reg_bodens_d.html>, accessed on 25 April 2003.

JACOBI, F. (1996), Ansatzpunkte zur Bewertung von Kooperationen im Tourismus am Beispiel ausgewählter Ferienorte des Alpenraums, (Bamberg: Universität Sankt Gallen, Dissertationen).

LIEBL, G. (2003), 'Umsetzung der Alpenkonvention – Ist der Durchbruch geschafft?', <http://www.naturschutz.at/service/akt_170103.htm - prot5>, accessed on 24 October 2003.

MINGHI, J.V. (1981), 'The Franco-Italian borderland: Sovereignty change and contemporary developments in the Alpes-Maritimes', *Regio Basiliensis* 22, 232–246.

MÜLLER-SCHNEGG, H. (1994), *Grenzüberschreitende Zusammenarbeit in der Bodenseeregion*, (St. Gallen: Hochschule, Dissertation).

PLATTNER, S. (2001), *Wieviel Staat braucht der Tourismus im Alpenraum? Ein Vergleich der Förderungsmöglichkeiten im Tourismus in den Arge-Alp Mitgliedern*, (Innsbruck: Diplomarbeit).

PLATZGUMMER, B. (2001), *Kooperationen im Tourismus: eine Analyse der Euregio "Zugspitze/ Wetterstein-Karwendel"*, (Innsbruck: Diplomarbeit).

RUMLEY, P.-A. (2003), 'Die Alpenkonvention – Für eine nachhaltige Berggebietsentwicklung', <http://www.are.admin.ch/imperia/md/content/are/are2/medienmitteilungen/2003/3. pdf>, accessed on 12 October 2003.

PROVINCIA AUTONOMA DI TRENTO (n.d.), 'Kriterienpool Alpenkonvention', <http://www. provincia.tn.it/agenda21/D/Hand-d/Ann0104.htm>, accessed on 30 October 2003.

SCHWEIZER BUNDESAMT FÜR RAUMENTWICKLUNG (2002), 'Protokoll zur Durchführung der Alpenkonvention von 1991 im Bereich der Raumplanung und nachhaltigen Entwicklung', <http://www.are.admin.ch/imperia/md/content/are/raumplanung/alpenkonvention/10. pdf>, accessed on 25 October 2003.

Selected cases of European cross-border co-operation with regard to tourism

Having introduced the various general issues underlying the diverse possibilities for co-operation across national boundaries on the European continent, the following section comprises literature that takes a closer look at specific case studies dealing with definite destinations or particular issues of co-operational efforts. An initial literature list introduces case studies that relate to regional peculiarities of cross-border co-operation within the EU. Studies and research findings listed below have investigated the geographically interesting situation of Baarle-Nassau-Hertog, an international exclave, itself hosting a different international exclave (Timothy, 2001), or the development of tourism in different international borderlands such as the ones between Germany and the Netherlands or the Anglo-French border. After having compiled generally interesting case studies dealing with various different destinations across Europe, subsequent literature compilations will then introduce literature examining the particular situations affecting tourism development on the divided islands of Cyprus and Ireland, the extensive literature available on the large European region Saar-Lor-Lux, and the condition in the formerly divided city of Berlin before and after its reunification.

Such case studies enable future researchers to grasp a first look at what has been examined before, and offer an easily accessible basis for comparing and contrasting further study findings realised through potentially consecutive research activities.

Table 9.16   Selected literature: Cross-border co-operation in selected European destinations

ARNOULD, E. AND PERRIN, S. (1993), 'Developpement touristique et dimension transfrontaliere: le cas de l'espace Gaume-Meuse du Nord', *Revue Geographique de l'Est* 33:3, 191–204.

AZIENDA TURISTICA (1994), *Campione d'Italia: Destined To Be Different*, (Campione: Azienda Turistica).

BAARLE-NASSAU TOURIST OFFICE (n.d.), *Baarle-Nassau-Hertog: A Remarkable Village in an Enchanting District*, (Baarle-Nassau: Tourist Office).

Boehm, N. (1981) 'Die Regio Aachen – Partner in der Euregio Maas-Rhein', *Informationen und Materialien zur Geographie der Euregio Maas-Rhein* 9, 37–41.

Boos, X. (1982), 'Economic aspects of a frontier situation: The case of Alsace', *West European Politics* 5:4, 81–97.

Briner, H.J. (1991), 'Region Basiliensis: une région, trois pays, un avenir européen', *Bulletin, Association de Géographes Français* 5, 377–382.

Brösse, U. (1992), 'Deutsch-Belgischer Grenzraum – Institutionen, Probleme und Stand des Erreichten', in Akademie für Raumforschung und Landesplanung (ed.), *Grenzübergreifende Raumplanung – Erfahrungen und Perspektiven der Zusammenarbeit mit den Nachbarstaaten Deutschlands*, (Hannover: Forschungs- und Sitzungsberichte der Akademie für Raumforschung und Landesplanung 188, 87–100).

Carmona-Schneider, J. (1987), 'EUREGIO: Über die deutsch-niederländische Zusammenarbeit zwischen Rhein, Ems und Ijssel', *Standort* 11:2, 3–9.

Cowell, A. (1997), 'Austrian border controls are slow to fall', *New York Times*, (published on 20 July), p. 3.

Economist Intelligence Unit (1995), *International Tourism Reports: France*, (London: Economist Intelligence Unit).

Essex, S.J. and Gibb, R.A. (1989), 'Tourism in the Anglo-French frontier zone', *Geography* 74:3, 222–231.

Euregio Maas-Rhein (1982), 'Literaturbericht (zur Euregio Maas-Rhein)', *Informationen und Materialien der Euregio Maas-Rhein* 10, 68 and *Informationen und Materialien der Euregio Maas-Rhein* 11, 63 – 64.

Euregio Maas-Rhein (1991), *Gebiet, Ziele, Organisation*, (Maastricht: EUREGIO).

Fichtner, U. (1988), 'Grenzüberschreitende Verflechtungen und regionales Bewußtsein in der Regio Basel', *Schriften der Regio* 10.

Graef, P. (1983), 'Funktionale Verflechtungen im deutsch-österreichischen Grenzraum: Grundlagen und mögliche Auswirkungen', in *43. Deutscher Geographentag Mannheim 1981*, (Wiesbaden, 330 – 334).

Gramm, M. (1979), *Das Belgisch-Niederländisch-Deutsche Dreiländereck*, (München).

Grenzregio Rhein-Maas-Nord (ed.) (1978), *Übereinkommen zur Bildung der Grenzregio Rhein-Maas-Nord gemäß Beschluß des Regio-Rates vom 13 Dezember 1978*, (Roermond).

Grenzregio Rhein-Maas-Nord (ed.) (1986), *Grenzüberschreitendes Aktionsprogramm für die Grenzregio Rhein-Maas-Nord*, (Mönchengladbach/Venlo).

Groben, K. (1983) 'Das Ourtal – landschaftliches Kleinod im "Grünen Herzen Europas"', *Eifeljahrbuch 1983*, 111–119.

Internationale Werbegemeinschaft Eifel-Ardennen (n.d.), 'CHARTA Eifel-Ardennen – Europa ohne Grenzen', *Arbeitspapier 1993-1995 für die internationale Werbegemeinschaft Eifel-Ardennen*.

Interreg.Ch (2004), 'Alpenrhein/Bodensee/Hochrhein: Transeuropäische Zusammenarbeit zwischen der Schweiz und der Europäischen Union', <http://www.interreg.ch/ir2reg_bodens_d.html>, accessed on 22 August 2004.

Interreg-Sekretariat PAMINA (n.d), 'Development of tourist products in the PAMINA region and expansion of co-operation in the tourist sector in PAMINA', <http://www.baden-wuerttemberg.de/interreg/e_interreg/interreg2/ pro_schw_01/Proj_bsp_06.html>, accessed on 3 April 2003.

KEOWN, I. (1991), 'Border towns: Stay in Holland's medieval Maastricht and see four countries in one unhurried day', *Travel and Leisure* 21:4, 112–121.

KUENZLER, L. (1982), 'Grenzüberschreitender Huckepackverkehr Deutschland – Frankreich', *Rationeller Transport* 24, (Frankfurt am Main: Studiengesellschaft für den kombinierten Verkehr).

LEIMGRUBER, W. (1991), 'Boundary, values and identity: The Swiss-Italian transborder region', in D. Rumley and J.V. Minghi (eds), *The Geography of Border Landscapes*, (London: Routledge, 43–62).

LOCHMANN, J.M. (1983), 'Basel – Zentrum einer europäischen Dreiländer-Region', in *18. Deutscher Schulgeographentag Basel 1982, Tagungsband*, 31–34 (Basel).

MINGHI, J.V. (1981), 'The Franco-Italian borderland: Sovereignty change and contemporary developments in the Alpes-Maritimes', *Regio Basiliensis* 22, 232–246.

MOHR, B. (1983), 'Elsässische Grenzgänger im Raum Freiburg i. Br.', in J. Maier (ed.), *Staatsgrenzen und ihr Einfluß auf Raumstrukturen 1*, (Bayreuth: Arbeitsmaterialien zur Raumordnung und Raumplanung 23, 69–94).

OHNE Verfasser (1999c), 'Grenzüberschreitende Projekte: EU fördert Mosel-Regionen', *Fremdenverkehrswirtschaft international* 19, 66.

PAGNINI, M.P. (1979), 'Friuli-Venetia Julia: A tourist border region', in G. Gruber and H. Lamping and W. Lutz' and J. Matznetter and K. Vorlaufer (eds), *Tourism and Borders: Proceedings of the Meeting of the IGU Working Group – Geography of Tourism and Recreation*, (Frankfurt a.M.: Insitut für Wirtschafts- und Sozialgeographie der Johann Wolfgang Goethe Universität, 205–213).

REICHART, T. (1988), 'Socio-economic difficulties in developing tourism in small Alpine countries: The case of Andorra', *Tourism Recreation Research* 13:1, 27–32.

RINSCHEDE, G. (1977), 'Andorra: vom abgeschlossenen Hochgebirgsstaat zum internationalen Touristenzentrum', *Erdkunde* 31, 307–314.

SANGUIN, A.L. (1991), 'L'Andorre ou la quintessence d'une economie transfrontaliere', *Revue Geographique des Pyrenees et du Sud-Ouest* 62:2, 169–186.

SEIDELMANN, E. (1983), 'Die Stadt Konstanz und die Sprachlandschaft am Bodensee', in H. Steger and E. Gabriel and V. Schupp (eds), *Forschungsbericht "Südwest-deutscher Sprachatlas"*, (Marburg: 156–234).

TAILLEFER, F. (1991), 'Le paradoxe Andorran', *Revue Geographique des Pyrenees et du Sud-Ouest* 62:2, 117–138.

VIERHAUS, R. and Busch, E. (1983), *Die Deutschen und ihre Nachbarn: Bemerkungen zur politischen Kultur einer Nation. Österreich – Deutschland – Freundschaft und Entfremdung*, (Marburg: Pressestelle der Phillipps-Universität, Marburger Universitätsreden 5).

Weber, P. and Schreiber, K.-F. (eds) (1983), *Westfalen und angrenzende Regionen. Festschrift zum 44. Deutschen Geographentag in Münster 1983*, I:15 (Paderborn: Schöningh-Verlag, Münstersche Geographische Arbeiten).

WIELENGA, F. (1983), 'Die Niederlande und Deutschland: Zwei unbekannte Nachbarn', *Internationale Schulbuchforschung* 5:2, 145–155.

WOLF, H.D. (1979), 'Der Einfluss der deutsch-österreichischen Staatsgrenze auf den Tourismus im Raum Salzburg, Bad Reichenhall, Berchtesgaden', in G. Gruber and H. Lamping and W. Lutz and J. Matznetter and K. Vorlaufer (eds), *Tourism and Borders: Proceedings of the Meeting of the IGU Working Group – Geography of Tourism and Recreation*, (Frankfurt a.M.: Institut für Wirtschafts- und Sozialgeographie der Johann Wolfgang Goethe Universität, 169-180).

ZEINER, M. (1986), 'Der grenzüberschreitende Ausflugsverkehr aus der Bundesrepublik Deutschland', *Jahrbuch für Fremdenverkehr* 34, 113–133.

Divided islands and cross-border co-operation: Ireland and Cyprus

Cross-border co-operation naturally starts with two nations willing to co-operate and work together to accomplish a common goal. Particularly in partitioned island-states, subject to limited natural and economic resources, this concept progressively becomes a necessity for further economic development, and has thus spurred the acceptance and will to co-operate across boundaries previously heavily fought over. In a European context, this development is most obvious with the currently partitioned islands of Ireland and Cyprus. Both islands are not only politically divided and shared between two different national sovereignties, but also signify a heavily contested separation between opposing beliefs and social norms, a situation that has, traditionally, severely impeded successful collaborations.

Considering the fact that tourists generally enjoy visiting such border landscapes and 'like to have themselves photographed straddling political boundaries because it is truly the only way they can be in two places at the same time' (Timothy, 2001: 44), such a situation does not necessarily have to be negative and hinder economic development. If employed carefully and communicated collaboratively, such a boundary might prove valuable as a tourism attraction and subsequently help overcome the various issues that originally led to strict separation.

In the case of Ireland, divided between the UK and the sovereign Republic of Ireland, this separation has historically created a duplication of tourism resources and promotional efforts, since each part of the island is individually engaged in tourism development and promotion. However, as Greer (2002), investigating the history of tourism related co-operation in Ireland, found out, this situation changed a few years ago, and initial partnerships were identified that jointly promoted Ireland as a single tourist destination. Supported by the European Union's various trans-boundary programs (Department of Agricultural and Rural Development, 2002), collaborative tourism development for the entire island as a whole was put higher on the political agenda and initial efforts were introduced to overcome the boundary and create a unified tourism destination.

The literature below concentrates on research and studies undertaken on this newly emerging idea of a, with regards to tourism, unified island and offers a first look at what has been accomplished so far as well as plans and projects for the near future.

Table 9.17   Selected literature: Ireland

Boyd, S.W. (1999), 'North-south divide: The role of the border in tourism to Northern Ireland', *Visions in Leisure and Business* 17:4, 50–71.
Campbell, H. (2000), 'Success story knows no bounds', *Belfast Telegraph Business*, (published on 26 September 2000), p. 5.
Central Statistics Office of Ireland (2002a), 'Tourism', <http://www.cso.ie/schools/tourism/tourism.html>, accessed on 3 April 2003.
Central Statistics Office of Ireland (2002b), 'Cross-Border Visits by Irish Residents', <http://www.cso.ie/schools/tourism/tourism.html#north>, accessed on 3 April 2003.
Clarke, W. and O'Cinneide, B. (1981), *Understanding and co-operation in Ireland: Tourism in the Republic of Ireland and Northern Ireland* 5, (Belfast and Dublin: Co-operation North).
Cullen, K. (1998), 'Britain and Ireland unveil draft for an Ulster accord', *Boston Globe*, (published on 13 January).
D'Arcy, M. and Dickson, T. (eds) (1995), *Border crossings developing Ireland's island economy*, (Dublin: Gill and Macmillian).
Deegan, J. and Dineen, D.A. (1997), *Tourism policy and performance – the Irish experience*, (London: International Thompson Business Press).
Department of Agriculture and Rural Development (2002), 'Rural Development supports cross-border activity tourism in Sliabh Beagh', <http://www.dardni.gov.uk/pr2002/pr020172.htm>, accessed on 3 April 2003.
Douglas, N. (1998), 'The politics of accommodation, social change and conflict resolution in Northern Ireland', *Political Geography* 17, 209–230.
Fitzpatrick, J. and McEniff, J. (1992),'Tourism', in Unknown author, *Ireland in Europe: A shared challenge – economic co-operation on the island of Ireland in an integrated Europe*, (Dublin: Stationary Office).
Greer, J. (2000), 'The local authority cross-border networks: lessons in partnership and North/South co-operation in Ireland', *Administration* 48:1, 52–68.
Greer, J. (2002), 'Developing trans-jurisdictional tourism partnerships – insights from the Island of Ireland', *Tourism Management* 23, 355-366.
Livingston, D.N. and Keane, M.C. and Boal, F.W. (1998), 'Space for religion: A Belfast case study?', *Political Geography* 17, 145–170.
O'Dowd, L. and Corrigan, J. and Moore, T. (1995), 'Border, national sovereignty and European integration: The British-Irish case', *International Journal of Urban and Regional Research* 19:2, 272–285.
Page, S.J. (1994), 'Perspectives on tourism and peripherality: Review of tourism in the Republic of Ireland', *Progress in Tourism, Recreation and Hospitality Management* 5, 26–53.
Scutt, C. (1998), 'Ireland plans to create islandwide tourism unit', *Travel Weekly* 57:45, 20.
Tansey, P. (1995), 'Tourism a product with big potential', in M. D'Arcy and T. Dickson (eds), *Border crossings developing Ireland's island economy*, (Dublin: Gill and Macmillian).

A similar situation arises on the divided island of Cyprus too, but appears to be in a rather infant stage and not yet developed as extensively as Ireland. In Cyprus, the border previously signified a very strict separation between the Greek and Turkish population and their respective ways of life, religious beliefs, and social composition. In this case, the border was actively used to restrict tourist flow across the island and encapsulate the enemy Turkish neighbour economically (Timothy, 2001). Given the solid fronts that built up on either side of the border, two distinctly different tourism industries developed, prospering the Greek side and leaving the Turkish side subject to economic decay. The historical atrocity of this fierce rivalry, therefore, requires potential co-operational partners to omit their previous hatred, overcome their fears, and commit to collaborative and sustainable future developments, making cross-border co-operation an unlikely and difficult undertaking. In addition, the Cypriot border dispute comprises a further dimension that is not existent in the Ireland example. Since the Greek part of the island is controlled by the sovereign power of an EU member state, and the Turkish part is controlled by the sovereign power of a non-member state, this situation additionally offers the possibility of investigating and studying the process of co-operating across the boundaries of the European Union.

The literature listed below has examined this unique situation and elaborated on the scope and potential of existing and future collaborative developments on the island.

Table 9.18   Selected literature: Cyprus

DIKOMITIS, L. (2004), 'A moving field: Greek Cypriot refugees returning "home"', *Durham Anthropology Journal* 12:1, 7–20.
GRUNDY-WARR, C. (1994), 'Peacekeeping lessons from divided Cyprus', in C. Grundy-Warr (ed.), *World Boundaries 3, Eurasia,* (London: Routledge, 71–88).
IOANNIDES, D. (1992), 'Tourism Development agents: The Cypriot resort cycle', *Annals of Tourism Research* 19, 711–731.
IOANNIDES, D. AND APOSTOLOPOULOS, Y. (1999), 'Political instability, war, and tourism in Cyprus: Effects, management, and prospects for recovery', *Journal of Travel Research* 38:1, 51–56.
KAMMAS, M. (1991), 'Tourism development in Cyprus', *The Cyprus Review* 3:2, 7–26.
KLIOT, N. AND MANSFELD, Y. (1997), 'The political landscape of partition: The case of Cyprus', *Political Geography* 16:6, 495–521.
LOCKHART, D.G. (1993), 'Tourism and Politics: The example of Cyprus', in D.G. Lockhart and D. Drakakis-Smith and J. Schembri (eds), *The Development Process in Small Island States,* (London: Routledge, 228–246).
LOCKHART, D.G. (1997b), 'Tourism in Malta and Cyprus', in D.G. Lockhart and D. Drakakis-Smith (eds), *Island Tourism: Trends and Prospectives*, (London: Pinter, 152–178).
LOCKHART, D.G. AND ASHTON, S. (1990), 'Tourism to Northern Cyprus', *Geography* 75:2, 163–167.

MANSFELD, Y. AND KLIOT, N. (1996), 'The tourism industry in the partitioned island of Cyprus', in A. Pizam and Y. Mansfeld (eds), *Tourism, Crime and International Security Issues*, (Chichester: Wiley, 187–202).

ROSSIDES, N. (1995), 'The conservation of the cultural heritage in Cyprus: A planner's perspective', *Regional Development Dialogue* 16:1, 110–125.

SÖNMEZ, S.F. AND APOSTOLOPOULOS, Y. (2000), 'Conflict resolution through tourism co-operation? The case of the partitioned island-state of Cyprus', *Journal of Travel and Tourism Marketing* 9:4, 35–48.

WITT, S.F. (1991), 'Tourism in Cyprus: Balancing the benefits and costs', *Tourism Management* 12, 37–46.

The formerly divided city of Berlin

Having discussed the particularly difficult situations prevailing on bisected islands, the following section introduces studies and research conducted on the formerly divided city of Berlin. For centuries, the Berlin Wall has been a role model of a concrete border, physically separating the communist from the capitalist world. The wall itself soon became a popular tourist destination and viewing points, such as the world-famous Checkpoint Charly, proved to be popular tourist hot spots. It was there that visitors could take a glimpse at the other side, gaze upon a foreign country, especially in this situation, and have a look at a completely different social and economic system. It has already been highlighted, that straddling a border enables visitors to somehow be in two places at the same time. Especially at Checkpoint Charly, this situation was even highlighted, due to the fact that visitors here were not only able to be on the verge of two different places, but of two distinctly different world views as well. It is therefore not surprising that even today, more than a decade after reunification and overcoming the differences, the very spot of Checkpoint Charly still attracts large amounts of visitors that want to catch an idea of what it must have been like in the days of the wall. A museum presents a reminder of this situation and provides education about the history of Berlin and its wall. In addition, the example of the Berlin Wall enables researchers to understand the phenomenon of borders as destinations since the memory of the wall is notoriously kept alive. A red line winds its way all across the city, today, indicating the previous course of the actual wall, a line that has already become a tourist attraction itself. Still, visitors enjoy being photographed standing next to or even across this line, supporting Ryden's (1993) idea of visitors longing to be in more than one place, and, in this case, more than one time zone, at the same time.

Hence, tourism at borders could be argued to incorporate a multidimensional facet similar to the one prevailing in historical or heritage tourism, a hypothesis deducted from studies conducted by Ryden (1993), Timothy (2001) or Vila (2003), but which requires substantial further research. Having reunified the German nation, the Berlin wall as a previous border destination is not exclusively of historical interest, though, but might support academic understanding of the difficult processes of overcoming the gap between two distinctly different economic and social systems, namely capitalism and socialism. The lessons learned from reunifying Germany might prove valuable for potential future reunification of the last remaining socialist/capitalist separations to be found, for example in Korea. The literature listed below might be of interest to researchers conducting studies on the Korean or Chinese cross-border relations as well.

Table 9.19   Selected literature: The formerly divided city of Berlin

BUCHHOLZ, H. (1994), 'The inner-German border: Consequences of its establishment and abolition', in C. Grundy-Warr (ed.), *World Boundaries, Vol.3, Eurasia*, (London: Routledge, 55–62).
ELKINS, T.H. AND HOFMEISTER, B. (1988), 'Berlin: The Spatial Structure of a Divided City', (London: Methuen).
ELLGER, C. (1992), 'Berlin: Legacies of division and problems of unification', *Geographical Journal* 158:1, 40–46.
FINN, P. (2000), 'Berlin seeking to keep memory of the Wall alive' *Arizona Republic*, (published on 27 August 2000), p. A24.
HARRIS, C.D. (1991), 'Unification of Germany in 1990', *Geographical Review* 81, 170–182.
KIEFER, F.S. (1989), 'Berliners rejoice at open gate', *Christian Science Monitor*, (published on 26 December 1989), p. 3.
KINZER, S. (1994), 'At Checkpoint Charlie, a museum remembers', *New York Times*, (published on 18 December), p. 3.
KOENIG, H. (1981), 'The two Berlins', *Travel Holiday* 156:4, 58–63 and 79–80.
RITTER, G. AND HAJDU, J.G. (1989), 'The east-west German boundary', *Geographical Review* 79:3, 326–344.
ROBERTS, G.K. (1991), 'Emigrants in their own country: German reunification and its political consequences', *Parliamentary Affairs* 44, 373–388.
SMITH, F. M. (1994), 'Politics, place and German reunification: A realignment approach', *Political Geography* 13:3, 228–244.

Saar-Lor-Lux Region

One of the best-researched European cross-border regions, apart from the Alpines and the German-Dutch border, is the so-called Saar-Lor-Lux Region, connecting Germany, France, and Luxembourg. Cross-border co-operation within this region comprises various aspects of economic, social, environmental and political issues influencing the everyday lives of its inhabitants by constantly improving living conditions. Profound studies of European cross-border co-operation are, therefore, likely to involve one or two observations or findings made from research conducted within this region. Especially in terms of tourism, numerous studies have been conducted within this region and contributed to the creation of a trans-boundary, tri-national tourism destination. One of the most comprehensive overviews about tourism related cross-border co-operation within the region could was found in the EURES-T

Saar-Lor-Lux study (Eures – Transfrontalier Saar-Lor-Lux-Rheinlandpfalz, 2000). A thorough description of this tri-national border region introduces each member, and lists, describes, and profoundly analyses its relative tourism potential. Discussing the differences in tourism perceptions and distribution among participating regions, it identifies and highlights the issues and challenges that prevail in all cross-border co-operation with regard to tourism. By presenting the differences in tourism policy formation between participating regions, it also indicates the diversity and heterogeneity of participating agencies that tremendously hinders any successful co-operation. Once again, the importance of collaborative efforts in the formulation of a common tourism strategy and implementation is stressed, and, therefore, should be of interest to general research within this academic field, as well as secondary research aimed specifically at the Saar-Lor-Lux.

Besides an overview of cross-border tourism activities within this region, the following literature compilation enables more focused research into particular activities. It introduces case studies conducted on social and economic developments relative to border issues, such as bilingualism and custom restrictions, or tourism activities related to hiking, shopping, appreciating culture, industrial accomplishments, or general nature, in a truly European context.

Table 9.20   Selected literature: Saar-Lor-Lux region

AUGUSTIN, CHR. (1978), *Die wirtschaftliche und soziale Entwicklung im Grenzraum Saar-Lor-Lux*, (Saarbrücken: Institut für empirische Wirtschaftforschung an der Universität des Saarlandes, Abteilung Stuktur – und Regionalforschung).
AUST, B. (1983), *Die staatliche Raumplanung im Gebiet Saar-Lor-Lux-Regionalkommission*, (Saarbrücken: Universität des Saarlandes, Arbeiten aus dem Geographischen Institut der Universität des Saarlandes, Sonderheft 4).
BECKER, Chr. and Moll, P. (1992a), 'Geographischer Wanderführer für den Raum Saar-Lor-Lux-Trier-Westpfalz', in Chr. Becker and W. Schertler and A. Steinecke (eds), *Perspektiven des Tourismus in Europa*, (Trier).
BECKER, CHR. (1991), 'Fremdenverkehrsentwicklung in der Region Saar-Lor-Lux-Trier-Westpfalz', in P. Herkert (ed.), *Saar-Lor-Lux-Westpfalz. Eine Region auf dem Weg nach Europa*, (Trier: 79–81).
BORCHERDT, CHR. (1964), 'Die Veränderungen in der Kulturlandschaft beiderseits der saarländisch-lothringischen Grenzen', in *Tagungsberichte und wissenschaftliche Abhandlungen des Deutschen Geographentages 1963*, (Wiesbaden, 335–350).
BRÜCHER, W. (ed.) (1982), Pilotstudie zu einem Saar-Lor-Lux-Atlas, (Saarbrücken: Schriftenreihe der Regionalkommission Saarland-Lothringen-Luxemburg-Rheinland-Pfalz 8).
BRÜCHER, W. (1983), 'Inhalte und Ziele eines grenzüberschreitenden Kartenwerkes in der Region Saar-Lor-Lux', in *43. Deutscher Geographentag Mannheim 1981*, (Wiesbaden, 324–327).

COMREGIO (1992), *Möglichkeiten und Notwendigkeiten eines grenzüberschreitenden Tourismuskonzeptes in der Großregion SAAR-LOR-LUX-RHEINLAND-PFALZ*, (Luxemburg: Akademie für Raumforschung Landesplanung, Protokoll der Fremdenverkehrstagung am 3.7.1992 in Trier).

ENDRUWEIT, G. (1983), 'Grenzlage als Bewußtseins- und Imageproblem – das Beispiel des Saarlandes', in LAG Hessen/Rheinland Pfalz/Saarland (ed.), *Probleme räumlicher Planung und Entwicklung in den Grenzräumen an der deutsch-französisch-luxemburgischen Staatsgrenze*, (Hannover: 137–168).

EURES – TRANSFRONTALIER SAAR-LOR-LUX-RHEINLANDPFALZ (2000), 'Tourismus und Arbeitsmarkt – Eine grenzüberschreitende Bestandsaufnahme EURES-T Saar-Lor-Lux-Rheinland-Pfalz (SLLR)', <http://www.eures-sllr.org/deutsch/info/Pdf-Downloads/Tourismus%20Studie%20Dt.pdf>, accessed on 24 April 2003.

GEIGER, K. (1999), *Tourismus im saarländisch-lothringischen Grenzraum – vergleichende Bewertung der grenzüberschreitenden Zusammenarbeit in drei ausgewählten Projekträumen*, (Diplomarbeit, Universität Saarbrücken).

GERNDT, F. (1988), 'Saarland – Europa – Chance für die Grenzregion', *Der Arbeitgeber* 12, 473–477.

GLOECKNER, Chr. (1982), 'Von den Grenzländern zu einer Europäischen Großregion', Saar-Lor-Lux, *Wirtschaft und Standort* 7, 2–6).

GUCKELMUS, Karl (1983), 'Einfluß der nationalen Grenze auf die Lebensbedingungen saarländischer Arbeitnehmer', in LAG Hessen/Rheinland Pfalz/Saarland (ed.), *Probleme räumlicher Planung und Entwicklung in den Grenzräumen an der deutsch-französisch-luxemburgischen Staatsgrenze*, (Hannover: 169–198).

INSTITUT FÜR REGIONALPOLITISCHE ZUSAMMENARBEIT IN INNERGEMEINSCHAFTLICHEN GRENZRÄUMEN DER EUROPÄISCHEN AKADEMIE OTZENHAUSEN E.V. (1975), *Freizeit, Erholung und Tourismus in der Großregion Lothringen-Luxemburg-Saar-Westpfalz-Trier*, (Saarbrücken: Institut für Regionalpolitische Zusammenarbeit in Innergemeinschaftlichen Grenzräumen).

LAFONTAINE, O. (1982), 'Saarbrücken – Dienstleistungszentrum in der Großregion Saar-Lor-Lux', *Wirtschaft und Standort* 7, 18–20.

MATTAR, M. (1983), *Die staats- und landesgrenzenüberschreitende kommunale Zusammenarbeit in der Großregion Saar-Lor-Lux-Trier*, (Frankfurt a. M.: Akademie für Raumforschung und Landesplanung Rheinland-Westpfalz-Lothringen-Luxemburg-Trier, Zusammenarbeit in Europäischen Grenzregionen 4).

MATTAR, M. (1983), *Die Staats- und Landesgrenzenüberschreitende kommunale Zusammenarbeit in der Großregion Saarland-Westpfalz-Lothringen-Luxemburg-Trier*, (Frankfurt a.M.: Verlag R.G.Fischer).

MOLL, P. (1990), 'Grundlagen und Ergebnisse der grenzüberschreitenden räumlichen Abstimmung und Planung im Raum Saar-Lor-Lux-Trier-Westpfalz', in *Saar-Lor-Lux-Trier-Westpfalz: Eine Region auf dem Weg nach Europa*, (Trier: 20. – 22. September 1990, Grenzüberschreitender Hochschulkongreß. Trier, 83-91).

MOLL, P. (1992), 'Stand und Probleme der grenzüberschreitenden Zusammenarbeit im Raum Saar-Lor-Lux-Trier. – Saarland/Lothringen/Luxemburg/westliches Rheinland-Pfalz', in Akademie für Raumforschung und Landesplanung (ed.), *Grenzübergreifende Raumplanung – Erfahrungen und Perspektiven der Zusammenarbeit mit den Nachbarstaaten Deutschlands*, (Hannover: ARL, 101–121, Forschungs- und Sitzungsberichte der Akademie für Raumforschung und Landesplanung 188).

MOLL, P. (1983), 'Der Beitrag der Raumordnung zur Überwindung der Grenzen im Gebiet Saarland / Rheinland-Pfalz – Lothringen – Luxemburg', in LAG Hessen/Rheinland Pfalz/ Saarland (ed.), *Probleme räumlicher Planung und Entwicklung in den Grenzräumen an der deutsch-französich-luxemburgischen Staatsgrenze*, (Hannover: 71–105).

MOLL, U. (1983), *Luxemburg. Entdeckungsfahrten zu den Burgen, Schlössern, Kirchen und Städten des Großherzogtums*, (Köln: DuMont Kunst-Reiseführer).

PROMOTECH (1991), *Erstellung eines gemeinsamen grenzüberschreitenden Entwicklungskonzeptes für die Großregion Saar-Lor-Lux-Trier/Westpfalz*, (Nancy: Brabois).

RAICH, S. (1995), *Grenzüberschreitende und interregionale Zusammenarbeit in einem 'Europa der Regionen', Dargestellt anhand der Fallbeispiele Großregion Saar-Lor-Lux, EUREGIO und 'Vier Motoren für Europa' – Ein Beitrag zum Europäischen Integrationsprozeß*, (Baden-Baden: Nomos, Schriftenreihe des Europäischen Zentrums für Föderalismus-Forschung 3).

REGIONALKOMMISSION LOTHRINGEN-LUXEMBURG-RHEINLAND-PFALZ-SAARLAND (1978), *Die wirtschaftliche und soziale Entwicklung im Grenzraum Saar-Lor-Lux*, (Mainz: Schriftenreihe der Regionalkommission Lothringen-Luxemburg-Rheinland-Pfalz-Saarland 6).

SCHILLING, H. (ed.) (1986), *Leben an der Grenze: Recherche in der Region Saarland/ Lorraine*, (Frankfurt a.M.: Institut für Kulturanthropologie und Europäische Ethnologieder Universität Frankfurt a. M.).

SOYEZ, D. (1985), 'Industrietourismus im Saar-Lor-Lux Raum – eine Chance für Industriegemeinden', in *Fremdenverkehr und Freizeit: Entwicklung ohne Expansion: Ergebnisse der 3. Fachtagung der Regionalen Arbeitsgruppe Saar/Mosel/Pfalz am 31. Juni in Trier*, (Bochum: Deutscher Verband für angewandte Geographie, 71–88, Materialien zur angewandeten Geographie 13:3).

SPEHL, H. (1983), 'Wirtschaftliche Auswirkungen von Grenzen – das Beispiel Saar-Lor-Lux', in J. Maier (ed.), *Staatsgrenzen und ihr Einfluß auf Raumstrukturen 1*, (Bayreuth: Arbeitsmaterialien zur Raumordnung und Raumplanung 23, 95–139).

SPEHL, H. (1983), 'Wirkungen der nationalen Grenzen auf Betriebe in peripheren Regionen, dargestellt am Beispiel des Saar-Lor-Lux-Raumes', in LAG Hessen/Rheinland Pfalz/ Saarland (ed.), *Probleme räumlicher Planung und Entwicklung in den Grenzräumen an der deutsch-französisch-luxemburgischen Staatsgrenze*, (Hannover, 199–224).

WACHOWIAK, H. (1994), 'Das Tourismuskonzept "Europäisches Tal der Mosel" – Ein Beitrag zur touristischen Entwicklung in der Region Saar-Lor-Lux-Trier/Westpfalz', in H.P. Burmeister (ed.), *Wohin die Reise geht – Perspektiven des Tourismus in Europa*, (Loccum: Loccumer Protokolle 2, 203–215).

WACHOWIAK, H. (1996), 'Denken über die Grenzen hinweg: Tourismus und Naherholung im deutsch-luxemburgischen Grenzgebiet', in Messe BerlinGmbH (ed.), *Brücke zwischen Wissenschaft und Praxis*, (Berlin: Dokumentation Wissenschaftszentrum ITB 1996).

ZIMMER, N. (2000), Ein kulturtouristisches Vermarktungskonzept für den saarländisch-lothringischen Grenzraum unter besonderer Berücksichtigung der Industriekultur, (Schriftenreihe der Regionalkommission Saarland-Lothringen-Luxemburg-Trier. Arbeitsgruppe Raumordnung 11).

Cross-border co-operation in Northern and Eastern Europe

Cross-border co-operation in Europe is not necessarily restricted to co-operation between EU member states but also comprises non members. Most of these are situated in the Eastern parts of the continent and co-operation often relates to Scandinavia or the Eastern block. The literature listed below discusses issues relative to tourism co-operation between Nordic countries, the Balkan states or the centrally-planned nations situated in Eastern Europe. One aspect of this is the formation of formerly communist nations to capitalist players on a global market and their adjustment processes to Western tourism industries. Social changes imposed by the breakdown of the former communist governments (Krätke, 1998) and the approach towards the European Union (Krätke, 2001) are discussed, and might support further research into these newly emerging but arguably potentially very powerful future tourist destinations. Cross-border co-operation has already begun to prosper regional development, as is the case within the newly created nature reserves in the Carpathian mountains (see especially Carpathian Foundation, 2002; Green Gate, 1998; Klimkiewicz, 2002) and indicates a willingness to engage in further trans-boundary development activities.

The literature below intends to give a preliminary look at what has previously been investigated and examined, and to create the basis of further studies researching the numerous aspects necessary for successful and sustainable regional tourism developments.

Table 9.21   Selected literature: Cross-border co-operation in Northern and Eastern Europe

BACHVAROV, M. (1979), 'The tourist traffic between the Balkan States and the role of the frontiers', in G. Gruber and H. Lamping and W. Lutz and J. Matznetter and K. Vorlaufer (eds), *Tourism and Borders: Proceedings of the Meeting of the IGU Working Group – Geography of Tourism and Recreation*, (Frankfurt a.M.: Institut für Wirtschafts- und Sozialgeographie der Johann Wolfgang Goethe Universität, 129–140).
BACHVAROV, M. (1997), 'End of the model? Tourism in post-communist Bulgaria' *Tourism Management* 18, 43–50.
BARSCHEL, U. (1983), 'Schleswig-Holstein und Dänemark, Perspektiven der Zusammenarbeit Vortrag' *Grenzfriedenshefte* 4, 213–221.
BATTISTI, G. (1979), 'Tourism on the border between Italy and Yugoslavia: Methods for research', in G. Gruber and H. Lamping and W. Lutz and J. Matznetter and K. Vorlaufer (eds), *Tourism and Borders: Proceedings of the Meeting of the IGU Working Group – Geography of Tourism and Recreation*, (Frankfurt a.M.: Institut für Wirtschafts- und Sozialgeographie der Johann Wolfgang Goethe Universität, 215–229).
BERG, E. (1999), 'National interests and local needs in a divided Setumaa: Behind the narratives', in H. Eskelinen and I. Liikanen and J. Oksa (eds), *Curtains of Iron and Gold: Reconstructing Borders and Scales of Interaction*, (Avebury: Aldershot, 167–177).

BERTRAM, H. (1998), 'Double transformation at the eastern border of the EU: The case of the Euroregion pro Europa Viadrina', *GeoJournal* 44:3, 215–224.

BALTIC SEA TOURISM COMMISSION (1996), 'Baltic Sea Tourism Commission', <http://www. balticsea.com>, accessed on 23 April 1998.

BUMBARU, D. (1992), 'Dubrovnik: Heritage and culture as targets', *Impact* 4:5, 1–2.

BYGVRÁ, S. (1990), 'Border shopping between Denmark and West Germany', *Contemporary Drug Problems*, 17:4, 595–611.

CARPATHIAN FOUNDATION (2002a), 'List of Programs' <http://www.carpathianfoundation.org/ languages/en/program.php?Program=2&>, accessed on 3 April 2003.

CARPATHIAN FOUNDATION (2002b), 'Cross-Border Co-operation in the Carpathian Euroregion', <http://www.carpahtianfoundation.org/download/programs/cbcen.doc>, accessed on 25 March 2003.

CARTER, F.W. (1991), 'Bulgaria', in D.R. Hall (ed.), *Tourism and Economic Development in Eastern Europe and the Soviet Union*, (London: Belhaven Press, 155–172).

CENTER FOR TRANSBOUNDARY CO-OPERATION (1999), 'Community Development and Cross-Border Co-Operation in the Estonian-Russian Border Area', <http://www. ctc.ee/old/lib/ pdf/cbc_2_eng.pdf>, accessed 1 February 2003.

DANISH CROSS BORDER INSTITUTE (2002), 'Cross-Border Shopping in the spring of 2001 – after the Oeresund Bridge and Schengen', <http://www.ifq.dk/en/ publications.html>, accessed on 3 April 2003.

EREG – EIGENSTÄNDIGE REGIONALENTWICKLUNG IM GRENZRAUM (1999), 'Projekt Retz-Znojmo-Pulkautal' <http://www.stb.tuwien.ac.at/FORSCHUNG/EREG/PROJEKT2/EREG_2_ KurzfassungDE(546KB).pdf>, accessed on 24 April 2003.

ERIKSSON, G.A. (1979), 'Tourism at the Finnish-Swedish-Norwegian borders', in G. Gruber and H. Lamping and W. Lutz and J. Matznetter and K. Vorlaufer (eds), *Tourism and Borders: Proceedings of the Meeting of the IGU Working Group – Geography of Tourism and Recreation*, (Frankfurt a.M.: Institut für Wirtschafts- und Sozialgeographie der Johann Wolfgang Goethe Universität, 151–162).

EURASIA FOUNDATION (2002), Regional Tourism Receives Boost from South Caucasus Co-operation Program', <http://www.eurasia.org/news/ SCCP_PR_02.01.html>, accessed on 3 April 2003.

GOSAR, A. (1999), *Recovering tourism in the Balkans*, (Honolulu: 23 March, Paper presented at the Annual Meetings of the Association of American Geographers).

GOSAR, A. AND KLEMENCIC, V. (1994), 'Current problems of border regions along the Slovene-Croatian border', in W.A. Gallusser (ed.), *Political Boundaries and Coexistence*, (Bern: Peter Lang, 30–42).

GREEN GATE (1998), *Creation of a model of transborder co-operation for sustainable development in the Carpathian Euroregion on an example of the International Biosphere Reserve "Eastern Carpathians"*, (Lviv-Sianky-Velykyi Bereznyi).

GREENWICH WATERFRONT DEVELOPMENT PARTNERSHIP (1995), *Greenwich 2000: Tourism Development*, (Greenwich: Waterfront Development Partnership).

GREENWICH WATERFRONT DEVELOPMENT PARTNERSHIP (1996), *Once in a Thousand Years: Maximising Opportunities in Greenwich*, (Greenwich: Waterfront Development Partnership).

HALL, D.R. (1984) 'Foreign tourism under socialism: The Albanian "Stalinist" model', *Annals of Tourism Research* 11, 539–555.

HALL, D.R. (1990a), 'Eastern Europe opens its doors', *Geographical Magazine* 62:4, 10–15.

HALL, D.R. (1990b), 'Stalinism and tourism: A study of Albania and North Korea', *Annals of Tourism Research* 17, 36–54.

HALL, D.R. (1991a), 'Evolutionary pattern of tourism development in Eastern Europe and the Soviet Union', in D.R. Hall (ed.), *Tourism and Economic Development in Eastern Europe and the Soviet Union*, (London: Belhaven Press, 79–115).

HALL, D.R. (1992), 'Albania's changing tourism environment', *Journal of Cultural Geography* 12:2, 35–44.

HALL, D.R. (1995), 'Tourism change in Central and Eastern Europe', in A. Montanari and A.M. Williams (eds), *European Tourism: Regions, Spaces and Restructuring*, (Chichester: Wiley, 221–244).

HAMILTON, J. AND BROUSTAS, M.V. (1996), 'Peace dividend: Tourists reclaim the Balkans', *Traveler* 37, November 1996.

HARRIS, M. (1997), 'Spirited, independent Slovenia' *New York Times*, (published on 11 May 1997), p. 12.

HELLE, R.K. (1979), 'Observations on tourism between Finland and the Soviet Union', in G. Gruber and H. Lamping and W. Lutz and J. Matznetter and K. Vorlaufer (eds), *Tourism and Borders: Proceedings of the Meeting of the IGU Working Group – Geography of Tourism and Recreation*, (Frankfurt a.M.: Institut für Wirtschafts- und Sozialgeographie der Johann Wolfgang Goethe Universität, 163–167).

INSTITUT FÜR REGIONALE FORSCHUNG UND INFORMATION IM DEUTSCHEN GRENZVEREIN E.V. (1982), *Grenzüberschreitende Reiseströme an der deutsch-dänischen Grenze*, (Flensburg: Deutscher Grenzverein e.V.).

KALUSKI, S. (1994), 'The "new" eastern Polish border in the face of environmental and socio-political problems', in W.A. Gallusser (ed.), *Political Boundaries and Coexistence*, (Bern: Peter Lang, 57–64).

KLEMENCIC, V. AND GOSAR, A. (1987), 'Grenzüberschreitende Raumwirksame Leitbilder dargestellt an Beispielen der Grenzräume Sloweniens in Jugoslawien', *Revue Géographique de l'Est* 27 :1/2, 27–38.

KLEMENCIC, V. AND BUFON, M. (1994), 'Cultural elements of integration and transformation of border regions: The case of Slovenia', *Political Geography* 13:1, 73–83.

KLIMKIEWICZ, M. (2000), 'Development of Trans-boundary Sustainable Tourism in the Carpathians: A Case of the Eastern Carpathians Biosphere Reserve', in J. Wyrzykowski, (ed.), *Conditions of the Foreign Tourism Development in Central and Eastern Europe: Changes in model of tourism in the last decade*, (Wrocaw: University of Wrocaw, Institute of Geography 6).

KLIMKIEWICZ, M. (2002), 'Tourism and Cross-Border Co-operation-Carpathains', <http://www.mtnforum.org/emaildiscuss/discuss02/031102269.htm>, accessed 3 February 2003.

KLIMKIEWICZ, M. (2002), *Sustainable Tourism Development in Mountain Regions: The Case of the International East Carpathians Biosphere Reserve*, (Prague: Research support scheme 1741/1999).

KORTELAINEN, J. (ed.) (1997), *Crossing the Russian Border, Regional Development and Cross-Border Co-operation in Karelia*, (TMR Course Report 3, University of Joensuu, Departement of Geography).

KOTER, M. (1994), 'Transborder "Euroregions" around Polish border zones as an example of a new form of political coexistence', in W.A. Gallusser (ed.), *Political Boundaries and Coexistence*, (Bern: Peter Lang, 77–87).

KOTKIN, S. AND WOLFF, D. (eds) (1995), *Rediscovering Russia in Asia: Serbia and the Russian Far East*, (Armonk, NY: M.E. Sharp).

KOVÁCS, Z. (1989), 'Border changes and their effect on the structure of Hungarian society', *Political Geography Quarterly* 8:1, 79–86.

KRÄTKE, S. (1998), 'Problems of cross-border regional integration: The case of the German-Polish border area', *European Urban and Regional Studies* 5:3, 249–262.

KRÄTKE, S. (2001), 'Cross-border co-operation in the German–Polish Border Area', in M.V. Geenhuizen and R. Ratti (eds), *Gaining advantage from open borders: An active space approach to regional development*, (Aldershot/Burlington: Ashgate, 213–232).

LAAR, M. AND BIRKAVS, V. AND SLEZEVICIUS, A. (1996), 'Free trade agreement between the Republic of Estonia, The Republic of Latvia, and the Republic of Lithuania', *Russian and East European Finance and Trade* 32:5, 39–48.

LIGHT, D. (2000), 'Gazing on communism: Heritage tourism and post-communist identities in Germany, Hungary and Romania', *Tourism Geographies* 2:2, 157–176.

MARIC, R. (1988), 'Osnovna prostorno-fizionomska I sadrzinska obelizja pogranicnog prostora SR Srbije van SAP', *Teorija i Praksa Turizma* 1, 26–31.

MIHALIC, T. (1996), 'Tourism and warfare: The case of Slovenia', in A. Pizam and Y. Mansfeld (eds), *Tourism, Crime and International Security Issues*, (Chichester: Wiley, pp. 231–246).

MINGHI, J.V. (1994b), 'The impact of Slovenian independence on the Italo-Slovene borderland: An assessment of the first three years', in W.A. Gallusser (ed.), *Political Boundaries and Coexistence*, (Bern: Peter Lang, 88–94).

NORDIC TRAVEL (2001), 'Hotel Complex "Matkaselka – Kirrkolahti"/executive summary of the investment project', <http://nordictravel.ru/p./matkaselka.html>, accessed 13 March 2003.

OFFICE OF THE GOVERNMENT OF KYUSTENDIL (ed.) (2000), 'Revised Version of the Regional Plan for Development of Region with Administrative Center Kyustendil', <http://www. kn. government.bg/kn/en/regional_development.htm>, accessed February 12 2003.

PAASI, A. (1994), 'The changing representations of the Finnish-Russian boundary', in W.A. Gallusser (ed.), *Political Boundaries and Coexistence*, (Bern: Peter Lang, 103–111).

PAASI, A. (1996), *Territories, Boundaries and Consciousness: The Changing Geographies of the Finnish-Russian Border*, (Chichester: Wiley).

PAASI, A. AND RAIVO, P.J (1998), 'Boundaries as barriers and promoters: Constructing the tourist landscapes of Finnish Karelia', *Visions in Leisure and Business* 17:3, 30–45.

PALOMÄKI, M. (1994), 'Transborder co-operation over Quarken Strait between Finland and Sweden', in W.A. Gallusser (ed.), *Political Boundaries and Coexistence*, (Bern: Peter Lang, 238–246).

PANIC-KOMBOL, T. (1996), 'The cultural heritage of Croatian cities as a tourism potential', *World Leisure and Recreation* 38:1, 21–25.

PEIPSI CENTRE FOR TRANSBOUNDARY CO-OPERATION (2002), 'Development of the Lake Peipsi region co-operation network in tourism', <http://www.ctc.ee/index.php?menu_id =224&lang_id=2>, accessed on 3 April 2003.

RAKOWSKI, G. (1993), *A system of cross-protected areas (TOCH) with leading tourist function in the eastern border region of Poland*, (Warsaw: Institute of Tourism – Institute of Environmental Protection).

SÁNDOR, J. (1990), 'A nyugati határszél idegenforgalma a vonzás és a keresletkinálat tükrében', *Idegenforgalmi Közlemények* 1, 25–34.

SEVODNYA (1994), 'Russia closes its borders with Azerbaijan and Georgia' *Sevodnya*, (published on 21 December 1994), p. 1.

STRYJAKIEWICZ, T. (1998), 'The changing role of border zones in the transforming economies of East-Central Europe: The case of Poland', *GeoJournal* 44:3, 203–213.

SUOMEN MATKAILULIITTO (1993), *Kalottireitti: Rajaton Tunturielämys Pohjoisessa Suomen, Ruotsin ja Norjan Halki*, (Helsinki: Suomen Matkailuliitto).

SWEEDLER, A. (1994), 'Conflict and co-operation border regions: An examination of the Russian-Finnish border', *Journal of Borderlands Studies* 9:1, 1–13.

VAAGT, G. (1983), 'Die deutsch-dänische Grenzregion, (Themen aus Geschichte und Gegenwart)', *Die Heimat* 90:4/5, 119–126.

VARDOMSKY, L. (1992), 'Blagoveshchensk and Heihe engage in across-the-border co-operation', *Far Eastern Affairs* 2, 81–86.

VIKEN, A. AND VOSTRYAKOV, L. AND DAVYDOV, A. (1995), 'Tourism in Northeast Russia', in C.M. Hall and M.E. Johnston (eds), *Polar Tourism: Tourism in the Arctic and Antarctic Region*, (Chichester: Wiley, 101–114).

VUORISTO, K.V. (1981), 'Tourism in Eastern Europe: Development and regional patterns', *Fennia* 159:1, 237–247.

WALL STREET JOURNAL (1996), 'Russia to tax travelers' *Wall Street Journal*, (published on 24 December 1996), p. A6.

WEIGAND, K. (1990), 'Drei Jahrzehnte Einkaufstourismus über die deutsch-dänische Grenze', *Geographische Rundschau* 42:5, 286–290.

WEIGAND, K. (1982), *Grenzüberschreitende Reiseströme an der deutsch-dänischen Grenze: Kontinuität und Wandel von 1957–1980/81*, (Flensburg: Institut für Regionale Forschung und Information im deutschen Grenzverein e.V.).

WEIGAND, K. (1983) 'Tourismus und Grenzhandel an der deutsch-dänischen Land – und Seegrenze. Bilanz 1981/83', *Landeskundliche Beiträge* 2, (Flensburg: Institut für Regionale Forschung und Information).

WEIGAND, K. (1983), 'Tourismus und Grenzhandel an der deutsch-dänischen Land – und Seegrenze. Eine weiterführende Untersuchung zur langfristigen Analyse des deutsch-dänischen Grenzverkehrs in der Nachkriegszeit', *Landeskundliche Beiträge* 2, (Flensburg: Forschungsstelle für Regionale Landeskunde).

WICHMANN MATTHIESSEN, C. (2000), 'Bridging the Öresund: potential regional dynamics Integration of Copenhagen (Denmark) and Malmö-Lund (Sweden): A cross-border project on the European metropolitan level', *Journal of Transport Geography* 8, 171–180.

WITT, S.F. (1998), 'Opening of the former communist countries of Europe to inbound tourism', in W.F. Theobald (ed.) *Global Tourism* – 2ⁿᵈ Edition, (Oxford: Butterworth-Heinemann, 380–390).

YOUNG, L. AND RABB, M. (1992), 'New park on the bloc: Hungary, Czechoslovakia, and Austria must overcome obstacles to create Eastern Europe's first trilateral park', *National Parks* 66:1, 35–40.

Peculiarities and special co-operation across national boundaries

Classifying cross-border co-operation with regard to tourism is not always a straightforward process. Having introduced the general nature of cross-border co-operation in Europe, having discussed its political requirements and geographic specialities and taken a closer look at popular and important destinations, the following list includes literature that could not easily be classified among the above sections.

The case studies compiled below take a very detailed approach to very particular aspects of the tourism industry such as golf tourism (Steinecke and Treinen, 1996), heritage tourism in a common European context (Ashworth, 1995) or general studies on tourism feasibility and its marketing potential (Becker, 1992; Wachowiak, 1994; Zimmer, 2000). They consequently deal with cross-border co-operation in European tourism and take a very focused approach, highlighting particular aspects in-depth, rather than examining and investigating the underlying rules and laws of European co-operation across national boundaries.

None the less, such studies are arguably capable of contributing tremendously to understanding these general principles and consequently deserve to be mentioned separately within this discussion on trans-boundary tourism co-operation on the European continent.

Table 9.22   Selected literature: Cross-border co-operation determined by specific interests

Ashworth, G.J. (1995), 'Heritage, tourism and Europe: A European future for a European past?', in D.T. Herbert (ed.), *Heritage, Tourism and Society*, (London: Mansell, 68–84).
Bachmann, U. (1982), *Der Wassersporttourismus am Bodensee*, (Innsbruck: Dissertation).
Barisic, G. and Wachowiak, H. (2002), 'Festungspartnerschaft Stevensweert – Zons: Kulturtourismus im deutsch-niederländischen Grenzraum', *Der Landkreis* 5, 368–369.
Becker, Chr. (1988), 'Kulturtourismus im Mosel-Ardennenraum: Zur Konzeption des Europäischen Tourismusinstitutes in Trier', in Chr. Becker (ed.), *Denkmalpflege und Tourismus III. Mißtrauische Distanz oder fruchtbare Partnerschaft, Vorträge und Diskussionsergebnisse des 3. Internationalen Symposiums vom 7. – 12.11.1990 in Trier, Materialien zur Fremdenverkehrsgeographie* 23, 33–40.
Becker, Chr. (1992), 'Kulturtourismus – Eine Zukunftsträchtige Entwicklungsstrategie für den Saar-Mosel-Ardennenraum', in Chr. Becker and W. Schertler and A. Steinecke (eds), *Perspektiven des Tourismus im Zentrum Europas*, (Trier: ETI-Studien 1, 21 – 25).
Becker, Chr. and Moll, P. (1992a), 'Geographischer Wanderführer für den Raum Saar-Lor-Lux-Trier-Westpfalz', in Chr. Becker and W. Schertler and A. Steinecke (eds), *Perspektiven des Tourismus in Europa*, (Trier: ETI-Studien 1).
Becker, Chr. (ed.) (1990), *Geographischer Wanderführer für den Saar-Mosel-Raum: 30 Rundwanderungen zu Fuß, mit dem Rad und per Boot*, (Saarbrücken: Regionale Arbeitsgruppe Saar / Mosel / Pfalz der DVAG's).

BENDERMACHER, J. (1983), 'Das alte Bauernhaus zwischen Mosel und Maas', *Heimatjahrbuch Kreis Daun* 90–96.

COSAERT, P. (1994), 'Frontiere et commerce de detail: la localisation des commerces de detail aux points de passage de la frontiére franco-belge au niveau de l'arrondissement de Lille', *Hommes et Terres du Nord* 2 :3, 134–141).

DANISH CROSS BORDER INSTITUTE (2002), 'Cross-Border Shopping in the spring of 2001 – after the Oeresund Bridge and Schengen', <http://www.ifq.dk/en/ publications.html>, accessed on 3 April 2003.

ESER, F. (1983), *BV – Radtourenblätter: Allgäu, Schwaben und angrenzendes Österreich,* (München: Bergverlag-Rother).

Euregio Rhein-Waal (n.d.), 'Grenzenlos genießen und erholen auf Bauernhöfen in der Euregio Rhein-Waal', <http://www.euregio.org/publications/publications.cfm?lang=2&id=57>, accessed on 16 October 2003.

EUROPÄISCHES TOURISMUS INSTITUT (ed.) (1993), *Tourismuskonzept 'Europäisches Tal der Mosel' – Grundlagenuntersuchungen im Auftrag der Arbeitsgruppe Fremdenverkehr der Saar-Lor-Lux-Trier/Westpfalz-Regionalkommission,* (Trier: ETI).

GIARDIN, A. (1983), *Kirrberg im Krummen Elsaß. Geschichte eines Hugenottendorfes im deutsch-französischen Grenzraum,* (Bad Neustadt: a.d.S, Schriften der Ewin –von Steinbach- Stiftung, 8).

GRAMM, M. (1983), 'Einkaufen im belgisch-niederländisch-deutschen Dreiländereck – ein Beispiel für grenzüberschreitende räumliche Interaktionen und ihrem Beitrag zur Entwicklung eines grenzüberschreitenden Nationalbewußtseins', in J. Maier (ed.), *Staatsgrenzen und ihr Einfluß auf Raumstrukturen 1,* (Bayreuth: Arbeitsmaterialien zur Raumordnung und Raumplanung 23, 51–69).

GRAMM, M. (1983), 'Die Euregio Maas-Rhein im Lufbild', *Informationen und Materialien zur Euregio Maas-Rhein* 12, 47–49.

GÜGEL, R. (1983), *Camping im deutsch-luxemburgischen Grenzgebieten – Struktur des Campings in einem Mittelgebirgserholungsraum unter Berücksichtigung der Belastungsprobleme und des aktionsräumlichen Verhaltens der Camper,* (Trier: Geographische Gesellschaft der Universität Trier).

HEPPERLE, I. (1982), *Mit dem Fahrrad zu Natur und Kunst, (am Oberrhein)* (Freiburg i.Br.).

JAENICKE, G. (1983), 'Die neue Großschifffahrtsstraße Rhein-Main-Donau: Eine völkerrechtliche Untersuchung über den rechtlichen Status', *Völkerrecht und Außenpolitik* 21, (Frankfurt a. M.).

KLOSS, H. (1982), 'Die niederländisch-deutsche Sprachgrenze, insbesondere in der Grafschaft Bentheim', *Das Bentheimer Land* 98, 145–156.

KRIER, J. AND WAGNER, R. (1983), 'Römisches Landgut bei Wasserbillig/Langsur "an der Freinen"', *Hemecht* 35:2, 211–276.

MESCHEDE, W. (1983), *Geschäftseinzugsbereiche und räumliches Einkaufsverhalten der niederländischen und deutschen Bevölkerung im Raum Enschede-Gronau,* (Münster/ Westfalen: Arbeitsgemeinschaft Angewandte Geographie Münster, Arbeitsberichte 2).

MINGHI, J.V. (1999), 'Borderland "day tourists" from the East: Trieste's transitory shopping fair', *Visions in Leisure and Business* 17:4, 32–49.

NEUWIRTH, H. (1983), *Die Traumstraßen Europas: Die schönsten Reiserouten planen und erleben,* (München: Süddeutscher Verlag).

PAUSE, W. (1983), 'Zwischen München und Salzburg: 40 Wandertips für kluge Autofahrer', in *Wer viel geht, fährt gut*, (München/Zürich: Verlag Schnell und Steiner).

REGIO+NET (2001), 'ViaSpluga – Kultur- und Weitwanderweg', <http://www.regioplus.ch/ViaSpluga/>, accessed on 24 April 2003.

RICHTER, G. (1983), *Mit Bacchus am Oberrhein und Bodensee: Die Badische Weinstraße, die Elsässische Weinstraße*, (Karlsruhe: Verlag Braun).

ROLLMANN, T. (1989), *Tagungsverkehr in der Region Trier – Auswirkungen auf den Raum, Standortwahl, Marketing von Tagungen*, (Trier: Diplomarbeit).

SOYEZ, D. (1985), 'Industrietourismus im Saar-Lor-Lux.Raum – eine Chance für Industriegemeinden', in DVAG (ed.), *Regionale Arbeitsgruppe Saar/Mosel/Pfalz: Fremdenverkehr und Freizeit: Entwicklung ohne Expansion: Ergebnisse der 3. Fachtagung der Regionalen Arbeitsgruppe Saar/Mosel/Pfalz am 31. Juni in Trier*, (Bochum: Materialien zur angewandten Geographie 13:3, Deutscher Verband für angewandte Geographie, 71–88).

STEINECKE, A. AND TREINEN, M. (eds.) (1996), 'Wachstumsmarkt Golftourismus: Chancen für die Nachbarn Luxemburg, Rheinland-Pfalz, Belgien und Lothringen' *ETI-Texte 9*, (Trier: Europäisches Tourismus Institut).

STEINECKE, A. AND WACHOWIAK, H. (1994), 'Kulturstraßen als innovative touristische Produkte – Das Beispiel der grenzüberschreitenden Kulturstraße "Straße der Römer" an der Mosel', in J. Maier (ed.), *Touristische Straßen – Beispiele und Bewertung*, (Bayreuth: Arbeitsmaterialien zur Raumordnung und Raumplanung 137, 5–33).

STERNER, S. (1982), *Drei-Länder-Weg: Vogesen-Jura-Schwarzwald Taschenbücher für Wanderfreunde*, (Neckartenzlingen: Verlag Walz).

VREULS, H.H.J. (1982), 'Grenzüberschreitende Pendlerwanderungen im Raum Aachen-Minjmstreek-Maastricht-Lüttich', *Informationen und Materialien zur Geographie der Euregio Maas-Rhein* 10, 1–42.

WACHOWIAK, H. (1994), 'Das Tourismuskonzept "Europäisches Tal der Mosel" – Ein Beitrag zur touristischen Entwicklung in der Region Saar-Lor-Lux-Trier/Westpfalz', in H.P. Burmeister (ed.), *Wohin die Reise geht – Perspektiven des Tourismus in Europa*, (Loccum: Loccumer Protokolle 2, 203–215).

WALKER, A.M. (1994), 'Euro Airport Basle-Mulhouse-Freiburg: Strengths and weaknesses of a bi-national airport', in W.A. Gallusser (ed.), *Political Boundaries and Coexistence*, (Bern: Peter Lang, 279–283).

ZIMMER, N. (2000), *Ein kulturtouristisches Vermarktungskonzept für den saarländisch-lothringischen Grenzraum unter besonderer Berücksichtigung der Industriekultur*, (Schriftenreihe der Regionalkommission Saarland-Lothringen-Luxemburg-Trier. Arbeitsgruppe Raumordnung 11).

Cross-border co-operation in international tourism industries situated in the Middle East

Operating successful tourism businesses in the Middle East has historically been a difficult task. Due to the internationally perceived high risk of terrorist attacks and the prevailing danger of becoming involved in warfare and military conflicts, tourism operators need to engage heavily in positive promotion and image creation.

Such efforts often relate to niche markets, such as Christian pilgrimages to the Holy Land, archaeological excursions to the ancient heritage of Jordan, or wellness related activities at the Dead Sea.

Considering that this chapter concentrates on cross-border co-operation between adjacent sovereignties, and most of the issues mentioned above tend to relate to attracting international visitors, rather than local engagement in trans-boundary co-operation, the literature listed below will omit these aspects, and instead concentrate on domestic and intra-regional tourism activities. These activities demand a higher level of co-operation within the entire region rather than unitarily national promotional efforts, and are capable of locally prospering citizens and regions economically. Such co-operation requires mutual agreements and common strategy formulation, though, which is often hindered dramatically by ongoing border disputes resulting in ferocious and unruly warfare and terrorist combats. Hence, any successful co-operation demands peace talks and friendly approaches which increasingly lead, in particular, to cease fires and peace agreements, thereby building the foundation for future co-operation and potential economic prosperity from newly developed tourism industries.

Despite the fact that such co-operation is momentarily hard to accomplish, some tourism businesses actually prosper from border disputes, as is specifically the case with Taba, situated between Egypt and Israel. Research conducted on tourism in border areas has indicated that much tourism development is deliberately undertaken at the border between opposing legislation in order to increase the geographical catchment area (Timothy, 2001). Gambling represents a tourism industry that highly prospers from the border and the different legislation on either side. Whereas gambling is legalised on Egyptian land, it is illegal in Israel. Casinos operating in Taba, right next to the Israeli border are, therefore, attractive tourist destinations for Israeli gambling tourists and serve a remarkable bi-national catchment area. However, such tourism industries also have severe social implications, and are therefore closely observed and opposed by Israel. Whereas Egypt prospers economically from this industry through "the import of tax income and the re-export of negative externalities, such as gambling addiction, bankruptcy and reduced labour productivity", Israel is left with all the externalities "that accompany the gamblers as they return to their homes on the other side of the border" as Felsenstein and Freeman (2001: 511) realised. Obviously, gambling tourism is arguably rather a prohibitive activity with regard to friendly approaches, but its negative implications might be overcome through engagement in further collaborative trans-boundary tourism developments.

None the less, gambling tourism is just one example of tourism activities currently operated within the Middle East. The literature below intends to introduce further developments within this region, which arguably might prove highly interesting for future tourism operators now that the previously exclusively oil-centred economies of the Middle East start to look out for further economic possibilities for regional development.

Table 9.23   Selected literature: Middle-East cross-border co-operation

CHRISTIAN SCIENCE MONITOR (1989), 'Israelis return last of captured land to Egypt', *Christian Science Monitor*, (published on 16 March 1989).

DRYSDALE, A. (1991), 'The Gulf of Aqaba coastline: An evolving border landscape', in D. Rumley and J.V. Minghi (eds), *The Geography of Border Landscapes*, (London: Routledge, 203–216).

FELSENSTEIN, D. AND FREEMAN, D. (1998), 'Simulating the impacts of gambling in a tourist location: Some evidence from Israel', *Journal of Travel Research* 37:2, 145–155.

FELSENSTEIN, D. AND FREEMAN, D. (2001), 'Estimating the impacts of crossborder competition: the case of gambling in Israel and Egypt', *Tourism Management* 22, 511–521.

FINEBERG, A. (1993) *Regional Co-operation in the Tourism Industry*, (Jerusalem: Israel/ Palestine Center for Research and Information).

FOSTER-CARTER, A. (1996), 'Monumental puzzle' *Far Eastern Economic Review*, (published on 26 June 1996), p. 36–39.

FRIEDMANN, J.S. (1988), 'Taba goes to Egypt: Settlement of border dispute with Israel boosts Egypt's leader', *Christian Science Monitor*, (published on 30 September 1988), p. 11.

GOVERNMENT OF ISRAEL (1997), *Programs for Regional Co-operation*, (Tel Aviv: Ministry of Foreign Affairs and Ministry of Finance).

GRADUS, Y. (1994), 'The Israel-Jordan Rift Valley: A border of co-operation and productive coexistence', in W.A. Gallusser (ed.), *Political Boundaries and Coexistence*, (Bern: Peter Lang, 315–321).

HAZAN, R. (1988), 'Peaceful conflict resolution in the Middle East: The Taba negotiations', *Journal of the Middle East Studies Society* 2:1, 39–65.

KEMP, A. AND BEN-ELIEZER, U. (2000), 'Dramatizing sovereignty: The construction of territorial dispute in the Israeli-Egyptian border at Taba', *Political Geography* 19, 315–344.

KLIOT, N. (1996), 'Turning desert to bloom: Israeli-Jordanian peace proposals for the Jordan rift valley', *Journal of Borderlands Studies* 11:1, 1–24.

KLIOT, N. (1997), 'Regional development for the peace era – Jordanian and Israeli Perspectives', in G. Blake and L. Chia and C. Grundy-Warr and M. Pratt and C.H. Schofield (eds), *International Boundaries and Environmental Security: Frameworks for Regional Co-operation*, (Amsterdam: Kluwer, 279–290).

LAPIDOTH, R. (1986), 'The Taba controversy', *The Jerusalem Quarterly* 37, 29–39.

MANSFELD, Y. (1996), 'Wars, tourism, and the "Middle East" factor', in A. Pizam and Y. Mansfeld (eds), *Tourism, Crime and International Security Issues*, (Chichester: Wiley, 265–278).

MOFFETT, G.D. (1988a), 'Disputed patch of desert a key to Egyptian-Israeli relations', *Christian Science Monitor*, (published on 8 June 1988), p. 11.

MOFFETT, G.D. (1988b), 'Taba verdict: Boost to Israel-Egypt ties?', *Christian Science Monitor*, (published on 28 September 1988), p. 7–8.

NEZAVISIMAYA GAZELA (1994), 'Special conditions for crossing the border are relaxed', *Nezavisimaya Gazela*, (published on 28 December 1994), p. 1.

RAAFAT, W. (1983), 'The Taba case between Egypt and Israel', *Revue Egyptienne de Droit International* 39, 1–22.

ROBERTS, J. (1995), 'Israel and Jordan: Bridges over the borderlands', *Boundary and Security Bulletin* 2:4, 81–84.

Tourism related cross-border co-operation in Asia

Similar to Europe, Asia offers a high potential for cross-border co-operation with regard to tourism. Especially in the Southeast, high clusters of nations, interconnected by natural similarities and cultural heritage, offer various opportunities for collaborative tourism development. In addition, the political systems of China and North Korea offer a unique chance to research the subject of potential cross-border co-operation between centrally planned communist countries and westernised capitalist nations. Cross-border co-operation, with regard to tourism, is widespread and comprises economic issues as well as social and cultural collaborative clusters engaging in regional economic development or environmental protection. Such co-operation often needs to consider safety and security issues resulting from the perceived danger arising from the severe clash between rich and highly developed nations and its less developed and rather poor neighbours as examples from China, Tibet, Buthan, Nepal or North Korea indicate (Timothy and Tosun, 2003).

Tourism development in Asia is often concerned with ensuring tourist security. However, it has progressively realised the benefits of supra-national co-operation even between nations of different economic development. Subsequently, it has seen the creation of the Association of South-East Asian Nations (ASEAN), an economic free trade area with supra-national sovereignty consisting of Singapore, Malaysia, Indonesia, Thailand, The Philippines, Vietnam, Brunei and the membership interested Cambodia and Myanmar (see especially Broinowski, 1982, and Var and Toh and Khan, 1998, for a detailed discussion of the history of ASEAN) which soon collaboratively created its own tourism policy for the ASEAN region. According to Timothy (2001), this strategy comprised six distinct aspects relating to marketing ASEAN as a single destination, encouraging tourism investment, developing a critical pool of tourism human resources, promoting environmentally sustainable tourism, and facilitating seamless intra-ASEAN travel, as well as the exchange of information and experiences. This common strategy symbolises a change in the function of ASEAN which is subsequently concerned with borderland integration and regional development. Such processes are more subtle and rather concerned with proximity, complementarity and shared history, aspects important for the tourism industry. They have improved ASEAN from simply being a trade block trying 'to open internal and seal external borders economically' (Ho and So, 1997: 242). Considering the broad acceptance of ASEAN among Asian nations and the Asian Development Bank's efforts undertaken in tourism development for Asia as a whole, investigating cross-border co-operation in Asia promises decent and abundant information on the mechanisms underlying successful collaborations across national boundaries in tourism.

The literature compiled below intends to give a profound insight into the creation and different functions of the Asian Development Bank and ASEAN. Both entities heavily influence any tourism activity pertinent in Asia.

Table 9.24   Selected literature: General structure and politics of Asian cross-border tourism co-operation

ASEAN SECRETARIAT (1999), 'Economic Co-operation', <http://www.aseansec.org/ economic/ eco.htm>, accessed on 24 November 1999.
ASIAN DEVELOPING BANK (2000), 'Co-operation in Tourism', <http://www.adb.org/ Documents/ Events/Privat_Sector_Forum/calcutta-tourism.pdf>, accessed 1 February 2003.
ASIAN DEVELOPMENT BANK (1998), 'Tourism', <http://www.adb.org/GMS/tour1.asp>, accessed 27 February 2003.
ASIAN DEVELOPMENT BANK (1999), 'Tenth Meeting of the Working Group on Tourism', <http://www.adb.org/GMS/wgt10.asp>, accessed 27 February 2003.
ASIAN DEVELOPMENT BANK (2002b), 'Tourism Sector could benefit from sub-regional co-operation', <http://www.adb.org/Documents/Speeches/2002/ sp2002011.asp>, accessed 27 February 2003.
ASIAN DEVELOPMENT BANK (2002c), 'Draft proposals for Infrastructure Development', <http://www.adb.org/Documents/Events/2002/RETA5936/Tourism/Infradev_alwis. pdf>, accessed on 25 March 2003.
ASIAN DEVELOPMENT BANK (2002d), 'A strategy for SASEC Tourism', <http://www.adb.org/ Documents/Events/2002/RETA5936/Tourism/Co-operation_alwis.pdf>, accessed on 25 March 2003.
ASIAN DEVELOPMENT BANK (2003a), 'Co-operation spurs tourism – Rich in environmental, cultural and historical treasures, the GMS has significant potential for becoming a major tourist destination-but infrastructure and co-operation must come first', <http://www.adb. org/Documents/Brochures/ GMS_Connecting_Nations/ co-operation_spurs_tourism. asp>, accessed 11 February 2003.
BROINOWSKI, A. (ed.) (1982), *Understanding ASEAN*, (New York: St Martin's Press).
BROWN, D. (1994), *The State and Ethnic Politics in Southeast Asia*, (London: Routledge).
CASTRO, A. (1982), 'ASEAN economic co-operation', in A. Broinowski (ed.), *Understanding ASEAN*, (New York: St Martin's Press, 70–91).
ESMARA, H. (ed.) (1988), *ASEAN Economic Co-operation: A New Perspective*, (Singapore: Chopmen Publishers).
GO, F.M. AND JENKINS, C.L. (eds) (1997), *Tourism and Economic Development in Asia and Australasia*, (London: Cassell).
HONG, M.S. (1987), 'Competition between NICs and ASEAN', *East Asia International Review of Economic, Political and Social Development* 4, 130–144.
HUSSEY, A. (1991), 'Regional development and co-operation through ASEAN', *Geographical Review* 81, 87–98.
NAIDU, G. (1988), 'ASEAN co-operation in transport', in H. Esmara (ed.), *ASEAN Economic Co-operation: A New Perspective*, (Singapore: Chopmen Publishers, 191–204).

PUNTASEN, A. (1988), *ASEAN co-operation in tourism*, (Bangkok: Faculty of Economics, Thammasat University).

REGNIER, R. AND NIU, Y. AND ZHANG, R. (1993), 'Towards a regional 'bloc' in East Asia: implications for Europe', *Issues and Studies* 29, 15–34.

RICHTER, L.K. AND RICHTER, W.L. (1985), 'Policy choices in South Asian tourism development', *Annals of Tourism Research* 12:2, 201–217.

SCALAPINO, R.A. (1992), 'Northeast Asia – prospects for co-operation', *The Pacific Review*, 5:2, 101–111.

SOMMERS, B.J. AND TIMOTHY, D.J. (1999), 'Economic development, tourism, and urbanization in the emerging markets of Northeast Asia', in G.P. Chapman and A.K. Dutt and R.W. Bradnock (eds), *Urban Growth and Development in Asia: Making the Cities*, (Aldershot: Ashgate, 111–131).

TASKER, R. AND SCHWARTZ, A. AND VATIKIOTIS, M. (1994),'ASEAN: Growing pains', *Far Eastern Economic Review*, (published on 28 July 2003), pp. 22–23.

TIMOTHY, D.J. (2003), 'Supranationalist alliances and tourism: Insights from ASEAN and SAARC', *Current Issues in Tourism* 6:3, 250–266.

TIMOTHY, D.J. AND WALL, G. (1995), 'Tourist accommodation in an Asian historic city', *Journal of Tourism Studies* 6:2, 63–73.

VAR, T. AND TOH, R. AND KHAN, H. (1998), 'Tourism and ASEAN Economic Development', *Annals of Tourism Research* 25:4, 195–197.

WU, C. (1998), 'Cross-border development in Europe and Asia', *GeoJournal* 44:3, 189–201.

The Koreas

One particularly interesting situation regarding potential cross-border co-operation across political boundaries can be investigated on the Korean peninsula. After centennials of severe cold warfare and strict separation of the two distinctly different Koreas, there has been a slight ease of tension in recent years. This has opened the gates for possible cross-border co-operation between the former enemies. Especially in terms of tourism, this co-operation has started to produce new ways of looking at the brother and might be considered a preliminary step in a future unification of the Korean peninsula. First attempts have been made to allow cross-border travel (Kirk, 2003) and have supported the idea of a potential reunification discussed within academia (see Eberstadt, 1997; Henriksen and Lho, 1994). Lessons learned from the reunification of Germany might assist in understanding the strenuous obstacles lying ahead of those willing to support Korea's reunification and contribute to further research conducted on the current and future potential of tourism on the Korean peninsula.

Table 9.25   Selected literature: The Koreas

AGENCE FRANCE PRESSE (2002), 'North Korea designates tax-free international tourism zone', <http://traveltax.msu.edu/news/Stories/afp10.htm>, accessed 13 March 2003.
CHINA RADIO INTERNATIONAL (1997), 'Republic of Korea reportedly to simplify visa procedures for PRC tourists', *China Radio International broadcast*, (Beijing, published on 5 June 1997).
EBERSTADT, N. (1995), *Korea Approaches Reunification*, (Armonk, NY: M.E. Sharpe).
FOSTER-CARTER, A. (1981), 'North Korea: Opening the door', *New Society*, (published on 27 August 1981), p. 224.
FOSTER-CARTER, A. (1994), 'Korea: Sociopolitical realities of reuniting a divided nation', in T.H. Henriksen and K. Lho (eds), *One Korea?: Challenges and Prospects for Reunification*, (Stanford, CA: Hoover Institution Press, 31–47).
HALL, D.R. (1986a), 'North Korea opens to tourism: A last resort', *Inside Asia* 9:4, 21–23.
HALL, D.R. (1990b), 'Stalinism and tourism: A study of Albania and North Korea', *Annals of Tourism Research* 17, 36–54.
HENRIKSEN, T.H. AND LHO, K. (1994a), 'Introduction', in T.H. Henriksen and K. Lho (eds), *One Korea?: Challenges and Prospects for Reunification*, (Stanford, CA: Hoover Institution Press, 1–11).
HENRIKSEN, T.H. AND LHO, K. (eds) (1994b), *One Korea?: Challenges and Prospects for Reunification*, (Stanford, CA: Hoover Institution Press).
KIM, S.S. AND PRIDEAUX, B. (2003), 'Tourism, peace and ideology: impacts of the Mt. Gumgang tour project in the Korean Peninsula', *Tourism Management* 24, 675–685.
KIRK, D. AT THE INTERNATIONAL HERALD TRIBUNE – THE IHT ONLINE (2003), 'Seoul leader deems mission 'a triumph': South Korean Busses cross border to North', <http://www.iht.com/articles/ 2003/02/06/talks.php>, accessed 15 February 2003.
KIM, Y.K. AND CROMPTON, J.L. (1990), 'Role of tourism in unifying the two Koreas', *Annals of Tourism Research* 17, 353–366.
MO, J. (1994), 'German lessons for managing the economic cost of Korean reunification', in T.H. Henriksen and K. Lho (eds), *One Korea?: Challenges and Prospects for Reunification*, (Stanford, CA: Hoover Institution Press, 48–67).
POLLACK, A. (1996b), 'Behind North Korea's barbed wire: Capitalism', *New York Times*, (published on 15 September 1996), p. A9.

Southeast Asian Cross-border Co-operation

Probably the most interesting Asian region in terms of cross-border potential lies in the Southeast and comprises nations such as Singapore, Thailand, Myanmar, Vietnam, Indonesia, or Cambodia, to name just a few popular examples. Large clusters of environmentally very similar nations, divided by geometric borderlines, have created a culturally compatible landscape predestined for collaboration across these man-made and imposed boundaries. This geographical and socio-economical distribution has led to the creation of unique 'growth triangles' concerned with reterritorialisation and cross-border region creation (see specifically Parsonage, 1992;

Perry, 1991). According to the Asian Development Bank, such 'growth triangles', uniquely restricted to Asia, are defined as the 'exploitation of complementarity among geographically contiguous countries to help them gain greater competitive advantage in export promotion' (Krongkaew, 2004: 979). Economically, such triangulations are signified by extensive economic cross-border co-operation in terms of land, capital and labour to avoid duplication, strengthen regional peculiarities and support local specialisation for the sake of sustainable and economically viable regional development (Sparke, Sidaway, Bunell and Grundy-Warr, 2004). Countries with high populations but little capital thereby send workers to neighbouring countries to work and create economic benefit for all participants. One of the most famous and best examined of these growth triangles is the Indonesia-Malaysia-Singapore growth triangle situated around their intersection.

The literature compiled below introduces the preliminary research conducted on the phenomenon of triangulation in South-East Asia, in addition to some general studies investigating the different aspects of cross-border co-operation with regard to tourism in the popular tourist destinations of South-East Asia.

Table 9.26   Selected literature: Southeast Asian cross-border co-operation

BUSZYNSKI, L. (1998), 'Thailand and Myanmar: The perils of "constructive engagement"', *The Pacific Review* 11:2, 290–305.
ECONOMIC AND BUSINESS REVIEW INDONESIA (1993), 'Cross-border shopping craze', *Economic and Business Review Indonesia* 57, 34.
GRUNDY-WARR, C. (2002), 'Cross-border regionalism through a "South-east Asian" looking-glass', *Space and Polity* 62, 215–225.
GRUNDY-WARR, C. AND PEACHEY, K. AND PERRY, M. (1999), 'Fragmented integration in the Singapore-Indonesian border zone: Southeast Asia's "growth triangle" against the global economy', *International Journal of Urban and Regional Research* 23:2, 304–328.
GRUNDY-WARR C. AND PERRY M. (2000), 'Tourism in an Inter-state Borderland: The Case of the Indonesian-Singapore Co-operation', in Chon K.S. (ed.) (2000), *Tourism in South-East Asia*, (Birmingham: Haworth Hospitality Press).
HAGIWARA, Y. (1973), 'Formation and development of the Association of Southeast Asian Nations', *The Developing Economies* 11:4, 443–465.
HALL, C.M. AND PAGE, S.J. (eds) (2000), *Tourism in South and Southeast Asia: Critical Perspectives*, (Oxford: Butterworth-Heinemann).
LEE, Y.L. (1980), *The razor's edge: Boundaries and boundary disputes in Southeast Asia*, Institute of Southeast Asian Studies Research Notes and Discussion 15.
LEIFER, M. (1962), 'Cambodia and her neighbours', *Pacific Affairs* 34:4, 361–374.
LIDDLE, R.W. AND MALLARANGENG, R. (1997), 'Indonesia in 1996: Pressures from above and below', *Asian Survey* 37, 167–174.
LOW, L. AND HENG, T.M. (1997), 'Singapore: Development of gateway tourism', in F.M. Go and C.L. Jenkins (eds), *Tourism and Economic Development in Asia and Australasia*, (London: Cassell, 236–254).

MOEDJANTO, G. (1986), *The Concept of Power in Javanese Culture*, (Yogyakarta: Gadjah Mada University Press).

NGAMSOM, B. (1998), 'Shopping tourism: A case study of Thailand', in K.S. Chon (ed.), *Proceedings, Tourism and Hotel Industry in Indo-China and Southeast Asia: Development, Marketing, and Sustainability*, (Houston, TX: University of Houston, 112–128).

PARSONAGE, J. (1992), 'Southeast Asia's growth triangle: a sub-regional response to global transformation', *International Journal of Urban and Regional Research* 16, 307–318.

PERRY, M. (1991), 'The Singapore Growth Triangle: State, capital and labour at a new frontier in the world economy', *Singapore Journal of Tropical Geography* 12:2, 138–151.

RIMMER, P.J. (1994), 'Regional economic integration in Pacific Asia', *Environment and Planning A* 26, 1731–1759.

SINGH, L.P. (1962), 'The Thai-Cambodian temple dispute', *Asian Survey* 2:8, 23–26.

ST JOHN, R.B. (1994), 'Preah Vihear and the Cambodia-Thailand borderland', *Boundary and Security Bulletin* 1:4, 64–68.

TIMOTHY, D.J. (1999c), 'Participatory planning: A view of tourism in Indonesia', *Annals of Tourism Research* 26, 371–391.

TIMOTHY, D.J. (2000c), 'Tourism planning in Southeast Asia: Bringing down borders through co-operation', in K.S. Chon (ed.), *Tourism in Southeast Asia: A New Direction*, (New York: The Haworth Hospitality Press, 21–35).

TURNER, P. (1994), *South-East Asia*, (Hawthorn: Lonely Planet).

Greater Mekong sub-region

Another regional exceptionality in Asian tourism is the Greater Mekong Sub-region (GMS). This regional entity was founded in 1992 by the six nations of China, Lao, Myanmar, Thailand, Cambodia, and Vietnam in co-operation with the Asian Development Bank (Krongkaew, 2004). The unique nature of the GMS results from its interaction between some of the world's least-developed and formerly strictly socialist oriented countries with decently developed and capitalist regional tourist magnets, such as Thailand. The overall aim of the GMS is to facilitate future co-operation by overcoming differences in economic development and even social systems. Like ASEAN, the GMS has realised the economic potential of successful tourism development and initiated a tourism promotion program based on eight priorities: efforts in destination marketing, the creation of sub-regional events, extensive training of tourism staff, state-of-the-art management of natural and cultural resources, the development of river-based tourism activities, the facilitation of sub-regional travel, the development of village-based tourism businesses and the control and monitoring of tourism flows within the GMS (Krongkaew, 2004).

The small list of literature below intends to impart the history of this already well-established tourism related cross-border co-operation in Asia.

Table 9.27   Selected literature: Greater Mekong sub-region

ASIAN DEVELOPMENT BANK (1996), 'Greater Mekong Sub-region: Sub-regional Programs in Tourism', <http://www.adb.org/GMS/proj-tour.pdf>, accessed 26 March 2003.
ASIAN DEVELOPMENT BANK (2000), 'Eleventh Meeting of the Working Group on the Greater Mekong Sub-region Tourism Sector', <http://www.adb.org/GMS/ wgt11.asp>, accessed 11 February 2003.
ASIAN DEVELOPMENT BANK (2002a), 'Developing Tourism in Lower Mekong River Basin Countries', <http://www.adb.org/Documents/News/2002/ nr2002250.asp>, accessed 27 February 2003.
ASIAN DEVELOPMENT BANK (2003b), 'ADB's role in the GMS Program – Background and History', <http://www.adb.org/GMS/gmsprogram10.asp#tourism>, accessed 11 February 2003.
KRONGKAEW, M. (2004), 'The development of the Greater Mekong Sub-region (GMS): Real promise or false hope?', *Journal of Asian Economics* 15, 977–998.

Chinese tourism and related cross-border co-operation

Discussion of tourism co-operation across national boundaries in Asia necessarily has to involve a discussion about China. As a mixed economy that opens up specific national areas and leaves them subject to capitalist developments, China is of specific interest when investigating cross-border relations and tourism developments. Complying with policies formulated under 'one country, two systems' China holds total control over economic, and consequently tourism, development on its mainland, but has submitted its executive powers in its special administrative regions (SAR) Hong Kong and Macao to regional and democratically elected governments. Obviously then, tourism policies and developments differ tremendously from those established in Mainland China and have spurred the creation of extensive trans-boundary economic regions. One region of specific interest for tourism related research into cross-border co-operation is the Pearl River Delta connecting Mainland China with Hong Kong. The relationship between Hong Kong and China in the Pearl River Delta is highly complementary with the People's Republic of China being the manufacturing and Hong Kong the service oriented location. This situation has created a trans-boundary economic region best described as 'front shop, back factory' and spurred social interactions across the border, hence tourism related activities, too (see Ho and So, 1997 and Yang, 2004 for a detailed description of the history of the Pearl River Delta).

Further studies conducted on the relationship between Hong Kong and China in modern times, as well as in the past, is compiled below and intends to give a general insight into the theory of running 'one country' under 'two systems'. But supra-national regional developments in China are not restricted to the Hong Kong area. Several further studies listed below have investigated China's tourism policies in Tibet, for example, its general decentralisation of tourism industries within its mainland, and introduce a unique form of tourist activities along the Sino-Russian borders. Given the strict visa procedures and immigration regulations opposed by China, several bilateral agreements have been arranged by the Russian and Chinese government to enable citizens along their common border to visit the neighbouring country. Considering the relatively poor economic conditions within these regions and the remaining severe controls on imports and currency exchanges, these visits have created a distinctive form of border trade, called barter tourism (Zhao, 1994) which is subsequently discussed in detail within some of the studies compiled below.

Table 9.28   Selected literature: Chinese cross-border co-operation

BEIJING REVIEW (1994), 'Tibet opens further to the outside world', *Beijing Review* 37:40, 31–32.

BELYY, N. (1996), 'Cross-border trade with China' *Rossiyskiye Vesti*, (published on 18 April 1996), p. 3.

BREITUNG, W. (2002), 'Transformation of a boundary regime: The Hong Kong and Mainland China case', *Environment and Planning A* 34:10, 1749–1762.

BRESLIN, S. (2000), 'Decentralisation, globalisation and China's partial re-engagement with the global economy', *New Political Economy* 5:2, 205–226.

CAO, X. AND GE, A. (1993), 'Shenzen: A new frontier', *China Tourism* 156, 7–15.

CHAN, R. C. K. (1998), 'Cross-border regional development in Southern China', *Geojournal* 44:3, 225–237.

CHEN, E.K.Y. AND HO, A. (1994), 'Southern China Growth Triangles: An Overview', in M. Thant and M. Tang and H. Kakazu (eds) (1994), *Growth Triangles in Asia: A new approach to regional economic co-operation*, (Hong Kong: Oxford University Press, 29–72).

CHENMING, R. (1992), 'Cross-border on-day trips along the Chinese boundary in Yunnan', *China Tourism* 145, 41–49.

CHEUNG, P. T. Y. (2002), 'Managing the Hong Kong–Guangdong relationship: Issues and challenges', in A. G. O. Yeh (ed.), *Building a competitive Pearl River Delta region: Co-operation, coordination and planning*, (Hong Kong: The Centre of Urban Planning and Environmental Management, The University of Hong Kong, 39–58).

DOBSON, W.J. AND FRAVEL, M.T. (1997), 'Red herring hegemon: China in the South China Sea', *Current History* 96, 258–263.

GENGXIN, J. (1997), 'Heilongjiang achieves gratifying results in border trade', *Heilongjiang Ribao*, (published on 15 May 1997), p. 2.

GORMSEN, E. (1995), 'International tourism in China: Its organization and socio-economic impact', in A.A. Lew and L.Yu (eds), *Tourism in China: Geographic, Political, and Economic Perspectives*, (Boulder, CO: Westview Press, 63–88).

HEUNG, V.C.S. (1997), 'Hong Kong: Political impact on tourism', in F.M. Gol and C.L. Jenkins (eds), *Tourism and Economic Development in Asia and Australasia*, (London: Cassell, 123–137).

HO, K.C. AND SO, A. (1997), 'Semi-periphery and borderland integration: Singapore and Hong Kong experiences', *Political Geography* 16:3, 241–259.

Hobson, J.S.P. (1995), 'Hong Kong: The transition to 1997', *Tourism Management* 16, 15–20.

Hu, Y. and Chan, R. (2002), 'Globablization, governance and development of the Pearl River Delta region', *The China Review* 2:1, 61–84.

Lintner, B. (1991b), 'Upstaging Macau: Casino at centre of border development plan?', *Far Eastern Economic Review*, (published on 16 May 1991), p. 24.

Macau Government Tourist Office (1994), *Bestway to Macau*, (Macau: Government Tourist Office).

McKercher, B. (2001), 'Cross-Border Tourism: An Empirical Study of Tourism into the Pearl River Delta', *Pacific Tourism Review* 5:1, 33–41.

Ng, E. (2002), 'Cross-boundary planning: The interface between Hong Kong and the mainland', in A.G.Yeh and Y.F. Lee and T. Lee and N.D. Sze (eds) (2002), *Building a competitive Pearl River Delta region: Co-operation, coordination and planning*, (Hong Kong: Centre of Urban Planning and Environmental Management, The University of Hong Kong, 271–281).

Nin, C.Y. (1994), 'Trade at the Sino-Kazakhstani border: A visit to Korgas and Yining', *China Tourism* 167, 60–65.

Nin, C.Y. (1998), 'A boundary waterfall', *China Tourism* 215, 10–15.

Porter, J. (1997), 'Macau 1999', *Current History* 96, 282–286.

Roehl, W.S. (1995), 'The June 4, 1989, Tiananmen Square incident and Chinese tourism', in A.A. Lew and L. Yu (eds), *Tourism in China: Geographic, Political, and Economic Perspectives*, (Boulder, CO: Westview Press, 19–39).

Rozman, G. (1995), 'Spontaneity and direction along the Russo-Chinese border', in S. Kotkin and D. Wolff (eds), *Rediscovering Russia in Asia: Siberia and the Russian Far East*, (Armonk, NY: M.E. Sharp, 275–289).

Shen, J. (2003), 'Cross-border connection between Hong Kong and mainland China under "two systems" before and beyond 1997', *Geografiska Annaler Series B, Human Geography* 85:1, 1–17.

Shen, J. (2004), 'Cross-border Urban Governance in Hong Kong: The Role of State in a Globalizing City-Region', *The Professional Geographer* 56:4, 530–544.

Sit, V.F.S. (1989), 'Industrial out-processing – Hong Kong's new relationship with the Pearl River Delta', *Asian Profile* 17, 1–13.

Smart, A. (2002), 'The Hong Kong/Pearl River Delta urban region: an emerging transnational mode of regulation or just muddling through?', in J.R. Logan (ed.), *The new Chinese city: Globalization and market reform*, (Oxford/Malden: Blackwell Publishers, 92–105).

Storey, R. (1992), *Hong Kong, Macau and Canton*, (Hawthorn: Lonely Planet).

Ting, Z. (1994), 'Hunchun: A trade city that's going places', *China Today* 43:1, 22–26.

Toops, S.W. (1992), 'Tourism in China and the impact of June 4, 1989' *Focus* 42:1, 3–7.

Ul-chul, Y. (1997), 'DPRK reportedly to designate Pidan Island free trade zone', *Hangyore*, (published on 10 June 1997), p. 1.

Walsh, J. (1992), 'Sea of troubles: China's offshore oil grab chills détente with Vietnam and rings wider Asian alarms', *Time*, (published on 27 July 1992), p. 40–41.

Yang, C. (2004), 'The Pearl River Delta and Hong Kong: An evolving cross-boundary region under "one country, two systems"', *Habitat International*, Article in Press.

YEH, A.G. AND LEE, YF. AND LEE, T. AND SZE, N.D. (eds) (2002), *Building a competitive Pearl River Delta region: Co-operation, coordination and planning*, (Hong Kong: Centre of Urban Planning and Environmental Management, The University of Hong Kong).

YU, L. (1992), 'Emerging markets for China's tourism industry', *Journal of Travel Research* 31:1, 10–13.

YU, L. (1997), 'Travel between politically divided China and Taiwan', *Asia Pacific Journal of Tourism Research* 2:1, 19–30.

ZHANG, G. (1993), 'Tourism crosses the Taiwan Straits', *Tourism Management* 14, 228–231.

ZHAO, X. (1994a), 'Barter tourism: A new phenomenon along the China-Russia border', *Journal of Travel Research* 32:3, 65–67.

ZHAO, X. (1994b), 'Barter tourism along the China-Russia border', *Annals of Tourism Research* 21, 401–403.

ZHENGE, P. (1993), 'Trade at the Sino-Russian border', *China Tourism* 159, 70–73.

Cross-border co-operation in tourism in the Americas

Looking at the political landscape of the Americas, contributions on border issues and tourism either focus on the northern or southern realm. Especially in North America, the possibilities for cross-border co-operation are rather effortlessly distinguishable. Tourism related trans-national co-operation either exists between the US and Canada, or the US and Mexico. A smaller amount of co-operation also appears on a subnational level within the US due to the general jurisdictional differences among the US-American states that sometimes differ tremendously. Co-operation across subnational and special area borders, such as the ones surrounding Indian reservations, is often subject to jurisdictional differences that have formed economic niches for specific tourism development, including gambling, cultural or nature-based tourism activities.

Before individual literature compilations introduce studies and research activities conducted on the individual borderlands of the USA and Canada in the North, and the USA and Mexico in the South, a general introduction to tourism related policies within the USA is listed below. In addition to introducing US tourism policies, research conducted on cross-border shopping activities in North America (McAllister, 1961; Timothy and Butler, 1995) is highlighted and the creation, formation and function of the North American Free Trade Association (NAFTA), comprising Canada, the USA and Mexico is discussed (see specifically Hufbauer and Schott, 1992), which arguably facilitated border-crossing procedures among these nations in general and spurred regional trans-national economic developments and tourism activities (Smith, 1994; Smith and Pizam, 1998).

Table 9.29   Selected literature: US general cross-border co-operation policies

BRANT, M. AND GALLAGHER, T.E. (1993), 'Tourism and the Old Order Amish', *Pennsylvania Folklife* 43:2, 71–75.

BREWTON, C. AND WITHIAM, G. (1998), 'United States tourism policy: Alive but not well', *Cornell Hotel and Restaurant Administration Quarterly* 39:1, 50–59.

CLEMENT, N. AND GANSTER, P. AND SWEEDLER, A. (1999), 'Development, environment, and security in asymmetrical border regions: European and North American perspectives', in H. Eskelinen and I. Liikanen and J. Oksa (eds), *Curtains of Iron and Gold: Reconstructing Borders and Scales of Interaction*, (Aldershot: Ashgate, 243–281).

CONZEN, M.P. (1990), 'Introduction', in M.P. Conzen (ed.), *The Making of the American Landscape*, (Boston: Unwin Hyman, 1–8).

DENVER POST (1995), 'Wan reception for US immigration proposal', *Denver Post*, (published on 30 August 1995), p. A9.

DEQUINE, J. (1989), 'Spring breakers head for the border', *USA Today*, (published on 13 March 1989), pp. 1, 14.

GERLACH, J. (1989), 'Spring Break at Padre Island: A new kind of tourism', *Focus* 39:1, 13–16, 29.

GRINSPUN, R. AND CAMERON, M. (eds) (1993), *The Political Economy of North American FreeTrade*, (New York: St Martin's Press).

HERRICK, T. (1997), 'Blurring the line' *Houston Chronicle*, (published on 10 August 1997), p. D1.

HUFBAUER, G.C. AND SCHOTT, J.J. (1992), *North American Free Trade: Issues and Recommendations*, (Washington D.C.: Institute for International Economics).

HUSBANDS, W. (1981), 'Centres, peripheries, tourism and socio-spatial development', *Ontario Geography* 17, 37–59.

INGRAM, H. AND MILICH, L. AND VARADY, R.G. (1994), 'Managing transboundary resources: Lessons from Ambos Nogales', *Environment* 36:4, 6–9 and 28–38.

INTERNATIONAL PEACE GARDEN (n.d.), *Like No Place on Earth!*, (Boissevain, MB: International Peace Garden).

KIY, R. AND WIRTH, J.D. (1998), 'Introduction', in R. Kiy and J.D. Wirth (eds), *Environmental Management on North America's Borders*, (Texas A&M University Press, 3–31).

KIY, R. AND WIRTH, J.D. (eds) (1998), *Environmental Management on North America's Borders*, (Texas: A&M University Press).

KJOS, K. (1986), 'Trans-boundary land-use planning: A view from San Diego County', in L.A. Herzog (ed.), *Planning the International Border Metropolis: Trans-Boundary Policy Options in the San Diego-Tijuana Region*, (San Diego: Center for US-Mexican Studies, University of California, 22–26).

LEWIS, P.F. (1979), 'Axioms for reading the landscape: Some guides to the American scene', in D.W. Meinig (ed.), *The Interpretation of Ordinary Landscapes: Geographical Essays*, (New York: Oxford University Press, 11–31).

LONG, V.H. (1993), 'Techniques for socially sustainable tourism development: Lessons from Mexico', in J.G. Nelson and R.W. Butler and G. Wall (eds), *Tourism and Sustainable Development: Monitoring, Planning, Managing*, (Waterloo, ON: University of Waterloo, Department of Geography, 201–218).

MARTINEZ, O.J. (1996) 'Introduction', in O.J. Martinez (ed.), *US-Mexico Borderlands: Historical and Contemporary Perspectives*, (Wilmington, DE: Scholarly Resources, xiii–xix).

MAYES, H.G. (1992), 'The International Peace Garden: A border of flowers', *The Beaver*, 72:4, 45–51.

MCALLISTER, H.E. (1961), 'The border tax problem in Washington', *National Tax Journal* 14:4, 362–374.

MCMILLAN, C. (1993), *Building Blocks or Trade Blocs: NAFTA, Japan and the New World Order*, (Ottawa: Canada-Japan Trade Council).

POLLACK, A. (1996a), 'At the DMZ, another invasion: Tourists', *New York Times*, (published on 10 April 1996), p. A10.

RANDALL, J. AND CONRAD, H.W. (eds) (1995), *NAFTA in Transition*, (Calgary: University of Calgary Press).

SCOTT, J. AND SWEEDLER, A. AND GANSTER, P. AND EBERWEIN, W. (eds) (1996), *Border Regions in Functional Transition. European and North American Perspectives on Transboundary Interaction*, (Erkner: Regio Series of the IRS 9, 171–191).

SCOTT, J. W. (1999), 'European and North American contexts for cross-border regionalism', *Regional Studies* 33:7, 605–618.

SMITH, G. (1994), 'Implications of the North American Free Trade Agreement for the US tourism industry', *Tourism Management* 15, 323–326.

SMITH, G. AND PIZAM, A. (1998), 'NAFTA and tourism development policy in North America', in E. Laws and B. Faulkner and G. Moscardo (eds), *Embracing and Managing Change in Tourism: International Case Studies*, (London: Routledge, 17–28).

STEFFENS, R. (1994), 'Bridging the border: As NAFTA goes into effect, US and Mexican officials must ensure that economics does not overshadow the need to protect parks and other public lands', *National Parks* 68:7, 36–41.

TIMOTHY, D.J. AND BUTLER, R.W. (1995), 'Cross-border shopping: A North American perspective', *Annals of Tourism Research* 22, 16–34.

TRAVEL AND TOURISM EXECUTIVE REPORT (1997), 'Sec. 110 "non" waiver rules for Canada, Mexico surprised travel industry', *Travel and Tourism Executive Report* 18:7, 1, 5, 8.

US MAYORS (n.d.), 'International Affairs – Facilitating Movement Across US International LandBorders'<http://www.usmayors.org/uscm/resolutions/66th_conference/facilitating_int.html>, accessed on 13 March 2003.

XX BORDER GOVENORS CONFERENCE (2002), 'Joint Declaration', <http://www.azmc.org/downloads/XX_BGC_Joint_Declaration-final.doc>, accessed on 3 April 2003.

YENCKEL, J.T. (1995), 'Big bargains across the borders: Deals abound in Mexico and Canada as currencies plunge', *Washington Post*, (published on 15 January 1995), p. E2.

US-Canada cross-border co-operation in tourism

Discussing the potential of cross-border co-operation in tourism in North America requires a general insight into the relationship between Canada and the USA. The following compilation of literature aims to introduce Canadian-US relations and specifies the difficulties and chances of cross-border travel between these nations. In particular, the North-Western trans-national region, Cascadia, is a focal point for tourism related research, since it has recently been identified as creating a 'new binationalism' (Schell and Hamer, 1995) which subsequently has a heavy impact on the will and capability for co-operation across the border and collaborative formulation of a unitary tourism development strategy. Various studies elaborate on the various aspects of this newly emerged bi-national region (Mazza, 1995; McCloskey, 1995; Rutan, 1985) that might prove to be a valuable example for further trans-boundary developments along this, the longest unfortified border in the world (Timothy, 2001).

Table 9.30   Selected literature: General considerations of US-Canada cross-border co-operation

ARTIBISE, A.F.J. (1996), *Redefining B.C.'s place in Canada: The emergence of Cascadia as a strategic alliance* (Vancouver: Manuscript).

BAKER FOX, A. AND HERO, A.O. AND NYE, J.S. JR. (eds) (1976), *Canada and the United States: Transnational and transgovernmental relations* (New York/London: Columbia Universtiy Press).

DI MATTEO, L. AND DI MATTEO, R. (1996), 'An analysis of Canadian cross-border travel', *Annals of Tourism Research* 23, 103–122.

DILLEY, R.S. AND HARTVIKSEN, K.R. AND NORD, D.C. (1991), 'Duluth and Thunder Bay: A study of mutual tourist attractions', *Operational Geographer* 9:4, 9–13.

DONALDSON, G. (1979), *Niagara! The Eternal Circus*, (Toronto: Doubleday Canada).

GODDARD, A.M. AND SMITH, P.J. (1993), *The development of subnational relations: The case of the Pacific Northwest and Western Canada*, (Manuscript).

GOLDBERG, M.A. AND LEVI M.D. (1992/93), 'The evolving experience along the Pacific North-West corridor called Cascadia', *The new pacific*, 28–32.

GRISWOLD, E.N. (1939), 'Hunting boundaries with car and camera in the Northeastern United States', *Geographical Review* 29, 353–382.

HENKEL, W.B. (1993), 'Cascadia: A state of (various) mind(s)', *Chicago Review* 39, 110–118.

HODGKINSON, T. (1992), 'Ottawa to launch travel campaign' *London Free Press*, (published on 5 September 1992.

KELLER, C.P. (1987), 'Stages of peripheral tourism development – Canada's Northwest Territories', *Tourism Management* 8, 20–32.

KENDALL, K.W. AND KRECK, L.A. (1992), 'The effect of the across-the-border travel of Canadian tourists on the city of Spokane: A replication', *Journal of Travel Research* 30:4, 53–58.

KEOHANE, R.O. AND NYE, J.S. (1976), 'Introduction', in A. Baker Fox and A.O. Hero and J.S. Nye, Jr. (eds) *Canada and the United States: Transnational and transgovernmental relations*, (New York/London: Columbia Universtiy Press, 3–17).

MACKAY, J.R. (1958), 'The interactance hypothesis and boundaries in Canada: A preliminary study', *Canadian Geographer* 11, 1–8.

MADDEN, K. (1995), 'The best of the Berkshires', *Travel and Leisure* 25:8, 76–83.

MAZZA, P. (1995), 'Amtrak rechristens Northwest line "The Cascadia" – Highspeed rail coming?' <http://www.tnews.com/text/the_cascadia.html>, accessed July 1996.

Mazza, P. (1995), 'Cascadia Emerging: The end and the beginning of the world', <http://www.tnews.com/text/emerge.html>, accessed July 1996.

McCLOSKEY, D.D. (1995), 'Cascadia: A great green land on the Northeast Pacific Rim', <http://www.tnews.com:80/text/mccloskey.html>, accessed in July 1996.

McGREEVY, P. (1988), 'The end of America: The beginning of Canada', *Canadian Geographer* 32:4, 307–318.

McKINSEY, L.S. AND KONRAD, V.A. (1989), *Borderlands Reflections: The United States and Canada*, (Orono, ME: University of Maine, Borderlands Project).

MERRETT, C. (1991), 'Crossing the Border: The Canada-United States Boundary', *Borderlands Monograph Series* 5, 19–54 (Orono, ME: University of Maine, The Canadian-American Center).

MINISTÈRE DU TOURISME (1993), *Estrie Eastern Townships*, (Quebec City: Ministère du Tourisme, Gouvernement du Quebec).

MURPHY, P.E. (ed) (1983), *Tourism in Canada: Selected Issues and Options*, (Victoria, BC: University of Victoria, Department of Geography).

QUINN, F. (1991), 'Canada – United States relations along the waterfront', *Zeitschrift für Kanada Studien* 11:½, 79–93.

RUTAN, G.F. (1985), 'British Columbia – Washington State governmental interrelations: Some findings upon the failure of structure', *American Review of Canadian Studies* 15:1, 97–110.

SCHELL, P. AND HAMER, J. (1995), 'Cascadia: The new binationalism of Western Canada and the US Pacific Northwest', in R.L. Earle and J.D. Wirth (eds), *Identities in North America. The Search for Community*, (Stanford: Stanford University Press, 140–156).

SCHWARTZ, F.D. (1997), 'Niagara Falls: For two hundred years it's been attracting tourists – and tourist traps', *American Heritage*, (published on September 1997), pp. 22–26.

VESILIND, P.J. (1990), 'Common ground, different dreams: The US-Canada border', *National Geographic* 177:2, 94–127.

Politics of US-Canada cross-border co-operation

Tourism related policies along the Canadian-US border mainly relate to regional diplomatic relations and border crossing procedures. Visa and entry regulations are discussed within the literature below that is compiled to offer an easy access to legislative and politically oriented information on cross-border relations between these adjacent super powers.

Table 9.31   Selected literature: Politics of US-Canada cross-border co-operation

ALPER, D. K. (1986), 'Recent trends in US-Canada regional diplomacy', in O.J. Martinez (ed.), *Across Boundaries – Transborder Interaction in Comparative Perspective*, (El Paso: Texas Western Press, 118–153).
ALPER, D. AND MONAHAN, R.L. (1986), 'Regional transboundary negotiations leading to the Skagit river treaty: analysis and future application', *Canadian Public Policy* 7:1, 163–174.
BELTRAME, J. (1997), 'US delays border-control law' *Kitchener-Waterloo Record*, (published on 10 November 1997).
BIDDLE, F.M. (1991), 'With "border war" over, councils from N.H., Mass. meets on tourism' *Boston Globe*, (published on 7 October 1991), p. 14.
CANADIAN PRESS (1997b), 'US visa law is "long way off", official says', *Kitchener-Waterloo Record*, (published on 14 October 1997), p. 2.
GRIFFITH, P. (1997), 'New law threatens gridlock on border' *The Blade*, (published on 28 September 1997), pp. 1, 4.
MCKENNA, B. (1997b), 'US border law attacked' *The Globe and Mail*, (published on 15 October 1997), pp. A1, A16.
MCKENNA, B. (1997a), 'Review of Canada-US border dispute months away', *The Globe and Mail*, (published on 8 October 1997), p. A17.
OFFICE OF THE GOVERNOR OF THE STATE OF NEW YORK (2001), 'Governor Pataki, Premier Harris discuss cross-border issues', <http://www.state.ny.us/governor/press/year01/ april4_1_01.htm>, accessed 13 March 2003.
RUTAN, G.F. (1981), 'Legislative interaction of a Canadian province and an American state – Thoughts upon sub-national cross-border relations', *American Review of Canadian Studies* 6:2, 67–79.

Economic aspects of US-Canada cross-border co-operation

Having introduced the political dimension to cross-border co-operation along this border, the literature listed below discusses the economic determinants of cross-border co-operation and their potential positive or negative influence on trans-national regional development. Since cross-border shopping has proven to be tremendously popular along the Canada-US borderline (see Bondi, 1997; Butler, 1991, Canadian Chamber of Commerce, 1992), various researchers set out to closely investigate this phenomenon and advise on potentially beneficial regulation mechanisms (see Boisvert and Thirsk, 1994; Chatterjee, 1991). Others simply examined the determinants and potential implications such developments might exert on cross-border tourism (see Bradbury and Turbeville, 1997; Chadee and Mieczkowski, 1987; DiMatteo and DiMatteo, 1993).

Table 9.32   Selected literature: Economic aspects of US-Canada cross-border co-operation

ASSOCIATED PRESS (1998), 'Drop in Canadian dollar draws US shoppers across border' *The Toledo Blade*, (published on 3 September 1998).
BAILEY, R. (1995), 'A new air transport agreement between the United States and Canada has made it easier to fly over the border', *Air Line Pilot* 64:7, 24–26.
BANK OF CANADA (1985–98), *Quarterly Report* (Ottawa: Bank of Canada).
BOISVERT, M. AND THIRSK, W. (1994), 'Border taxes, cross-border shopping, and the differential incidence of the GST', *Canadian Tax Journal* 42:5, 1276–1293.
BONDI, N. (1997), 'Shoppers head to Windsor for deals: High exchange rate on American dollar fuels big-ticket buys' Detroit News, (published on 30 December 1997).
BONDI, N. (1998), 'Bargain hunters hit Canada: Loonie's record fall an economic windfall for Americans shopping across the border' *Detroit News*, (published on 30 January 1998).
BRADBURY, S.L. AND TURBEVILLE III, D.E. (1997), 'Communities in transition: the experience of towns on the Washington State/British Columbia border since the implementation of Free Trade' *Small Town* 28:2, 10–15.
BUTLER, R.W. (1991), 'West Edmonton Mall as a Tourist Attraction', *The Canadian Geographer* 35, 287–295.
CANADIAN CHAMBER OF COMMERCE (1992), *The Cross Border Shopping Issue* (Ottawa: Canadian Chamber of Commerce).
CHADEE, D. AND MIECZKOWSKI, Z. (1987), 'An Empirical Analysis of the Exchange Rate on Canadian Tourism', *Journal of Travel Research* 26:1, 13–17.
CHAMBERLAIN, L. (1991), *Small Business Ontario Report No.44: Cross Border Shopping* (Toronto: Ministry of Industry, Trade and Technology).
CHATTERJEE, A. (1991), 'Cross-border shopping: Searching for a solution', *Canadian Business Review* 18, 26–31.
DILLEY, R.S. AND HARTVIKSEN, K.R. (1993), 'Duluth and Thunder Bay tourism after the Free Trade Agreement' *Operational Geographer* 11:3, 15–18.
DI MATTEO, L. (1993), 'Determinants of cross-border trips and spending by Canadians in the United States: 1979–1991', *Canadian Business Economics* 1:3, 51–61.
DI MATTEO, L. AND DI MATTEO, R. (1993), 'The determinants of expenditures by Canadian visitors to the United States', *Journal of Travel Research* 31:4, 34–42.
FISHER, M. (1990), 'Canadians give in to shopping drive' *The Globe and Mail*, (published on 26 October 1990), p. A4.
GOODMAN, L.R. (1992), 'A working paper on crossborder shopping: The Canadian impact on North Dakota', in H.J. Selwood and J.C. Lehr (eds) *Reflections from the Prairies: Geographical Essays* (Winnipeg: University of Winnipeg, Department of Geography, 80–89).
GOVERNMENT OF ONTARIO (1991), 'Report on Cross-Border Shopping' (Toronto: Standing Committee on Finance and Economic Affairs).
GRINSPUN, R. (1993), 'The economics of free trade in Canada', in R. Grinspun and M. Cameron (eds) The Political Economy of North American FreeTrade (New York: St Martin's Press, 105–124).
KEMP, K. (1992), 'Cross-border shopping: Trends and measurement issues', *Canadian Economic Observer* 5, 1–13.

KRECK, L.A. (1985), 'The effect of the across-the-border commerce of Canadian tourists on the city of Spokane', *Journal of Travel Research* 24:1, 27–31.

KRESL, P.K. (1993), 'The impact of free trade on Canadian-American border cities', *Canadian-American Public Policy* 16, 1–44.

LEWIS, K. (1990), 'Buying across the border', *Canadian Consumer* 20:3, 9–14.

RITCHIE, K.D. (1993), *Spatial analysis of cross-border shopping in Southern Ontario* (Unpublished B.A. honor's thesis, Department of Geography, University of Waterloo).

SCANIAN, D. (1991b), 'The recession: Canadian dollars helping Massena ride it out' *The Ottawa Citizen*, (published on 18 May 1991), p. B1.

SCHNEIDER, H. (1998), 'Canadian dollar hits record low' *Washington Post,* (published on 23 January 1998).

SLOAN, G. (1998), 'Now's a great time to drop a dime in Canada' *USA Today*, (published on 14 August 1998).

STEVENSON, D. (1991), 'Cross-border dispute', *Canadian Consumer* 21:7, 8–15.

STINSON, M. AND BOURETTE, S. (1998), 'Dollar sinks to lowest ever: Americans flock to Canadian border towns to wine and dine at a discount', *The Globe and Mail*, (published on 3 January 1998), p. 1.

TAYLOR, G.D. (1994), 'The implications of free trade agreements for tourism in Canada', *Tourism Management* 15, 315–356.

TIMOTHY, D.J. (1999b), 'Cross-border shopping. Tourism in the Canada-United States borderlands', *Visions in Leisure and Business* 17:4, 4–18.

TRAVEL WEEKLY (1998), 'Windsor Group opens permanent facility in Ontario' *Travel Weekly*, (published on 5 October 1998), p. 92.

Social aspects of US-Canada cross-border co-operation

Besides the various economic considerations demanded by cross-border activities, they also create social impacts that require profound investigation to enable researchers to elaborate on their outcome. The literature listed below, therefore, intends to offer an overview about what has been examined and evaluated to date, in terms of cross-border region building and territorial integration (Blatter, 2000). Questions of identity formation and the potential emergence of a distinct borderland population are addressed and investigated by numerous studies that conclude that trans-border regionalism does not necessarily have to lead to diminishing national identity since 'cross-cultural events and attractions can also serve as a tool for affirming the positive aspects of differences among peoples' (Association for Canadian Studies in the United States, 2001: 6) and hence rather lead to an increased awareness of national identity as opposed to being absorbed entirely by a trans-national borderland identity.

Table 9.33   Selected literature: Social aspects of US-Canada cross-border co-operation

ASSOCIATION FOR CANADIAN STUDIES IN THE UNITED STATES (2001), 'Cross-border cultural tourism–a two way street – Facts and Figures about cultural tourism across the Canada – USA border', <http://www.theniagarasguide.com/partners/tou.pdf>, accessed 2 February 2003.
ALPER, D. K. (1996), 'The idea of Cascadia: Emergent transborder regionalism in the Pacific-Northwest-Western Canada', *Journal of Borderland Studies* 11:2, 1–22.
APPLEBY, T. (1995), 'Bordering on a reversal of fortune', *The Globe and Mail*, (published on 10 November 1995).
CANADIAN PRESS (1997a), 'Canadians to avoid border checks', *The Globe and Mail*, (published on 18 October 1997).
DI MATTEO, L. (1999a), 'Cross-border trips by Canadians and Americans and the differential impact of the border', *Visions in Leisure and Business* 17:4, 72–92.
MCALLISTER, B. (1996), 'Canadians find shelter in US border enclave', *Washington Post*, (published on 14 May 1996).
MCGREEVY, P. (1991), 'The Wall of Mirrors: Nationalism and Perceptions of the Border at Niagara Falls', *Borderlands Monograph Series* 5, (Orono, ME: The Canadian-American Center, University of Maine, 1–18).
MCKENNA, B. (1998), 'Crossing US border still easy – for now', *The Globe and Mai,* (published on 29 September), p. A5.
REZA, H.G. (1995), 'US eyes open border with Canada', *The Toronto Star*, (published on 27 August 1995), p. 3.
RINEHART, D. (1992), 'Canadians making fewer trips south', *London Free Press,* (published on 15 August 1992), p. B1.
SANGUIN, A.L. (1974), 'La frontière Québec–Maine: quelques aspects limologiques et socio-économiques', *Cahiers de Géographie de Québec* 18:43, 159–185.
SCANIAN, D. (1991a), 'Hard times on the line', *The Ottawa Citizen*, (published on 18 May 1991), p. B2.

Geographic aspects of US-Canada cross-border co-operation

The literature incorporated below takes a rather geographical approach to potential cross-border collaborations. As such, it discusses the unique situation of the Canadian exclave of Point Roberts (see German, 1984; Minghi and Rumley, 1972), as well as the spatial similarities or differences either prospering or hindering potential co-operation across existing boundaries.

Table 9.34   Selected literature: Geographic aspects of US–Canada cross-border co-operation

ARTIBISE, L.F.J. (1995), 'Achieving sustainability in Cascadia: An emerging model of urban growth management in the Vancouver-Seattle-Portland corridor', in P.K. Kresl and G.Gappert (eds), *North American Cities and the Global Economy: Challenges and Opportunities*, (Thousand Oaks, CA: Sage Publications, 221–250).
DYKSTRA, T.L. AND IRONSIDE, R.G. (1972), 'The effects of the division of the city of Lloydminster by the Alberta-Saskatchewan inter-provincial boundary', *Cahiers de Géographie de Québec* 16, 263–283.
GERMAN, A.L. (1984), 'Point Roberts: A tiny borderline anomaly', *Canadian Geographic* 104:5, 72–74.
MINGHI, J.V. AND RUMLEY, D. (1972), 'Integration and system stress in an international enclave community: Point Roberts, Washington. B.C.', *Geographical Series* 15, 213–229.
REITSMA, H.J. (1971), 'Crop and livestock production in the vicinity of the United States-Canada border', *The Professional Geographer* 23:3, 216–223.
REITSMA, H.J. (1972), 'Areal differentiation along the United States-Canada border', *Tijdschrift voor Economische and Sociale Geografie* 63, 2–10.
SLOWE, P.M. (1991), 'The geography of borderlands: The case of the Quebec-US borderlands', *Geographical Journal* 157:2, 191–198.
SLOWE, P.M. (1994), 'The geography of borderlands: The case of the Quebec-US borderlands', in P.O. Girot (ed.) *World Boundaries 4, The Americas* (London: Routledge, 3–17).

US-Canada environmental cross-border co-operation

In terms of environmental cross-border co-operation, the borderline between the US and Canada offers two specialities that are especially worthy of notice. For one thing, these two nations both border the world famous Niagara Falls. Even though a tourist attraction on either side of the border, it appeared that the Canadian side appeals more to tourists than the American side, as research conducted by McGreevy (1988) reveals. Obviously then, tourism development differs drastically on either side of the border and could benefit positively from further cross-border co-operation. This assumption is supported by the fabulous success story of the International Peace Garden (see Lieff and Lusk, 1990; Timothy and Tosun, 2003 for a detailed discussion on the International Peace Garden) which enables visitors to freely cross the border within the park area without having to go through extensive entry procedures and visa regulations, thereby tremendously spurring tourist satisfaction and benefiting attendance levels. Collaborative planning and human resource management within the park enable such borderline exceptions and support the notion of cross-border collaboration leading to bilateral economic improvements.

Table 9.35   Selected literature: US–Canada environmental cross-border co-operation

ALLEY, J. (1995); *The British Columbia/Washington Environmental Co-operation Council: An Evolving model of Canada/US interjurisdictional co-operation* (Manuscript).
GEORGIA STRAIT ALLIANCE (ed.) (2001), 'Putting Orca Pass on the Map. – Quarterly News Letter March 2001, 8, 1, 12' <http://www.georgiastrait.org/ Newsletters/news0103-12.php>, accessed 15 February 2003.
GREEN, S. (1991), 'A garden for peace: A beautiful idea blooms on the border', *North Dakota Horizons* 21:3, 8–15.
LIEFF, B.C. AND LUSK, G. (1990), 'Transfrontier co-operation between Canada and the USA: Wateron-Glacier International Peace Park', in J. Thorsell (ed.), *Parks on the Borderline: Experience in Transfrontier Conservation*, (Gland: IUCN, 39–49).
NIAGARA PARKS COMMISSION (1992), *Niagara Falls: The Wonder of the World*, Niagara Falls, (ON: Niagara Parks Commission).
TIMOTHY, D.J. (1999), 'Cross-border partnership in tourism resource management: international parks along the US-Canada border', *Journal of Sustainable Tourism* 7:3, 182–205).
TIMOTHY, D.J. (2000a), 'Cross-Border Partnership in Tourism Resource Management: International Parks along the US-Canada Border', in B. Bramwell and B. Lane (eds), *Tourism Collaboration and Partnerships: Politics, Practice and Sustainability*, (Clevedon: Channel View Publications).
TIMOTHY, D.J. AND TOSUN, C. (2003), 'Tourists' perceptions of the Canada–US border as a barrier to tourism at the International Peace Garden', *Tourism Management* 24, 411–421.

Cross-border co-operation in tourism along the US-Mexico border

Whereas the co-operation across boundaries between the USA and Canada are initiated between similarly developed nations and equal partners, this situation is shown to be entirely different when investigating the relationships that exist between the USA and its southern neighbour, Mexico. The considerable distinction between these neighbours, although obvious when considering traditions and cultural characteristics, becomes even more so when investigating the economic situation pertinent to each country. Mexico is not only one of the world's most important tourist destinations among the less developed nations (Demler, 2004), but worldwide the only economically less developed nation that borders directly to a highly developed nation such as the US. Consequently, it offers vast potential for establishing economically related binational interrelations with specific regard to tourism on its northern borderlines (Demler, 2004; Vorlaufer, 1996).

Many issues already discussed within a US-Canadian context above are relevant to cross-border co-operation between Mexico and the USA, too, and subsequently appear to be duplicated. However, a closer look reveals that co-operation, despite the high level of binational interaction, is not as diversified along the US southern border than in the North. Whereas co-operation across the US–Canada border is signified by the high degree of organisation and profound structure evident in intentionally introduced and carefully monitored projects, the situation with most of the co-operation and cross-border activities at the US-Mexican border appears to have developed somewhat naturally out of cultural necessity, desire or concern. This holds specifically true for border tourism in the region which, according to Demler (2004), evolved naturally and rather unplanned. Such unplanned, evolutionary processes have distinctively transformed either side of the border and created a distinct border landscape which rates among the top tourist destinations within Mexico (Arreola and Curtis, 1993; Demler, 2004). Nowadays, this border zone holds the highest degree of Mexican urbanisation, thus indicating its economic and social importance to the country, and is mainly signified by the six most important twin-cities of Tijuana-San Diego, Mexicali-Calexico, Ciudad Juárez-El Paso, Nuevo Laredo-Laredo, Reynosa-McAllan and Matamoras-Brownsville that have grown across the international border. Functioning as a conjunction between distinctively different economies, cultures, and traditional ways of life, and profoundly employed in tourism industries, these cities soon realised the economic value of serving popular clichés, sometimes providing a rather devious and misleading image. Hence, it should not be surprising that 'To Americans and Mexicans from the interior of each nation, [border] cities (…) evoke images of gaudy tourist districts, unsavory bars and nightclubs, loud discotheques, tasteless curio shops, liquor stores, bargain dentists, and 'hustlers' of many types'. (Arreola and Curtis, 1993: XIV). Despite this image, the region has been extremely successful in attracting international tourists and has consequently seen the emergence of globally unique tourist districts (as discussed in detail by Arreola, 1996; Arreola and Curtis, 1993; Timothy, 2001). These districts serve as a city within a city, profoundly saturated with perceived traditional Mexican images and values, catering for a US American lifestyle which highly influences the formation of cultural identity within the regional populations by creating an artificial "Mexicoland", solely serving the needs of international tourists (Demler, 2004).

Literature discussing the general nature of US-Mexican cross-border relations, therefore, mainly focuses on the social implications of interaction across these national boundaries, due to the distinctly different cultures that meet at the borderline. The impact of tourism development on either side of the border are investigated and examined in relation to the relatively strict border controls and immigration regulations. Investigations also try to answer the emerging questions about how this region managed to exploit its scarce historical and culturally attractive attributes in a way that enabled it to become one of Mexico's oldest and, in terms of quantities, most important tourism location.

Table 9.36   Selected literature: General considerations of cross-border co-operation between Mexico and the USA

ALARCÓN CANTÚ, E. (1990), Evolucíon y Dependencia en el Noreste. Las Ciudades Fronterizas de Tamaulipas, (Tijuana).

ALVAREZ, R., JR. (1995), 'The Mexican-US border: The making of an anthropology of borderlands', *Annual Review of Anthropology* 24, 447–470.

ANDERSON, J. B. (2003), 'The US-Mexico Border. A Half Century of Change', *The Social Sciences Journal* 40:4, 535–554.

ARREOLA, D.D. (1996), 'Border-City ideé Fixe', *The Geographical Review* 86:3, 356–369.

ARREOLA, D.D. (1999), 'Across the street is Mexico: Invention and Persistence of the Border Town Curio Landscape', *Yearbook of the Association of Pacific Coast Geographers* 61, 9–41.

ARREOLA, D.D. (2001), 'Curio Consumerism and Kitsch culture in the Mexican-American borderland', *Journal of the West* 40:2, 24–31.

ARREOLA, D.D. AND CURTIS, J.R. (1993), *The Mexican Border Cities: Landscape Anatomy and Place Personality*, (Tucson: University of Arizona Press).

ARREOLA, D.D. AND MADSEN, K. (1999), 'Variability of tourist attractiveness along an international boundary: Sonora, Mexico Border Towns', *Visions in Leisure and Business* 17:4, 19–31.

BURDACK, J. (1996), 'Doppelstädte an der Grenze USA-Mexiko: Von der Border Town zur grenzübergreifenden Agglomeration', in Steinecke, A. (ed.) (1996), *Stadt und Wirtschaftsraum*, (Berlin: Berliner Geographische Studien 44, 271–282).

BUREAU OF TRANSPORTATION STATISTICS (BTS) (ed.) (2004), 'Border Crossing. US-Mexico Border Crossing Data', <http://www.bts.gov/programs/international/border_crossing_entry_data/ us_mexico/entire.pdf>, accessed on 12 February 2004.

BYRD, B. AND BYRD, S.M. (eds) (1996), *The Late Great Mexican Border: Reports from a Dissappearing Line*, (El Paso).

DEMARIS, O. (1970), *Poso del Mundo: Inside the Mexican-american Border, from Tijuana to Matamoros*, (Boston/Toronto).

DEMLER, D. (2004), *Der Internationale Tourismus in Nordmexikanischen Grenzstädten am Beispiel von Nuevo Laredo*, (Trier: Geographische Gesellschaft der Universität Trier. Materialien zur Fremdenverkehrsgeographie 62).

FERNANDEZ, R.A. (1977), *The United States-Mexico Border: A Politico-Economic Profile*, (Notre Dame: University of Notre Dame Press).

FOX, B. (2002), 'Mexican Border Towns want raucous crowds', *Marketing News* 36:8, 16.

GANSTER, P. (1996), 'On the road to interdependence? The United States-Mexico border region', in J. Scott and A. Sweedler and P. Ganster and W. Eberwein (eds) (1996), *Border Regions in Functional Transition. European and North American Perspectives on Transboundary Interaction*, (Erkner: Regio Series of the IRS 9, 171–191).

GORMSEN, E. (1979), 'The impact of tourism on the development of Mexican cities along the US border: The example of Tijuana', in G. Gruber and H. Lamping and W. Lutz and J. Matznetter and K. Vorlaufer (eds), *Tourism and Borders: Proceedings of the Meeting of the IGU Working Group – Geography of Tourism and Recreation*, (Frankfurt a.M.: Institut für Wirtschafts- und Sozialgeographie der Johann Wolfgang Goethe Universität, 345).

HANSEN, N. (1986), 'Border region development and co-operation: Western Europe and the US-Mexico borderlands in comparative perspective', in O.J. Martinez (ed.), *Across Boundaries: Transborder Interaction in Comparative Perspective*, (El Paso: Center for Inter-American and Border Studies, University of Texas, 31–44).

HERZOG, L.A. (1985), 'Tijuana', *Cities* 2, 297–306.

HERZOG, L.A. (1986a), 'Overview', in L.A. Herzog (ed.), *Planning the International Border Metropolis: Trans-Boundary Policy Options in the San Diego-Tijuana Region*, (San Diego: Center for US-Mexican Studies, University of California, 67–71).

HERZOG, L.A. (ed.) (1986b), *Planning the International Border Metropolis: Trans-Boundary Policy Options in the San Diego-Tijuana Region*, (San Diego: Center for US-Mexican Studies, University of California).

HERZOG, L.A. (1990), *Where North Meets South: Cities, Space, and Politics on the US-Mexico Border*, (Austin: Center for Mexican American Studies, University of Texas).

HERZOG, L.A. (1991c), 'USA-Mexico border cities: A clash of two cultures', *Habitat International* 15:1/2, 261–273.

HERZOG, L.A. (1992), 'The US-Mexico transfrontier metropolis', *Business Mexico* 2, 14–17.

LORREY, D.E. (ed.) (1990), *United States-Mexico Border Statistics since 1900*, (Los Angeles).

LORREY, D.E. (ed.) (1999), *The US-Mexican Border in the Twentieth Century: A History of Economic and Social Transformation*, (Wilmington).

MARTÍNEZ, O. (1978), *Border Boom Town. Ciudad Juárez since 1848*, (Austin/London: University of Texas Press).

MARTÍNEZ, O.J. (1986), *Across boundaries: Transborder interactions in comparative perspective*, (El Paso: University of Texas Press).

ROSS, S.R. (1978), *Views across the border: The United States and Mexico*, (Albuquerque: University of New Mexico Press).

SPARROW, G. (2001), 'San Diego-Tijuana: Not quite a binational city or region', *GeoJournal* 54:1, 73–83.

STODDARD, E. R. (2001), *US-Mexico Borderland Issues: The Binational Boundary, Immigration and Economic Policies*, (El Paso: Borderlands Trilogy, Vol. 1).

STODDARD, E.R. AND NOSTRAND, R.L. AND WEST, J.P. (eds) (1983), *Borderlands Sourcebook: A Guide to the Literature on Northern Mexico and the American Southwest*, (Norman: University of Oklahoma Press).

WEISMAN, A. (1986), *La Frontera. The United States Border with Mexico* (Tucson).

Politics of cross-border co-operation between the USA and Mexico

Politically, the trans-boundary relations along the borderline deal with tourism imperatives and the formulation of a common tourism strategy. Border controls, crossing procedures and most importantly, the avoidance of illegal immigration into the US, are major political issues that are addressed comprehensively and attempted to be solved collaboratively in order to gain mutually beneficial results. The generally high political involvement in Mexican tourism planning and development, with specific emphasis on developing and promoting peripheral and rural areas, has prospered the remote border regions in the North by creating an attractive climate for international privately operated investments (Demler, 2004). These private operations have consequently spurred the interest in tourism across the border and initiated various incidences of co-operation across the border. One of the most important examples is the collaborative development of tourism products by tourism boards on either side and greater attention being paid to the avoidance of overlapping important events and festivals. Such political involvement is necessary to improve the overall tourism infrastructure and economic climate required for further improvements and hence economic prosperity.

The three items listed below, therefore, intended to offer a convenient starting point, suggesting the vast potential for further research into the field of cross-border tourism policy co-operation.

Table 9.37   Selected literature: Politics of Mexican-US cross-border co-operation

BUDD, J. (1990), 'Mexico's border towns to be focus of new tourism imperative', *Travel Weekly* 49:49, 15.
DILLMAN, C.D. (1970a), 'Recent developments in Mexico's National Border Program', *Professional Geographer* 22:5, 243–247.
TEXAS DEPARTMENT OF AGRICULTURE (2002), 'XX Border Governors' Conference United States-Mexico June 21–22, 2002, Phoenix, Arizona – Joint Declaration', <http://www.agr.state.tx.us/border/activities/iga_declaration_eng_2002.htm>, accessed 25 February 2003.

Economic aspects of US-Mexico cross-border co-operation

The economic aspects of US-Mexico cross-border interactions and their impact have resulted in some researchers identifying the borderland, or 'franja fronteriza' (Demler, 2004), as a single economic region in its own right (see especially Brown, 1997). One reason why binational relations were positively influenced within the region was identified through its historic peripherality that, by its very nature, demanded creativity in economic development and hence might explain the early connectivity and economic interrelations with its northern neighbours in the US. It should not be surprising then, that most co-operation aims at economically prospering both sides. Especially on the Mexican side, these efforts resulted in the creation of distinctly recognisable tourism business districts unique to those Mexican towns and cities along the actual borderlines (see Curtis, 1993; Hansen, 1981; Holden, 1984; Mikus, 1986). Such economic prosperity along the border has been specifically spurred by increasing shopping related tourism activities (see particularly Timothy, 2001; Timothy and Butler, 1995) in recent years, after having prospered from vice industries over the past few decades. Historically, the importance of tourism as one of the most influencial economic factors in the region is rooted in the cities of Juárez and Tijuana and their development of specific vice industries (Bowman, 1994). Early on, these cities realised the economic potential of attracting US visitors by offering gambling, alcohol and prostitution which was, and largely remains, prohibited across the US.

It should not be surprising then, that the golden years of tourism in Mexican border towns began in the 1920's, when the US increasingly introduced prohibition acts that disapproved any production and consumption of alcoholic beverages (Arreola, 1999; Arreola and Curtis, 1993; Bowman, 1994). Consequently, in the years to come, tourism industries in and around Mexican border towns peaked and declined with war times and the tightening or loosening of prohibitions of vice industries in the US, where Mexican border cities soon became known as 'playgrounds for the rich, famous, and gambling crowds' (Arreola, 1999: 18). Having realised that such vice industries, even though highly profitable, created a relatively shady image of the region's tourism products across its main generating region, the US, the industry needed to reconsider, and devise new strategies (Arreola and Curtis, 1993). The improved standard of living initiated by the increased economic prosperity of the residents, gained through the engagement in tourism businesses, facilitated further diversification of product offerings. Nowadays, US-Mexican border tourism directives are keen on further enhancing the image of Mexican border town tourism as being more family oriented. Particularly prostitution, therefore, was strictly removed and relocated to alleged 'zonas de tolerancia' on the cities' peripheries, where such industries can continue to generate significant economic contributions without significantly harming the region's overall image (Curtis and Arreola, 1991). Motivations for visiting the region today, therefore, are observably more diversified than in the early years and have come to comprise medical reasons (cheaper drugs and medicine), general shopping tourism, visiting friends and relatives, partying and, increasingly, stop-overs on the way to further destinations in mainland Mexico (Demler, 2004). Such activities and their economic results offer significant potential for further academic research into trans-boundary economic relations with regard to tourism industries.

Table 9.38   Selected literature: Economic aspects of Mexico–US cross-border co-operation

Asgary, N. and de Los Santos, G. and Vincent, V. and Davila, V. (1997), 'The determinants of expenditures by Mexican visitors to the border cities of Texas', *Tourism Economics* 3:4, 319–328.
Austin, J.P. (1979), 'Laredo: Trade Center on the border', *Texas Business Review* 53:2, 57–60.
Baerresen, D.W. (1983), 'The economy', in E.R. Stoddard and R.L. Nostrand and J.P. West (eds), *Borderlands Sourcebook: A Guide to the Literature on Northern Mexico and the American Southwest*, (Norman: University of Oklahoma Press, 121–124).

BERRUETO, E.M. (ed.) (1982), *Proceedings of the First Conference on Regional Impacts of United States-Mexico Economic Relations*, (Guanajuato, Mexico: Conference on Regional Impacts of United States-Mexico Economic Relations).

BROWN, T.C. (1997), 'The fourth member of NAFTA: The US-Mexico border', *Annals of the American Academy of Political and Social Science* 550, 105–121.

CURTIS, J.R. (1993), 'Central business districts of the two Loredos', *Geographical Review* 83:1, 54–65.

DIEHL, P.N. (1983), 'The effects of the peso devaluation on Texas border cities', *Texas Business Review* 57, 120–125.

DILLMAN, C.D. (1976), 'Maquiladoras in Mexico's northern border industrialization program', *Tijdschrift voor Economische en Sociale Geografie* 67, 138–150.

GIBBONS, J.D. AND FISH, M. (1985a), 'Mexico's balance of payments 1970–1983: Contributions of international tourism and border transactions', *Tourism Management* 6:2, 106–112.

GIBBONS, J.D. AND FISH, M. (1985b), 'Devaluation and US tourism expenditure in Mexico', *Annals of Tourism Research* 12:4, 547–561.

GIBBONS, J.D. AND FISH, M. (1987), 'Market sensitivity of US and Mexican border travel', *Journal of Travel Research* 26:1, 2–6.

HANSEN, N. (1981), *The Border Economy: Regional Development in the Southwest* (Austin: University of Texas Press).

HOLDEN, R.J. (1984), '"Maquiladoras" employment and retail sales effects on four Texas border communities, 1978–1983', *Southwest Journal of Business and Economics* 2:1, 16–26.

MELDMAN, M. (1995), 'Four faces of Mexico: Along the US border, boundless bargains' *Washington Post*, (published on 24 September 1995), p. E1.

MIKUS, W. (1986), 'Grenzüberschreitende Verflechtungen im tertiären Sektor zwischen USA und Mexiko: Das Beispiel Kaliforniens', *Geographica Helvetica* 36, 207–217.

PARFIT, M. (1996), 'Tijuana and the border: Magnet of opportunity', *National Geographic* 190:2, 94–107.

PATRICK, J.M. AND RENFORTH, W. (1996), 'The effects of the peso devaluation on cross-border retailing', *Journal of Borderlands Studies* 11:1, 25–41.

PAVLAKOVIC, V.K. AND KIM, H.H. (1990), 'Outshopping by maquila employees: Implications for Arizona's border communities' *Arizona Review*, (published in Spring 1990), pp. 9–16.

PERTMAN, A. (1995), 'California trolley enhances border appeal' *Boston Globe*, (published on 14 August), p. 3.

PROCK, J. (1983), 'The peso devaluations and their effect on Texas border economies', *Inter-American Economic Affairs* 37:3, 83–92.

SAINT-GERMAIN, M.A. (1995), 'Problems and opportunities for co-operation among public managers on the US-Mexico border', *American Review of Public Administration* 25:2, 93–117.

STODDARD, E.R. (1987), *Maquila Assembly Plants in Northern Mexico*, (El Paso: Texas Western Press).

SVERDLIK, A. (1994), 'Fast track to bargains in Tijuana' *Atlanta Journal Constitution*, (published on 3 April 1994), p. K5.

SZABO, J. (1996), 'Bonanza on the border: A loco buying spree along the Tex-Mex trail', *Travel and Leisure* 26:11, 58–64.

TAYLOR, J.C. and Robideaux, D.R. and Jackson, G.C. (2004), 'Costs of the US-Canada border', *Research in Global Strategic Management*, 10, 283–297.

YOSKOWITZ, D.W. and Giermanski, J.R. and Pena-Sánchez, R. (2002), 'The Influence of NAFTA on Socio-economic Variables for the US-Mexico Border Region', *Regional Studies* 36:1, 25–31.

Social aspects of Mexico-US cross-border co-operation

The improvement of living conditions along the borderlines has subsequently attracted numerous migrants from central Mexico to the border towns and regions searching for their own individual prosperity and economic improvements. Naturally, such migrations create social conflicts and change a region's social composition over time. These issues are investigated within the literature that is listed below. Security issues are addressed with regards to increasing crime rates within the border regions (see Jud, 1975; Lin and Loeb, 1977), as well as the unique culture that has evolved within these border landscapes (see Bustamante, 1988; Monsivais, 1978). Vila (2003), in a discussion on processes of identity formation in border regions, proposes that pluralistic cultural personalities will potentially result, juggling cultures to fit current needs. According to him, such 'border people' are individuals with numerous cultures at the same time, and are capable of quickly adjusting to a situationally required or appropriate cultural expression. Such observation led him to discuss these borderlands as being a 'third country', a hybrid cultural mixture of both sides with a living claim in itself.

    This argument is supported by the notion of recognising the US-Mexico border area as a unitary economic region and is further elaborated on within the literature assembled below. Such realisations contribute to current and future investigations of gradually established remedies intended to address potential fears of crossing borders and entering unknown territories pertinent among latent visitors. Subtle efforts are aimed at further enlarging the potential visitor base. Additionally, such culturally ambiguous individuals, capable of switching roles according to current desires and needs, might positively influence the excitement felt when pondering foreign and exotic lands, by portraying and fulfilling clichés. Both aspects are not only restricted to being overcome by the social adaptability of the regions' inhabitants, but are additionally actively addressed by employing very physical means as well. For one thing, the fear of having to deal with lengthy bureaucratic procedures on crossing the border, or falling victim to crimes in foreign and unfamiliar places on the other side, is minimised by facilitating border crossing procedures, reaching as far as enabling visitors to cross the border on foot. In addition, vistas are gradually changed from bilingual directions and advertisements towards the more authentic Mexican views with increasing distance from the actual borderline. On the other hand, the excitement of pondering in exotic places is actively promoted by serving popular clichés about Mexico held by the majority of US American visitors through the creation of stereotypical 'Mexicolands' (Demler, 2004). Such effects have not only created a unique, socially distinctive border zone, but have also managed to keep visitor numbers quantitatively the highest among Mexican incoming tourism. According to Timothy (2001), the increasing fear of further hispanisation of the US might negatively influence the endless flow of US Americans to the Mexican border cities in the near future. In the long run, such minimised tourism flows might not only harmfully influence the economic situation, but, over time, may have an impact on the overall behaviour and social characteristics that have evolved around the visitors' needs, and subsequently dramatically change the social pattern currently pertinent within this unique border region.

Researchers are, therefore, offered various possibilities for investigating processes potentially influencing formative social aspects prevailing and evolving around this specific border landscape.

Table 9.39   Selected literature: Social aspects in Mexico-US cross-border co-operation

ACKLESON, J. (2005), 'Constructing security on the US-Mexico border', *Political Geography* 24, 165–184.
ARREOLA, D.D. (2002), *Tejano South Texas. A Mexican American Cultural Province*, (Austin).
ARREOLA, D.D. AND CURTIS, J.R. (1996), 'Cultural Landscapes of the Mexican Border Cities', *Aztlán, A Journal of Chicano Studies* 21:1–2, (special issue: Borders), 1–47.
BARRERA, E. (1995), 'Apropiación y tutelaje de la frontera norte', *Puente Libre: Revista de Cultura* 4, 13–17.
BUSTAMANTE, J. (1988), 'Identidad, cultura nacional y frontera', in A. Malagamba (ed.), *Encuentros. Los festivales internacionales de la raza*, (Tijuana, Baja California: El Colegio de la Frontera Norte).
CAHILL, R. (1987), *Border Towns of the Southwest: Shopping, Dining, Fun and Adventure from Tijuana to Juarez*, (Boulder, CO: Pruett Publishing Co).
CURTIS, J.R. AND ARREOLA, D.D. (1989), 'Through Gringo eyes: Tourist districts in the Mexican border cities as other-directed places', *North American Culture* 5:2, 19–32.
CURTIS, J.R. AND ARREOLA, D.D. (1991), 'Zonas de tolerancia on the northern Mexican border', *Geographical Review* 81:3, 333–346.
HACKENBERG, R. AND ALVAREZ, R.R. (2001), 'Close-Ups of Postnationalism: Reports from the US-Mexico Borderlands', *Human Organization* 60:2, 97–104.
HERZOG, L.A. (1982), 'Cross cultural barriers to planned urban development in the US-Mexico border zone: A case study of the San Diego/Tijuana metropolitan region', in E.M. Berrueto (ed.), *Proceedings of the First Conference on Regional Impacts of United States-Mexico Economic Relations*, (Guanajuato, Mexico: Conference on Regional Impacts of United States-Mexico Economic Relations, 836–867).
JONES, K. (1997), 'Slipping in and out of Mexico', *New York Times*, (published on 6 January 1997), p. 15.
JUD, G.D. (1975), 'Tourism and crime in Mexico', *Social Science Quarterly* 56:2, 324–330.
LIN, V.L. AND LOEB, P.D. (1977), 'Tourism and crime in Mexico: Some comments', *Social Science Quarterly* 58:1, 164–167.

MARTÍNEZ, O. (1994), *Border people. Life and society in the US-Mexico borderlands*, (Tucson: The University of Arizona Press).

MONSIVÁIS, C. (1978), 'The culture of the frontier: The Mexican side', in S.R. Ross, *Views across the border, The United States and Mexico*, (Albuquerque: University of New Mexico Press).

NORTH, D.S. (1970), *The Border Crossers: People who live in Mexico and work in the United States*, (Washington, D.C.).

PUENTE, M. (1996), 'So close, yet so far: San Diego, Tijuana bridging gap', in O.J. Martinez (ed.), *US-Mexico Borderlands: Historical and Contemporary Perspectives*, (Wilmington: Scholarly Resources, 249–255).

SAN DIEGO UNION TRIBUNE (1999), 'Discouraging Tourism', <http://traveltax.msu.edu/ news/ Stories/sandiego_ut4.htm>, accessed on 3 April 2003.

TEEGEN, H.J. AND DOH, J.P. (2002), 'US-Mexican alliance negotiations: Impact of culture on authority, trust, and performance', *Thunderbird International Business Review* 44:6, November/December, 749–775.

VILA, P. (2003), 'Processes of identification on the US-Mexico border', *The Social Science Journal* 40, 607–625.

Geographic considerations of cross-border co-operation along the US-Mexico border

Geographic considerations of cross-border interactions along the US-Mexico border often relate to land-use planning and urban growth. With ever increasing economic and social trans-boundary relations, the originally strict separation between Mexican and US territories blurs into a bi-national urban conglomeration of different aspects recognisable in both cultures. Cross-border metropolises emerge that, even though obviously bisected between the sovereign nations, do not reveal the cultural identity of either side, but rather belong to a mixture of both. This is already recognisable in the architecture and infrastructure prevalent in cities such as El Paso/Juares or San Diego/Tijuana.

Table 9.40   Selected literature: Geographic aspects in Mexico-US cross-border co-operation

ALARCÓN CANTÚ, E. (1997), *Interpretación de la Estructura Urbana de Laredo y Nuevo Laredo*, (Tijuana).

ALARCÓN CANTÚ, E. (2000), *Estructura Urbana en Ciudades Fronterizas. Nuevo Laredo-Laredo, Reynosa-McAllen, Matamoros-Brownsville*, (Tijuana).

DILLMAN, C.D. (1970b), 'Urban growth along Mexico's northern border and the Mexican National Border Program', *Journal of Developing Areas* 4, 487–507.

GILDERSLEEVE, C.R. (1978), *The International Border City. Urban Spatial Organization in a Context of Two Cultures along the United States-Mexico Boundary*, (Lincoln: University of Nebraska Dissertation).

GRAIZBORD, C. (1986), 'Trans-boundary land-use planning: A Mexican perspective', in L.A. Herzog, (ed.), *Planning the International Border Metropolis: Trans-Boundary Policy Options in the San Diego-Tijuana Region*, (San Diego: Center for US-Mexican Studies, University of California, 13–20).

HERZOG, L.A. (1991a), 'Cross-national urban structure in the era of global cities: The US-Mexico transfrontier metropolis', *Urban Studies* 28:4, 519–533.

HERZOG, L.A. (1991b), 'The transfrontier organization of space along the US-Mexico border', *Geoforum* 22:3, 255–269.

HOFFMANN, P.R. (1983), *The Internal Structure of Mexican Border Cities*, (Los Angeles: University of California Dissertation).

Cross-border co-operation in environmental protection along the US-Mexican border

In contrast to the environmental protection employed on the northern US border, cross-border co-operation along the US-Mexico border rather relates to specific and spatially segregated environmental problems such as waste water disposal and the care of a common ecosystem rather than the creation of extensive trans-boundary nature parks. This might be due to the relative strength of the border which in contrast to the Canadian-US border, is severely fortified and physically demarcated in order to prevent undesired Mexican citizens from immigrating illegally. None the less, some efforts are being undertaken to collaboratively improve the environmental conditions on both sides of the border and potentially diminish existing mutual suspicions, which could successively spur further collaboration and cross-border co-operation within this unique border landscape.

Table 9.41   Selected literature: Environmental aspects of Mexico-US cross-border co-operation

ELLMAN, E. AND ROBBINS, D. (1998), 'Merging sustainable development with wastewater infrastructure improvement on the US-Mexico border', *Journal of Environmental Health* 60:7, 8–13.

GAINES, S.E. (1995), 'Bridges to a better environment: Building cross-border institutions for environmental improvement in the US-Mexico border area', *Arizona Journal of International and Comparative Law* 12:2, 429–471.

HERZOG, L.A. (1986b), 'San Diego-Tijuana: The emergence of a trans-boundary metropolitan ecosystem', in L.A. Herzog (ed.), *Planning the International Border Metropolis: Trans-Boundary Policy Options in the San Diego-Tijuana Region*, (San Diego: Center for US-Mexican Studies, University of California, 1–10).

JOHNSTONE, N. (1995), 'International trade, transfrontier pollution, and environmental co-operation: A case study of the Mexican-American border region', *Natural Resources Journal* 35:1, 33–62.

PARENT, L. (1990), 'Tex-Mex Park: Making Mexico's Sierra del Carmen a sister park to Big Bend', *National Parks* 64:7, 30–36.

RIOGRANDE INSTITUTE (2002), 'Projects' <http://www.riogrande.org/Projects.html>, accessed on 3 April 2003.
VARADY, R. G. AND COLNIC, D. AND MERIDETH, R. AND SPROUSE, T. (1996), 'The US-Mexican border environment co-operation commission: Collected perspectives on the first two years', *Journal of Borderlands Studies* 11:2, 89–113.

Sub-national cross-border co-operation between US federal states

An interesting individual category among cross-border interactions in tourism relates to trans-boundary co-operation across sub-national borders. Such co-operation between federal states is more easily manageable since it is mainly initiated and operated on a local rather than national level, but incorporates the same positive and negative potential as any other cross-border co-operation. Often co-operation within the US relates to historic border disputes or artificial separations dating back to the War of Independence or the Civil War. Other issues investigated previously relate to the use of highway welcome centres and the joint promotion of tourist attractions close to interstate borders. Crossing borders, according to Timothy (2001) has always had a degree of mystical impact upon travellers who feel they have accomplished something great by crossing a border. They have entered a distant and foreign land that requires exploration and feel they have overcome the fear of the new by leaving their acquainted surroundings behind. Such psychological considerations might help to understand the fact that tourists enjoy being photographed at border demarcations, and also the importance of such demarcations as tourism attractions themselves, as is the case with the 'Four Corners Monument', the physical demarcation of the intersection between the US states of New Mexico, Utah, Arizona and Colorado (Timothy, 2001). Crossing sub-national borders within one's own country hence facilitates such psychologically perceived achievements but minimises the risks incorporated with actually travelling to remote and unfamiliar places that are culturally and linguistically different from home.

Accordingly, this phenomenon deserves further attention within tourism research and the literature listed below, therefore, introduces some preliminary studies that have explored the phenomenon of sub-national border interactions.

Table 9.42   Selected literature: US-American sub-national cross-border co-operation

BOWDEN, J.J. (1959), 'The Texas-New Mexico boundary dispute along the Rio Grande', *Southwestern Historical Quarterly* 63, 221–237.

BILLINGTON, M. (1959), 'The Red River boundary controversy', *Southwestern Historical Quarterly* 62, 356–363.

CARPENTER, W.C. (1925), 'The Red River boundary dispute (Oklahoma-Texas)', *American Journal of International Law* 19, 517–529.

DE VORSEY JR., L. (1982), *The Georgia-South Carolina Boundary: A Problem in Historical Geography*, (Athens, GA: University of Georgia Press).

DELAWARE TOURISM OFFICE (1987), *Delaware: Small Wonder*, (Dover: Delaware Tourism Office).

FESENMAIER, D.R. AND VOGT, C.A. AND STEWART, W.P. (1993), 'Investigating the influence of welcome center information on travel behavior', *Journal of Travel Research* 31:3, 47–52.

FESENMAIER, D.R. AND VOGT, C.A. (1993), 'Evaluating the economic impact of travel information provided at Indiana welcome centers', *Journal of Travel Research* 31:3, 33–39.

FOX, W.F. (1986), 'Tax structure and the location of economic activity along state borders', *National Tax Journal* 39:4, 387–401.

GITELSON, R. AND PERDUE, R.R: (1987), 'Evaluating the role of state welcome centers in disseminating travel related information in North Carolina', *Journal of Travel Research* 25:4, 15–19.

GONZALES ASSOCIATES (1970), *Proposed General Plan, City of Nogales, Arizona, Nogales Downtown Area*, (Phoenix: Gonzales Associates).

GREENHOUSE, L. (1998), 'High court awards New Jersey sovereignty over most of Ellis Island' *New York Times*, (published on 27 May 1998), 1, 21.

HOVINEN, G.R. (1995), 'Heritage issues in urban tourism: An assessment of new trends in Lancaster County', *Tourism Management* 16, 381–388.

HOWARD, D. AND GITELSON, R. (1989), 'An analysis of the differences between state welcome center users and nonusers: A profile of Oregon vacationers', *Journal of Travel Research* 28:4, 38–40.

ISEMINGER, G. (1991), 'Stone border', *North Dakota Horizons* 21:1, 14–21.

JACKSON, J.B. (1984), *Discovering the Vernaculas Landscape*, (New Haven: Yale University Press).

JOHNSTON, R.J. (1982), *Geography and the State: An Essay in Political Geography*, (New York: St Martin's Press).

LA GANGA, M.L. (1995), 'Pit stop on Nevada border now a hot spot', *Los Angeles Times*, (published on 11 April 1995), p. A1.

LEW, A.A. (1996), 'Tourism management on American Indian lands in the USA', *Tourism Management* 17, 355–365.

LONDON FREE PRESS (1992), 'Yes Michigan' *London Free Press*, (published on 7 March 1992), p. F10.

MARTIN, L. (1938), 'The second Wisconsin-Michigan boundary case in the supreme court of the United States, 1930–1936', *Annals of the Association of American Geographers* 28, 77–126).

MARTINEZ, O.J. (1988), *Troublesome Border*, (Tucson: University of Arizona Press).

MARYLAND OFFICE OF TOURISM DEVELOPMENT (1989), *Maryland Travel and Outdoor Guide*, (Baltimore, MD: Maryland Office of Tourism Development).

MOREHOUSE, B.J. (1996), 'Conflict, space, and resource management at Grand Canyon', *Professional Geographer* 48:1, 46–57.

MUHA, S. (1977), 'Who uses highway welcome centers?', *Journal of Travel Research* 15:3, 1–4.

NAVAJO PARKS AND RECREATION DEPARTMENT (n.d.), *Four Corners Monument Navajo Tribal Park*, (Window Rock, AZ: Navajo Parks and Recreation Department).

NEW BRUNSWICK TOURISM (1997), *Oh, Say Can You Save!*, (Fredericton: New Brunswick Tourism).

NORTH DAKOTA PARKS AND TOURISM (1992), *Discover the Spirit!*, (Bismark: North Dakota Parks and Tourism).

OUTSIDE (1994), 'Big Bend National Park', *Outside* 19:7, 63–68.

RENNICKE, J. (1995), 'Rafting the Rio', *National Geographic Traveler* 12:2, 106–115.

SAN DIEGO CONVENTION AND VISITORS BUREAU (1999), *Official Visitors Planning Guide*, (San Diego: Convention and Visitors Bureau).

SÁNCHEZ, M.L. (ed.) (1994), *A Shared Experience: The History, Architecture and Historic Designations of the Lower Rio Grande Heritage Corridor*, (Austin, TX: Los Caminos del Rio Heritage Project, Texas Historical Commission).

SIMMONS, T. AND TURBEVILLE III, D.E. (1984), 'Blaine, Washington: Tijuana of the north?', *B.C. Geographical Series* 41, 47–57.

SOMMERS, L.M. AND LOUNSBURY, J.F. (1991), 'Border boom towns of Nevada', *Focus* 41:4, 12–18.

SONDEREGGER, J. (1996), 'Border war' *St Louis Post-Dispatch*, (published on 25 October 1996).

STEWART, W.P. AND LUE, C. AND FESENMAIER, D.R. AND ANDERSON, B.S. (1993), 'A comparison between welcome center visitors and general highway auto travelers', *Journal of Travel Research* 31:3, 40–46.

TEXARKANA CHAMBER OF COMMERCE (n.d.), *Texarkana's Trail of Two Cities*, (Texarkana: Chamber of Commerce).

TIERNEY, P.T. (1993), 'The influence of state traveler information centers on tourist length of stay and expenditures', *Journal of Travel Research* 31:3, 28–32.

TIMOTHY, D.J. (1998d), 'Tourism development in a small, isolated community: The case of Northwest Angle, Minnesota', *Small Town* 29:1, 20–23.

WESTERN ARCTIC TOURISM ASSOCIATION (n.d.), *The Dempster Highway: A Road Less Traveled, A Land Unspoiled* (Inuvik, NWT: Western Arctic Tourism Association).

WILSON, J.R. AND MATHER, C. (1990), 'Photo essay: The Rio Grande borderland', *Journal of Cultural Geography* 10:2, 66–98.

ZELINSKI, W. (1988), 'Where every town is above average: Welcoming signs along America's highways', *Landscape* 30:1, 1–10.

Gambling tourism within the US

The reasons for casino developments along political boundaries and the prerequisites for gambling tourism have already been profoundly discussed in the context of the Taba dispute between Israel and Egypt. The situation within the US does not differ significantly from these prerequisites, and hence does not require a repeat discussion. What is of particular interest for the gambling industry in the US, though, is that gaming is exclusively restricted to Indian reservation lands and very few other places, including Reno, Las Vegas, or Atlantic City. It follows, that gambling in the US requires the crossing of legal boundaries to reach jurisdictions in which gambling is legally authorised, and thus closely relates to cross-border relations. In terms of Indian reservations, the gambling industry offers decent economic benefits and, therefore, is often highly endorsed and actively promoted to attract visitors from all over the US. Such situations obviously spur tourism related activities too, and thus relate to the particular topic of this chapter. The literature compiled below closely investigates the gambling and gaming tourism industry in the US and highlights legal requirements as well as social implications alike, thereby offering a profound preliminary insight into this highly volatile and contested tourism related activity in the US.

Table 9.43   Selected literature: Gambling

BIXBY, L. (1996), 'A new player at the table, the Sun Casino: Foxwoods gets a rival', *Hartford Courant*, (published on 5 October 1996), pp. 1 and 10–11.
CARMICHAEL, B.A. AND PEPPARD JR., D.M. AND BOUDREAU, F.A. (1996), 'Megaresort on my doorstep: Local resident attitudes toward Foxwoods Casino and casino gambling on nearby Indian reservation land', *Journal of Travel Research* 34:3, 9–16.
DETROIT NEWS (1997), 'Fun at your fingertips! Casino Windsor', *Detroit News*, (published on 31 August 1997), p. 15A.
EADINGTON, W.R. (1996), 'The legalization of casinos: Policy objectives, regulatory alternatives, and cost/benefit considerations', *Journal of Travel Research* 34:3, 3–8.
EADINGTON, W.R. (ed.) (1990), *Indian Gaming and the Law*, (Reno: Institute for the Study of Gambling and Commercial Gaming, University of Nevada).
GABE, T. AND KINSEY, J. AND LOVERIDGE, S. (1996), 'Local economic impacts of tribal casinos: The Minnesota case', *Journal of Travel Research* 34:4, 81–88.
GREENE, B.M. (1996), 'The reservation gambling fury: Modern Indian uprising or unfair restraint on tribal sovereignty?', *BYU Journal of Public Law* 10:1, 93–116.
JACKSON, R.H. AND HUDMAN, L.E. (1987), 'Border towns, gambling and the Mormon culture region', *Journal of Cultural Geography* 8:1, 35–48.
LEW, A.A. AND VAN OTTEN, G.A. (eds) (1998), *Tourism and Gaming on American Indian Lands* (New York: Cognizant).
NIEVES, E. (1996), 'Casino envy gnaws at Falls on US side' *New York Time* , (published on 15 December 1996), p. 49.

RANDERSON, M. (1994), 'There's a casino just across the border' *Houston Post* 30, (published on January 1994), p. F1.
SMITH, G. AND HINCH, T.D. (1996), 'Canadian casinos as tourist attractions, Chasing the pot of gold', *Journal of Travel Research* 34:3, 37–45.
STANSFIELD, C. (1996), 'Reservations and gambling: Native Americans and the diffusion of legalized gaming', in R.W. Butler and T. Hinch (eds), *Tourism and Indigenous Peoples*, (London: Routledge, 129–147).
SWANSON, E.J. (1992), 'The reservation gaming craze: Casino gambling under the Indian Gaming and Regulatory Act of 1988', *Hamline Law Review* 15, 471–496.
TRUITT, L.J. (1996), 'Casino gambling in Illinois: Riverboats, revenues, and economic development', *Journal of Travel Research* 34:3, 89–96.
WEAVER, G.D. (1966), *Some spatial aspects of Nevada's gambling economy*, (Paper presented at the annual meeting of the Association of American Geographers, August, Toronto).

Cross-border co-operation in tourism situated in the Caribbean

Islands offer uncountable opportunities for engaging in cross-border co-operation, since leaving an island often means leaving a national territory and, subsequently, crossing a national border. The specific general implications, threats and opportunities of islands engaging in cross-border co-operation in tourism have already been introduced in the context of co-operation customary in microstates. The literature below takes an even closer look at the particular state of affairs on small islands situated within the Caribbean. Hosting one of the smallest islands bisected between two sovereign European nations, the island of St. Maarten/St. Martin, the Caribbean has become of particular interest to researchers studying tourism development on bisected island nations. It should not be surprising then, that most of the literature found below investigates cross-border interactions relating to the French-Dutch co-operation established on this peculiar island. Such preliminary investigations and research findings might prove to be valuable in examining further aspects or different situations emerging on other Caribbean island nations willing to engage in cross-border co-operation to enhance local tourism industries.

Table 9.44   Selected literature: Caribbean islands engaging in cross-border co-operation

CHARDON, J.P. (1995), 'Saint-Martin ou l'implacable logique touristique', *Cahiers d'Outre-Mer* 48:189, 21–33.
CHASE, H. (1996), 'A road sign of good times: Sint Maarten/Saint Martin', *American Visions* 11:6, 42.
DE ALBUQUERQUE, K. AND MCELROY, J.L. (1995), 'Tourism development in small islands: St Maarten/St Martin and Bermuda', in D. Barker and D.F.M. McGregor (eds), *Environment and Development in the Caribbean: Geographical Perspectives*, (Kingston: University of the West Indies Press, 70–89).

KERSELL, J.E. AND BROOKSON, A. AND DUZANSON, L.L. AND GROENEVELDT, R.A. AND ARTS, X. (1993), 'Small-scale administration in St Martin: Two governments of one people', *Public Administration and Development* 13, 49–64.

LANGFORD, D.L. (1998), 'It's party time in St Maarten' *The Toronto Sun*, (published on 12 April), p. T14.

NO AUTHOR (N.D.), 'St Martin/Sint Maarten', *Islands* 11:1, 100–110.

O'NEIL, D. (1996), 'St Maarten: Double your pleasure', *Leisure World* 8:1, 16–18.

OFFICE DU TOURISME (1996), *Reflets, ile de Saint Martin*, (Marigot: Office du Tourisme de l'ile de Saint Martin).

REYNOLDS, C. (1997), 'US ban aside, Americans visit Cuba easily' *Los Angeles Times*, (published on 15 June 1997), pp. F1–2.

ROYLE, S.A. (1997), 'Tourism in the South Atlantic islands', in D.G. Lockhart and D. Drakakis-Smith (eds), *Island Tourism: Trends and Prospects*, (London: Pinter, 323–344).

SUCCESSFUL MEETINGS (1992), 'St Maarten/St Martin', *Successful Meetings* 41:12, 35–36.

TIMOTHY, D.J. (2004), 'Tourism and supranationalism in the Caribbean', in D.T. Duval (ed.) *Tourism in the Caribbean: Trends, Development, Prospects*, (London: Routledge, 119–135).

TRAVEL WEEKLY (1991), *Vendome Guide: St Martin/St Maarten*, (Toronto: Travel Weekly).

WEAVER, D.B. (1998), 'Peripheries of the periphery: Tourism in Tobago and Barbuda', *Annals of Tourism Research* 25, 292–313.

Cross-border co-operation in South America

Unfortunately, only very little research has been conducted on the cross-border co-operation in South America and, therefore, the following list of literature is rather diminutive. Such previous disinterest in the cross-border state of affairs of South American nations might be explained by considering that most of these nations are considered to be in a development stage and hence have not largely attracted tourism industries, yet. Considering the unique natural resources available in South America and the increasing need for environmental protection of such magnificent landscapes as the Amazon rainforest or the Attacama Desert, the need for collaborative multi-national co-operation surfaces and indicates a potential future research field for tourism related studies.

The literature listed below discusses the current situation of some of these states and the preliminary efforts of engaging in supra-national activities, and might prove to be a convenient starting point for investigating future South American potential in cross-border co-operation with regard to international tourism.

Table 9.45   Selected literature: South American cross-border co-operation

CÁRDENAS, E.J. (1992), 'The Treaty of Asunción: A Southern Cone Common Market (Mercosur) begins to take shape', *World Competition* 15:4, 65–77.
CAVIEDES, C.N. (1994), 'Argentine-Chilean co-operation and disagreement along the southern Patagonian border', in W.A. Gallusser (ed.), *Political Boundaries and Coexistence*, (Bern: Peter Lang, 135–143).
FRIEDMANN, J. (1966), *Regional Development policy: A Case of Venezuela*, (Cambridge: MIT Press).
GRIFFIN, E.C. AND FORD, L.R. (1980), 'Model of Latin American city structure', *Geographical Review* 70, 397–422.
LAMPMANN, J. (1997), 'Argentina side trips open the door to wonder and adventure' *Christian Science Monitor*, (published on 17 July 1997), p. 13.
MIKUS, W. (1994), 'Research methods in border studies: Results for Latin America', in W.A. Gallusser (ed.), *Political Boundaries and Coexistence*, (Bern: Peter Lang, 441–449).
NORTON, P. (1989), 'Archaeological rescue and conservation in the north Andean area', in H. Cleere (ed.), *Archaeological Heritage Management in the Modern World*, (London: Unwin Hyman, 142–145).
SALOMON, J.N. (1992), 'Le complexe touristico-industriel d'Iguacu-Itaipu (Argentine-Brésil-Paraguay)', *Cahiers d'Outre-Mer* 45, 5–20.
TIMOTHY, D.J. AND WHITE, K. (1999), 'Community-based ecotourism development on the periphery of Belize', *Current Issues in Tourism* 2:2/3, 226–242.

Cross-border co-operation on the African continent

The African continent, due to its exotic and adventurous flair, has proven highly attractive for international tourism development and hence indicates high potential for cross-border co-operation in terms of tourism promotion and attraction management. This potential has recently been realised by tourism operators and attraction managers that are increasingly beginning to get involved in trans-national collaborative planning and development projects for future tourism businesses.

Probably the most popular example of African trans-boundary co-operation is the Kruger National Park that reaches across several national borders and has become a famous international tourism attraction. Since any collaborative activities, particularly if they involve a supra-national dimension, require a decent and well organised political framework enabling the formulation of a common strategic plan, the literature within the following section will introduce studies already undertaken on the political state of affairs prevalent in some African nations and the requirements for engaging in supra-national tourism development. A major issue to be addressed within this context is the South African history of being an apartheid state, bisecting its populations, and granting privileges to the Whites only. Overcoming these issues is a first prerequisite towards engagement in future sustainable development that prospers all participants economically, while at the same time adhering to requirements imposed by environmental protection. The literature below describes recent improvements and positive developments.

Table 9.46 Selected literature: Political requirements and co-operation

COWLEY, J. AND LEMON, A. (1986), 'Bophuthatswana: Dependent development in a black "homeland"', *Geography* 71:3, 252–255.
DENTLINGER, L. (2003), 'Cross-border tourism takes off: The Namibian' <http://www.namibian.com.na/2003/june/national/03E417D6C8.html>, accessed on 19 October 2003.
EGERÖ, B. (1991), *South Africa's Bantustans: From Dumping Grounds to Battlefronts*, (Uppsala: Nordiska Afrikainstitutet).
GRAF, W.D. (1992), 'Sustainable ideologies and interests: Beyond Brundtland', *Third World Quarterly* 13, 553–559.
GRUBER, G.R. (1979), 'The influence of national borders on tourism in Africa, (the Zambian example)', in G. Gruber and H. Lamping and W. Lutz and J. Matznetter and K. Vorlaufer (eds), *Tourism and Borders: Proceedings of the Meeting of the IGU Working Group – Geography of Tourism and Recreation*, (Frankfurt a.M.: Institut für Wirtschafts- und Sozialgeographie der Johann Wolfgang Goethe Universität, 181–194).
LUBOMBO TRANSFRONTIER CONSERVATION AREA (2000), 'Transfrontier protocol paves the way for cross-border conservation and tourism development', <http://www.environment. gov. za/NewsMedia/MedStat/2000jun22_3/Lubombo_22062000.htm>, accessed on 3 April 2003.
MERRETT, C. (1984), 'The significance of the political boundary in the apartheid state, with particular reference to Transkei', *South African Geographical Journal* 66:1, 79–93.
ROGERSON, C.M. (1990), 'Sun International: The making of a South African tourism multinational', *GeoJournal* 22:3, 345–354.
SOUTH AFRICAN TOURISM BOARD (1991), *The Splendor of Five African States*, (Pretoria: South African Tourism Board).
SOUTH AFRICAN TOURISM BOARD (1999), 'Tourism', <http://www.southafrica.net/ government/ tourism.html>, accessed on 10 December 1999.
SOUTHALL, R.J. (1983), *South Africa's Transkei: The Political Economy of an 'Independent' Bantustan*, (New York: Monthly Review Press).
TEYE, V.B. (2000), 'Regional co-operation and tourism development in Africa', in P.U.C. Dieke (ed.), *The Political Economy of Tourism Development in Africa*, (New York: Cognizant, 217–227).

Cross-border co-operation of tourism businesses in particular African destinations

Not much research has been conducted on the level of cross-border co-operation of tourism on the African continent and the literature compiled below intends to grant a preliminary overview of the shape and nature such projects might take. Discussion is concerned with the potential negative aspects of unplanned and uncontrolled tourism development, the creation and emergence of vices, possible increasing crime rates, the politically unstable situation still existing in some African nations, as well as the fear of losing traditional values and identities when engaging in tourism businesses. Such issues become particularly important when considering the case of Lesotho and Swaziland, Africa's gambling nations, which have employed gambling tourism as a means of developing and establishing incomparable economic prosperity which subsequently supports the manifestation of their new states' independence and sovereignty (Timothy, 2001). None the less, having discussed the negative implications of gambling tourism and the negative image attached to such activities, discussions on the value and convenience of such vice developments is likely to last some time. Further research and related studies might help overcome the fears of tourism development and contribute to sustainable developments across the entire continent.

Table 9.47   Selected literature: Cross-border tourism with regard to particular African destinations

CRUSH, J. AND WELLINGS, P. (1983), 'The southern African pleasure periphery, 1966–1983', *Journal of Modern African Studies* 21:4, 673–698.
CRUSH, J. AND WELLINGS, P. (1987), 'Forbidden fruit and the export of vice: Tourism in Lesotho and Swaziland', in S. Britton and W.C. Clarke (eds), *Ambiguous Alternative: Tourism in Small Developing Countries*, (Suva, Fiji: University of the South Pacific, 91–112).
DIEKE, P.U.C. (1998), 'Regional tourism in Africa: Scope and critical issues', in E. Laws and B. Faulkner and G. Moscardo (eds), *Embracing and Managing Change in Tourism: International Case Studies*, (London: Routledge, 29–48).
HARRISON, D. (1992), 'Tradition, modernity and tourism in Swaziland' in D. Harrison (ed.), *Tourism and the Less Developed Countries*, (Chichester: Wiley, 148–162).
LOVE, T. (1998), 'Uganda: Beautiful but AIDS stricken', *Tennessee Alumnus* 78:2, 40–41.
McNEIL, D.G. (1997), 'Out of Pretoria by luxury train', *New York Times*, (published on 12 October 1997), p. 14.
TEYE, V. (1988), 'Coup d'Etat and African tourism: A study of Ghana', *Annals of Tourism Research* 15, 315–356.

Transfrontier conservation areas in Africa

Proving supporters right and showing how cross-border collaboration and close supra-national interaction might prosper tourism developments and support environmental protection are the large transfrontier nature reserves and biosphere reserves. The compilation below introduces literature written on the creation, planning, managing, monitoring, and controlling of such large areas multi-nationally.

Table 9.48   Selected literature: Transfrontier conservation areas in Africa

FERREIRA, S. (2004), 'Problems associated with tourism development in Southern Africa: The case of transfrontier conservation areas', *GeoJournal* 60:3, 301–310.
GREAT LIMPOPO TRANSFRONTIER PARK (2002), *GLTP Joint Management Plan*, (January 2002 Draft Document, Unpublished).
GREAT LIMPOPO TRANSFRONTIER PARK (2003), 'The Great Limpopo Transfrontier Park', <http://www.gkgpark.com>, accessed 27 January 2003.
JOINT WORKING GROUP AND TECHNICAL COMMITTEE (2002), *Great Limpopo Transfrontier Park: Background Information Document*, (Unpublished).
PEACE PARKS FOUNDATION (2003), 'Great Limpopo Transfrontier Park – current status', <http://www.peaceparks.org/content/newsroom/news_pop.php?id=55>, accessed 6 May 2003.
WOLLMER W. (2003), 'Transboundary Conservation: The Politics of Ecological Integrity in the Great Limpopo Transfrontier Park', *Journal of Southern African Studies* 29, 261–278.
ZIMBABWE TOURISM AUTHORITY (2003), *Gonarezhou – Greater Limpopo Transfrontier Park: Tourism Development Framework*, (Harare: Unpublished Draft Working Document).

Cross-border issues of international tourism in Microstates

Considering the relatively small size in terms of geographical territory, small or microstates are highly capable of engaging in cross-border co-operation in tourism. Being the smallest national entities after the geographic peculiarities of international exclaves and enclaves (see Timothy, 2001 for a detailed discussion), tourism industries in microstates appear to be tremendously dependent on international arrivals and hence could prosper extraordinarily from industry related co-operation across national boundaries. The literature listed below discusses research conducted on the scale of the tourism industry in such microstates, and elaborates upon chances, possibilities, threats, and opportunities for microstates engaging in international co-operation to attract or reject international tourism development within their small sovereign national territories. Some literature examples listed below comprise European microstates such as Monaco, prospering from Gambling Tourism and its image as a high-class destination for the rich and famous, Liechtenstein, San Marino and Andorra, as well as the specific state of affairs pertinent in the island microstates situated within the Pacific Ocean. Considering that numerous islands around the world, in terms of size, are easily comparable to microstates, general literature on tourism co-operation on islands is incorporated below too. In addition, the situation of divided islands, for example St. Martin/St. Maarten, the smallest island territory bisected between two sovereign national entities, namely France and the Netherlands (Timothy, 2001) is also considered. Even though they belong to bigger national sovereignties, the islands' remoteness creates a situation similar to that of microstates and justifies classifying them in the literature relevant to tourism cross-border co-operation in microstates.

The small amount of literature available on tourism in microstates and bi-national islands indicates the infancy of research within this highly specialised segment and might prove valuable for further exploratory studies and research undertakings evolving from and building upon previous study findings which have been assembled below.

Table 9.49   Selected literature: Cross-border co-operation in Microstates

BALDACCHINO, G. (1993), 'Bursting the bubble: The pseudo-development strategies of microstates', *Development and Change* 24:1, 29–51.
BALDACCHINO, G. (1994), 'Peculiar human resource management practices? A case study of a microstate hotel', *Tourism Management* 15, 46–52.
BAUM, T. (1997), 'The fascination of islands: A tourist perspective'?, in D.G. Lockhart and D. Drakakis-Smith (eds), *Island Tourism: Trends and Prospects*, (London: Pinter, 21–35).
BUTLER, R.W. (1993), 'Tourism development in small islands: Past influences and future directions', in D.G. Lockhart and D. Drakakis-Smith and J. Schembri (eds), *Change in Tourism: People, Places, Processes*, (London: Routledge, 92–113).
ECONOMIST (1993a), 'Coming of age in Andorra' *The Economist*, (published on 13 March 1993), p. 60.
FAGENCE, M. (1997), 'An uncertain future for tourism in microstates: The case of Nauru', *Tourism Management* 18, 385–392.
JENNER, P. AND SMITH, C. (1993), 'Europe's microstates: Andorra, Monaco, Liechtenstein and San Marino', *EIU International Tourism Reports* 1, 69–89.
LOCKHART, D.G. (1997a), 'Islands and tourism: An overview', in D.G. Lockhart and D. Drakakis-Smith (eds), *Island Tourism: Trends and Prospectives*, (London: Pinter, 3–20).
MILNE, S. (1992), 'Tourism and development in South Pacific microstates', *Annals of Tourism Research* 19, 191–212.
REICHART, T. (1988), 'Socio-economic difficulties in developing tourism in small Alpine countries: The case of Andorra', *Tourism Recreation Research* 13:1, 27–32.
RINSCHEDE, G. (1977), 'Andorra: vom abgeschlossenen Hochgebirgsstaat zum internationalen Touristenzentrum', *Erdkunde* 31, 307–314.
SANGUIN, A.L. (1991), 'L'Andorre ou la quintessence d'une economie transfrontaliere', *Revue Geographique des Pyrenees et du Sud-Ouest* 62:2, 169–186.
TAILLEFER, F. (1991), 'Le paradoxe Andorran', *Revue Geographique des Pyrenees et du Sud-Ouest* 62:2, 117–138.
WILKINSON, P.F. (1987), 'Tourism in small island nations: A fragile dependence', *Leisure Studies* 6:2, 127–146.
WILKINSON, P.F. (1989), 'Strategies for tourism in island microstates', *Annals of Tourism Research* 16, 153–177.

Note

This bibliographical approach definitely lacks completeness. Although the literature listed above might cover substantial areas, not all publications made in the field of 'tourism and borders' could be listed.

Further literature on the topical issues presented in chapters one to eight of this book are available from the reference lists at the end of each chapter.

An alphabetized bibliography of all 1,280 literature items employed for this chapter is available at http://www.ashgate.com/subject_area/downloads/sample_chapters/Tourism_ and_Borders_Bibliography.pdf.

References

Arreola, D.D. (1996), 'Border-City ideé Fixe', *The Geographical Review*, Vol. 86, No. 3, pp. 356–369.

Arreola, D.D. (1999), 'Across the street is Mexico: Invention and Persistence of the Border Town Curio Landscape', *Yearbook of the Association of Pacific Coast Geographers*, Vol. 61, pp. 9–41.

Arreola, D.D. and Curtis, J.R. (1993), The Mexican Border Cities: Landscape Anatomy and Place Personality, Tucson: University of Arizona Press.

Ashworth, G.J. (1995), 'Heritage, Tourism and Europe: A European future for a European Past?', In: D.T. Herbert (ed.), *Heritage, Tourism and Society*, London: Mansell, pp. 68–84.

Association for Canadian Studies in the United States (2001), Cross-border cultural tourism – A two way street – Facts and Figures about Cultural Tourism across the Canada-USA Border [online], available at URL: http://www.theniagarasguide.com/partners/tou.pdf, accessed 2 February 2003.

Bar-On, R. (1988), 'International day trip, including cruise passenger excursions', *Revue de Tourisme*, Vol. 43, No. 4, pp. 12–17.

Becker, Chr. (1992), 'Kulturtourismus – Eine Zukunftsträchtige Entwicklungsstrategie für den Saar-Mosel-Ardennenraum', In: Chr. Becker and W. Schertler and A. Steinecke (eds), *Perspektiven des Tourismus im Zentrum Europas*, Trier: ETI-Studien, Vol. 1, pp. 21–25.

Blatter, J. (2000), 'Emerging Cross-Border regions as a step towards sustainable development? Experiences and Considerations from examples in Europe and North America', International Journal of Economic Development [online]. Available at URL: www.spaef. com/IJED_PUB/v2n3/ v2n3_4_blatter.pdf [Accessed on 25 March 2003].

Boal, F.W. (1994), 'Encapsulation: Urban dimensions of national conflict', In: S. Dunn (ed.), *Managing Divided Cities*, Keele: Ryburn, pp. 30–40.

Boisvert, M. and Thirsk, W. (1994), 'Border taxes, cross-border shopping, and the differential incidence of the GST', *Canadian Tax Journal*, Vol. 42, No. 5, pp. 1276–1293.

Bondi, N. (1997), 'Shoppers head to Windsor for deals: High exchange rate on American dollar fuels big-ticket buys', *Detroit News*, 30 December.

Bowman, K.S. (1994). 'The border as locator and innovator of vice', *Journal of Borderlands Studies*, Vol. 9, No. 1, pp. 51–67.

Bramwell, B. and Lane, B. (eds) (2000), *Tourism Collaboration and Partnerships: Politics, Practice and Sustainability*, Clevedon: Channel View Publications.

Braun-Moser, U. (1991), *Europäische Tourismuspolitik*. Sindelfingen: Libertas Verlag.

Broinowski, A. (ed.) (1982), *Understanding ASEAN*. New York: St Martin's Press.

Brown, T.C. (1997), 'The fourth member of NAFTA: The US-Mexico border', *Annals of the American Academy of Political and Social Science*, Vol. 550, pp. 105–121.

Bustamante, J. (1988), 'Identidad, cultura nacional y frontera', In: A. Malagamba (ed.), *Encuentros. Los festivales internacionales de la raza*, Tijuana, Baja California: El Colegio de la Frontera Norte.

Butler, R.W. (1991), 'West Edmonton Mall as a tourist attraction', *The Canadian Geographer*, Vol. 35, pp. 287–295.

Butler, R.W. (1996), 'The development of tourism in frontier regions: Issues and approaches', In: Y. Gradus and H. Lithwick (eds), *Frontiers in Regional Development*, Lanham, MD: Rowman & Littlefield, pp. 213–229.

Butler, R.W. and Mao, B. (1995), 'Tourism between quasi-states: International, domestic or what?', In: R.W. Butler and D. Pearce (eds), *Change in Tourism: People, Places, Processes*, London: Routledge, pp. 92–113.

Butler, R.W. and Mao, B. (1996), 'Conceptual and theoretical implications of tourism between partitioned states', *Asia Pacific Journal of Tourism Research*, Vol. 1, No. 1, pp. 25–34.

Bygvra, S. (1990), 'Border shopping between Denmark and West Germany', *Contemporary Drug Problems*, Vol. 17, No. 4, pp. 595–611.

Canadian Chamber of Commerce (1992), *The Cross-Border Shopping Issue*, Ottawa: Canadian Chamber of Commerce.

Carpathian Foundation (2002), Cross-Border Co-operation in the Carpathian Euroregion [online], Available at URL: http:// www.carpahtianfoundation.org/download/programs/ cbcen.doc [Accessed on 25 March 2003].

Chadee, D. and Mieczkowski, Z. (1987), 'An empirical analysis of the exchange rate on Canadian tourism', *Journal of Travel Research*. Vol. 26, No. 1, pp. 13–17.

Chatterjee, A. (1991), 'Cross-border shopping: Searching for a solution', *Canadian Business Review*. Vol. 18, pp. 26–31.

Curtis, J.R. (1993), 'Central business districts of the two Loredos', *Geographical Review*, Vol. 83, No. 1, pp. 54–65.

Curtis, J.R. and Arreola, D.D. (1991), 'Zonas de tolerancia on the northern Mexican border', *Geographical Review*, Vol. 81, No. 3, pp. 333–346.

Davis, G. and Guma, G. (1992), *Passport to Freedom: A Guide for World Citizens*, Washington: Seven Locks Press.

Demler, D. (2004), *Der Internationale Tourismus in Nordmexikanischen Grenzstädten am Beispiel von Nuevo Laredo*, Trier: Geographische Gesellschaft der Universität Trier, Materialien zur Fremdenverkehrsgeographie, No. 62.

Department of Agriculture and Rural Development (2002), Rural Development supports cross-border activity tourism in Sliabh Beagh (online). Available at URL: http://www. dardni.gov.uk/pr2002/ pr020172.htm, accessed on 3 April 2003.

DiMatteo, L. (1993), 'Determinants of cross-border trips and spending by Canadians in the United States: 1979-1991', *Canadian Business Economics*. Vol. 1, No. 3, pp. 51–61.

DiMatteo, L. and DiMatteo, R. (1993), 'The determinants of expenditures by Canadian visitors to the United States', *Journal of Travel Research*, Vol. 31, No. 4, pp. 34–42.

Dunn, S. (ed.) (1994), *Managing Divided Cities*, Ryburn: Keele.

Eberstadt, N. (1995), *Korea Approaches Reunification*, Armonk, NY: M.E. Sharpe.

EURES – TRANSFRONTALIER Saar-Lor-Lux-Rheinland-Pfalz (2000), Tourismus und Arbeitsmarkt – Eine grenzüberschreitende Bestandsaufnahme EURES-T Saar-Lor-Lux-Rheinland-Pfalz (SLLR) (online). Available at URL: http://www.eures-sllr.org/deutsch/

info/Pdf-Downloads/Tourismus%20Studie%20Dt.pdf, accessed on 24 April 2003.

Felsenstein, D. and Freeman, D. (2001), 'Estimating the impacts of crossborder competition: the case of gambling in Israel and Egypt', *Tourism Management*, Vol. 22, pp. 511–521.

Franklin, A. (2003), *Tourism – An Introduction*, London: SAGE Publications.

German, A.L. (1984), 'Point Roberts: A tiny borderline anomaly', *Canadian Geographic*, Vol. 104, No. 5, pp. 72–74.

Getz, D. (1993), 'Tourist Shopping Villages: Development and Planning Strategies', *Tourism Management*, Vol. 14, pp. 15–26.

Gramm, M. (1983), 'Einkaufen im belgisch-niederländisch-deutschen Dreiländereck – ein Beispiel für grenzüberschreitende räumliche Interaktionen und ihrem Beitrag zur Entwicklungeines grenzüberschreitenden Nationalbewußtseins', In: J. Maier (ed.), *Staatsgrenzen und ihr Einfluß auf Raumstrukturen*, Teil 1. Arbeitsmaterialien zur Raumordnung und Raumplanung, Heft 23, Bayreuth, pp. 51–69.

Greer, J. (2002), 'Developing trans-jurisdictional tourism partnerships – insights from the island of Ireland', *Tourism Management*, Vol. 23, pp. 355–366.

Green Gate (1998), *Creation of a model of transborder co-operation for sustainable development in the Carpathian Euroregion on an example of the International Biosphere Reserve 'Eastern Carpathians'*, Lviv-Sianky-Velykyi Bereznyi.

Hansen, N. (1981), *The border economy: Regional development in the Southwest*, Austin: University of Texas Press.

Henriksen, T.H. and Lho, K. (eds) (1994), *One Korea?: Challenges and Prospects for Reunification*, Stanford, CA: Hoover Institution Press.

Herzog, L.A. (1990), *Where North Meets South: Cities, Space, and Politics on the US–Mexico Border*, Austin: Center for Mexican American Studies, University of Texas.

Hidalgo, L. (1993), 'British shops suffer as "booze cruise" bargain hunters flock to France', *The Times*, 22 November, p. 5.

Ho, K.C. and So, A. (1997), 'Semi-periphery and borderland integration: Singapore and Hong Kong experiences', *Political Geography*, Vol. 16, No. 3, pp. 241–259.

Holden, R.J. (1984). '"Maquiladoras" employment and retail sales effects on four Texas border communities, 1978–1983', *Southwest Journal of Business and Economics*, Vol. 2, No. 1, pp. 16–26.

House, J.W. (1980), 'The frontier zone: A conceptual problem for policy makers', *International Political Science Review*, Vol. 1, No. 4, pp. 456–477.

Hufbauer, G.C. and Schott, J.J. (1992), *North American Free Trade: Issues and Recommendations*, Washington D.C.: Institute for International Economics.

Jud, G.D. (1975), 'Tourism and crime in Mexico', *Social Science Quarterly*, Vol. 56, No. 2, pp. 324–330.

Kirk, D. at the International Herald Tribune – The IHT online (2003), Seoul Leader deems mission 'a triumph': South Korean busses cross border to North (online), available at URL: http://www.iht.com/articles/ 2003/02/06/talks.php, accessed on 15 February 2003.

Kliemkiewicz, M. (2002), Tourism and Cross-Border Co-operation-Carpathains (online). Available at URL: http://www.mtnforum.org/ emaildiscuss/discuss02/031102269.htm, accessed 3 February 2003.

Krakover, S. (1985), 'Development of tourism resort areas in arid regions', In: Y. Gradus (ed.), *Desert Development: Man and Technology in Sparselands*, Dordrecht: D. Reidel Publishing, pp. 271–284.

Krätke, S. (1998), 'Problems of cross-border regional integration: The case of the German-Polish border area', *European Urban and Regional Studies*, Vol. 5, No. 3, pp. 249–262.

Krätke, S. (2001), 'Cross-border co-operation in the German–Polish Border Area', In: M.V. Geenhuizen and R. Ratti (eds), *Gaining advantage from open borders: an active space approach to regional development*, Aldershot/Burlington: Ashgate, pp. 213–232.

Krongkaew, M. (2004), 'The development of the Greater Mekong Sub-region (GMS): Real promise or false hope?', *Journal of Asian Economics*. Vol. 15, pp. 977–998.

Leimgruber, W. (1980), 'Die Grenze als Forschungsobjekt der Geographie', *Regio Basiliensis*, Vol. 21, pp. 67–78.

Leimgruber, W. (1998), 'Defying political boundaries: Transborder tourism in a regional context', *Visions in Leisure and Business*, Vol. 17, No. 3, pp. 8–29.

Lieff, B.C. and Lusk, G. (1990), 'Transfrontier co-operation between Canada and the USA: Wateron-Glacier International Peace Park', In: J. Thorsell (ed.), *Parks on the Borderline: Experience in Transfrontier Conservation*, Gland: IUCN, pp. 39–49.

Lin, V.L. and Loeb, P.D. (1977), 'Tourism and crime in Mexico: Some comments', *Social Science Quarterly*, Vol. 58, No. 1, pp. 164–167.

Lorenz, T. and Stoklosa, K. (n.d.), Bibliographie zur Grenzregion (online). Available at URL: http://www.wsgn.uni-ffo.de/bibliographie.pdf, accessed on 23 February 2005.

Lucas, V. (2004), 'Cross-border shopping in a federal economy', *Regional Science and Urban Economics*, Vol. 34, pp. 365–385.

Martinez, O.J. (1994), 'The dynamics of border interaction: New approaches to border analysis', In: C.H. Schofield (ed.), *World Boundaries, Vol. 1, Global Boundaries*, London: Routledge, pp. 1–15.

Matznetter, J. (1979), 'Border and tourism: Fundamental relations', In: G. Gruber and H. Lamping and W. Lutz and J. Matznetter and K. Vorlaufer (eds), *Tourism and Borders: Proceedings of the Meeting of the IGU Working Group – Geography of Tourism and Recreation*, Frankfurt a.M.: Institut für Wirtschafts- und Sozialgeographie der Johann Wolfgang Goethe Universität, pp. 61–75.

Mazza, P. (1995), Cascadia Emerging: The end and the beginning of the world (online). Available at URL: http://www.tnews.com/text/emerge.html, accessed in July 1996.

McAllister, H.E. (1961), 'The border tax problem in Washington', *National Tax Journal*, Vol. 14, No. 4, pp. 362–374.

McCloskey, D.D. (1995), 'Cascadia: A great green land on the Northeast Pacific Rim', (online). Available at URL: http://www.tnews.com:80/ text/mccloskey.html, accessed in July 1996.

McGreevy, P. (1988), 'The end of America: The beginning of Canada', *Canadian Geographer*. Vol. 32, No. 4, pp. 307–318.

Mikus, W. (1986), 'Grenzüberschreitende Verflechtungen im tertiären Sektor zwischen USA und Mexiko: Das Beispiel Kaliforniens', *Geographica Helvetica*, Vol. 36, pp. 207–217.

Minghi, J.V. (1994), 'European borderlands: International harmony, landscape change and new conflict', In: C. Grundy-Warr (ed.), *World Boundaries, Vol. 3, Eurasia*. London: Routledge, pp. 89–98.

Minghi, J.V. and Rumley, D. (1972), 'Integration and system stress in an international enclave community: Point Roberts, Washington', *B.C. Geographical Series*, Vol. 15, pp. 213–229.

Monsiváis, C. (1978), 'The culture of the frontier: The Mexican side', In: S.R. Ross, *Views across the border: The United States and Mexico*, Albuquerque: University of New Mexico Press.

O'Byrne, D.J. (2001), 'On passports and border controls', *Annals of Tourism Research*, Vol. 28, No. 2, pp. 399–416.

Palomäki, M. (1994), 'Transborder co-operation over Quarken Strait between Finland and

Sweden', In: W.A. Gallusser (ed.) *Political Boundaries and Coexistence*, Bern: Peter Lang, pp. 238–246.

Parsonage, J. (1992), 'Southeast Asia's growth triangle: a sub-regional response to global transformation', *International Journal of Urban and Regional Research*, Vol. 16, pp. 307–318.

Pearcy, G.E. (1965), 'Boundary Functions', *Journal of Geography*, Vol. 64, No. 8, pp. 346–349.

Perry, M. (1991), 'The Singapore Growth Triangle: State, capital and labour at a new frontier in the world economy', *Singapore Journal of Tropical Geography*, Vol. 12, No. 2, pp. 138–151.

Pezolli, K. (1997a), 'Sustainable Development Literature: A Transdisciplinary Bibliography', *Journal of Environmental Planning and Management*, Vol. 40, No. 5, pp. 575–601.

Pezolli, K. (1997b), 'Sustainable Development: A Transdisciplinary Overview of the Literature', *Journal of Environmental Planning and Management*, Vol. 40, No. 5, pp. 549–574.

Prescott, J.R.V. (1987), *Political Frontiers and Boundaries*, London: Allen and Unwin.

Richard, W. E. (1996), Cross Border Tourism and Shopping: The Policy Alternatives (online). Available at URL: http://www.Usm.maine.edu/cber/ mbi/winter96/tourism.htm (Accessed on 3 April 2003).

Rojek, C. (1998), 'Cybertourism and the phantasmagoria of place', In: G. Ringer (ed.), *Destinations: Cultural Landscapes of Tourism*, London: Routledge, pp. 33–48.

Rumley, D. and Minghi, J.V. (eds) (1991), *The Geography of Border Landscapes*, London: Routledge.

Rutan, G.F. (1981), 'Legislative interaction of a Canadian province and an American state – Thoughts upon sub-national cross-border relations', *American Review of Canadian Studies*, Vol. 6, No. 2, pp. 67–79.

Ryden, K.C. (1993), *Mapping the Invisible Landscape: Folklore, Writing, and the Sense of Place*, Iowa City: University of Iowa Press.

Schell, P. and Hamer, J. (1995), 'Cascadia: The new binationalism of Western Canada and the US Pacific Northwest', In: R.L. Earle and J.D. Wirth (eds), *Identities in North America, The search for community*, Stanford: Stanford University Press, pp. 140–156.

Scott, J. (1995), 'Sexual and national boundaries in tourism', *Annals of Tourism Research*, Vol. 22, pp. 385–403.

Scott, J. W. (1999), 'European and North American contexts for cross-border regionalism', *Regional Studies*, Vol. 33, No. 7, pp. 605–618.

Smith, G. (1994), 'Implications of the North American Free Trade Agreement for the US tourism industry', *Tourism Management*, Vol. 15, pp. 323–326.

Smith, V.L. (1996), Foreword, In: Price, M.F. (ed.) (1996), *People and Tourism in Fragile Environments*, Wiley: Chichester.

Smith, V.L. (2000), 'Space tourism: The 21st century "frontier"', *Tourism Recreation Research*, Vol. 25, No. 3, pp. 5–15.

Smith, G. and Pizam, A. (1998), 'NAFTA and tourism development policy in North America', In: E. Laws and B. Faulkner and G. Moscardo (eds), *Embracing and Managing Change in Tourism: International Case Studies*, London: Routledge, pp. 17–28.

Sparke, M. and Sidaway, J.D. and Bunell, T. and Grundy-Warr, C. (2004), *Triangulating the borderless world: geographies of power in the Indonesia-Malaysia-Singapore Growth Triangle*, Transnational Institute of British Geographers: Royal Geographical Society.

Stansfield, C.A. AND Rickert, J.E. (1970), 'The Recreational Business District'. *Journal of

Leisure Research, Vol. 2, No. 4, pp. 213–225.

Steinecke, A. and Treinen, M. (eds) (1996), *Wachstumsmarkt Golftourismus: Chancen für die Nachbarn Luxemburg, Rheinland-Pfalz, Belgien und Lothringen*, Trier: ETI-Texte, Vol. 9, Europäisches Tourismus Institut.

Steinecke, A. and Wachowiak, H. (1994b), 'Kulturstraßen als innovative touristische Produkte – Das Beispiel der grenzüberschreitenden Kulturstraße "Straße der Römer" an der Mosel', in: J. Maier (ed.), *Touristische Straßen – Beispiele und Bewertung*, Bayreuth: Arbeitsmaterialien zur Raumordnung und Raumplanung, No. 137, pp. 5 – 33.

Tenhiälä, H. (1994), 'Cross-border co-operation: Key to international ties', *International Affairs*, Vol. 6, pp. 21–23.

Timothy, D.J. (1995), 'Political boundaries and tourism: Borders as tourist attractions', *Tourism Management*, Vol. 16, pp. 525–532.

Timothy, D.J. (1999), 'Cross-border partnership in tourism resource management: International parks along the US-Canada border', *Journal of Sustainable Tourism*, Vol. 7, No. 3/4, pp. 182–205.

Timothy, D.J. (2000), 'Borderlands: An unlikely tourist destination?', *Boundary and Security Bulletin*, Vol. 8, No. 1, pp. 57–65.

Timothy, D.J. (2001), *Tourism and Political Boundaries*, London: Routledge.

Timothy, D.J. (2002), 'Tourism in borderlands: Competition, complementarity, and cross-frontier co-operation', in: S. Krakover and Y. Gradus (eds), *Tourism in frontier areas*, Lanham, MD: Lexington Books, pp. 233–258.

Timothy, D.J. and Butler, R.W. (1995), 'Cross-border shopping – A North American Perspective', *Annals of Tourism Research*, Vol. 22, No. 1, pp. 16–34.

Timothy, D.J. and Tosun, C. (2003), 'Tourists' perceptions of the Canada-US border as a barrier to tourism at the International Peace Garden', *Tourism Management*, Vol. 24, pp. 411–421.

Turbeville III, D.E. and Bradbury, S.L. (1997), *Borderlines, border towns: Cultural landscapes of the post-free trade 49th parallel*, Paper presented at the annual meeting of the Association of American Geographers, Forth Worth, Texas, April.

Urry, J. (1992), 'The Tourist Gaze and the 'Environment'', *Exploration in Critical Social Science*, Vol. 9, No. 3, pp. 1–26.

Var, T. and Toh, R. and Khan, H. (1998), 'Tourism and ASEAN Economic Development', *Annals of Tourism Research*, Vol. 25, No. 4, pp. 195–197.

Vila, P. (2003), 'Processes of identification on the US – Mexico border', *The Social Science Journal*, Vol. 40, pp. 607–625.

Vorlaufer, K. (1996), 'Mexiko: Regionale Disparitäten, Staat und Tourismus', *Zeitschrift für Wirtschaftsgeographie*, Vol. 40, No. 4, pp. 193–223.

Wachowiak, H. (1994a), 'Das Tourismuskonzept "Europäisches Tal der Mosel" – Ein Beitrag zur touristischen Entwicklung in der Region Saar-Lor-Lux-Trier/Westpfalz', in: H.P. Burmeister (ed.), *Wohin die Reise geht – Perspektiven des Tourismus in Europa*, Loccum: Loccumer Protokolle, Vol. 2, pp. 203–215.

Wachowiak, H. (1994c), 'Grenzüberschreitende Zusammenarbeit im Tourismus auf der Ebene der öffentlichen Hand entlang der westdeutschen Staatsgrenze', *Raumforschung und Raumordnung*, Vol. 52, No. 6, pp. 397–405.

Wachowiak, H. (1994d), *Grenzüberschreitende Zusammenarbeit im Tourismus – Eine Analyse grenzüberschreitender Maßnahmen an der westlichen Staatsgrenze der Bundesrepublik Deutschland zwischen Ems-Dollart und Baden-Nordelsaß-Südpfalz*, Trier: Europäisches Tourismus Institut.

Wachowiak, H. (1997), 'Tourismus im Grenzraum – Touristische Nachfragestrukturen unter dem Einfluß von Staatsgrenzen am Beispiel der Grenzregion Deutschland-Luxemburg', *Materialien zur Fremdenverkehrsgeographie* 38, (Trier).

Warszynska, J. and Jackowski, A. (1979), 'Impact of passport facilities in the passenger traffic between Poland and the German Democratic Republic (GDR) on the development of touristic phenomena', In: G. Gruber and H. Lamping and W. Lutz and J. Matznetter and K. Vorlaufer (eds), Tourism and borders: Proceedings of the meeting of the IGU Working Group – Geography of Tourism and Recreation, Frankfurt A.M.: Institut für Wirtschafts- und Sozialgeographie der Johann Wolfgang Goethe Universität, p. 353.

White, M. (1999), 'Entrepreneurs study space tourism with "reality" 15–20 years in future', *Sentinal Tribune*, 23 September, p. 10.

Wilson, J. (2000), 'Postcards from the moon: A lunar vacation isn't as far-out an adventure as you think', *Popular Mechanics*, June, pp. 97–99.

Wilson, T.M. and Donnan, H. (eds) (1998), *Border Identities: Nation and State at International Frontiers*, Cambridge: Cambridge University Press.

World Tourism Organization (2005), Why Tourism? (online). Available at URL: http://www. world-tourism.org/aboutwto/eng/menu.html, accessed on 23 February 2005.

Yang, C. (2004), 'The Pearl River Delta and Hong Kong: an evolving cross-boundary region under "one country, two systems"', *Habitat International*, Article in Press.

Zhao, X. (1994a), 'Barter tourism: A new phenomenon along the China-Russia border', *Journal of Travel Research*, Vol. 32, No. 3, pp. 65–67.

Zhao, X. (1994b), 'Barter tourism along the China-Russia border', *Annals of Tourism Research*, Vol. 21, pp. 401–403.

Zimmer, N. (2000), 'Ein kulturtouristisches Vermarktungskonzept für den Saarländisch-Lothringischen Grenzraum unter besonderer Berücksichtigung der Industriekultur', *Schriftenreihe der Regionalkommission Saarland-Lothringen-Luxemburg-Trier*, Arbeitsgruppe Raumordnung, Band 11.

Final word

The growth of cross-border tourism operations and the increasing foundations of supra-national tourism attractions impose various management challenges due to different managerial approaches, value systems, and cultural perceptions on either side of any border. Such challenges can only be overcome by engaging in collaborative planning and co-operative developments, indicating the importance of the principles of sustainability and stakeholder collaboration theory for cross-border interactions. Increasingly, borders are changing in their nature from being barriers towards being 'lines of contact'. Understanding the nature and various functions of borders and the related mechanisms and procedures of crossing them, might consequently prove highly valuable to tourism planners, operators, or destination marketing organisations.

According to several authors in the field, many more empirical case studies are needed on cross-border co-operation. Pioneers such as *Martinez* and *Timothy* have already proposed a preliminary agenda for future research activities. The consideration that the world is constantly changing and globalisation keeps on shrinking the world towards a 'global village', indicates the growing importance of understanding these phenomena for successful future tourism developments. Arguably, border studies are likely to increase in importance in coming years.

Understanding border relations and investigating the mechanisms underlying successful co-operation, therefore, appears to be a major task in future tourism related research activities. The bibliography chapter of this book intended to address this task and offers an initial convenient access point to the literature already published on this academic topic. The chapters of this book have introduced the concepts that are currently interesting. Perhaps they have inspired more ideas for further research on *Tourism and Borders*.

Index